MW00635480

THE ART OF DRUG SYNTHESIS

The Wiley Bicentennial–Knowledge for Generations

Each generation has its unique needs and aspirations. When Charles Wiley first opened his small printing shop in lower Manhattan in 1807, it was a generation of boundless potential searching for an identity. And we were there, helping to define a new American literary tradition. Over half a century later, in the midst of the Second Industrial Revolution, it was a generation focused on building the future. Once again, we were there, supplying the critical scientific, technical, and engineering knowledge that helped frame the world. Throughout the 20th Century, and into the new millennium, nations began to reach out beyond their own borders and a new international community was born. Wiley was there, expanding its operations around the world to enable a global exchange of ideas, opinions, and know-how.

For 200 years, Wiley has been an integral part of each generation's journey, enabling the flow of information and understanding necessary to meet their needs and fulfill their aspirations. Today, bold new technologies are changing the way we live and learn. Wiley will be there, providing you the must-have knowledge you need to imagine new worlds, new possibilities, and new opportunities.

Generations come and go, but you can always count on Wiley to provide you the knowledge you need, when and where you need it!

William J. Pesce
President and Chief Executive Officer

Peter Booth Wiley
Chairman of the Board

THE ART OF DRUG SYNTHESIS

Edited by

Douglas S. Johnson

Jie Jack Li

Pfizer Global Research and Development

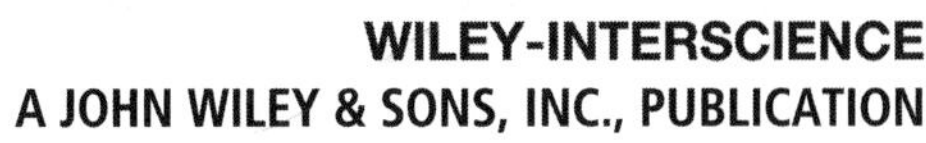

Published by John Wiley & Sons, Inc., Hoboken, New Jersey
Published simultaneously in Canada

Wiley Bicentennial Logo: Richard J. Pacifico

***Library of Congress Cataloging-in-Publication Data*:**

The art of drug synthesis / edited by Douglas S. Johnson and Jie Jack Li.
p. cm.
Includes bibliographical references and index.
ISBN 978-0-471-75215-8 (cloth)
1. Drugs—Design. 2. Pharmaceutical chemistry. I. Johnson, Douglas S. (Douglas Scott), 1968- II. Li, Jie Jack.
[DNLM: 1. Drug Design. 2. Chemistry, Pharmaceutical—methods. QV 744 A784 2007]
RS420.A79 2007
615′.19--dc22

2007017891

Printed in the United States of America

10 9 8 7 6 5 4 3 2 1

CONTENTS

II CARDIOVASCULAR AND METABOLIC DISEASES

FOREWORD

The discovery of efficacious new human therapeutic agents is one of humanity's most vital tasks. It is an enormously demanding activity that requires creativity, a vast range of scientific knowledge, and great persistence. It is also an exceedingly expensive activity. In an ideal world, no education would be complete without some exposure to the ways in which new medicines are discovered and developed. For those young people interested in science or medicine, such knowledge is arguably mandatory.

In this book, Douglas Johnson, Jie Jack Li, and their colleagues present a glimpse into the realities and demands of drug discovery. It is both penetrating and authoritative. The intended audience, practitioners and students of medicinal and synthetic chemistry, can gain perspective, wisdom, and valuable factual knowledge from this volume. The first two chapters of the book provide a clear view of the many complexities of drug discovery, the numerous stringent requirements that any potential therapeutic molecule must meet, the challenges and approaches involved in finding molecular structures that "hit" a biological target, and the many facets of chemical synthesis that connect initial small-scale laboratory synthesis with the evolution of a process for successful commercial production. The remaining 15 chapters provide a wealth of interesting synthetic chemistry as applied to the real world of the molecular medicine of cancer, infectious, cardiovascular, and metabolic diseases. At the same time, each of these chapters illuminates the way in which a first-generation therapeutic agent is refined and improved by the application of medicinal chemistry to the discovery of second- and third-generation medicines.

The authors have produced a valuable work for which they deserve much credit. It is another step in the odyssey of drug finders; a hardy breed that accepts the high-risk nature of their prospecting task, the uncertainties at the frontier, and the need for good fortune, as well as focus and sustained hard work. My ability to predict the future is no better than that of others, but I think it is possible that a highly productive age of medicinal discovery lies ahead, for three reasons: (1) the discovery of numerous important new targets for effective disease therapy, (2) the increasing power of high-throughput screening and bio-target structure-guided drug design in identifying lead molecules, and (3) the ever-increasing sophistication of synthetic and computational chemistry.

E. J. Corey

PREFACE

Our first book on drug synthesis, *Contemporary Drug Synthesis*, was published in 2004 and was well received by the chemistry community. Due to time and space constraints, we only covered 14 classes of top-selling drugs, leaving many important drugs out. In preparing *The Art of Drug Synthesis*, the second volume in our series on "*Drug Synthesis*," we have enlisted 16 chemists in both medicinal and process chemistry, encompassing nine pharmaceutical companies. Some authors were even intimately involved with the discovery of the drugs that they reviewed. Their perspectives are invaluable to the reader with regard to the drug discovery process.

In Chapter 1, John Lowe details "The Role of Medicinal Chemistry in Drug Discovery" in the twenty first century. The overview should prove invaluable to novice medicinal chemists and process chemists who are interested in appreciating what medicinal chemists do. In Chapter 2, Neal Anderson summarizes his experience in process chemistry. The perspectives provide a great insight for medicinal chemists who are not familiar with what process chemistry entails. Their contributions afford a big picture of both medicinal chemistry and process chemistry, where most of the readers are employed. Following two introductory chapters, the remainder of the book is divided into three major therapeutic areas: I. Cancer and Infectious Diseases (five chapters); II. Cardiovascular and Metabolic Diseases (six chapters); and III. Central Nervous System Diseases (four chapters).

We are grateful to Susan Hagen and Derek Pflum at Pfizer, and Professor John Montgomery of the University of Michigan and his students Ryan Baxter, Christa Chrovian, and Hasnain A. Malik for proofreading portions of the manuscript. Jared Milbank helped in collating the subject index.

We welcome your critique.

Douglas S. Johnson
Jie Jack Li

Ann Arbor, Michigan
April 2007

CONTRIBUTORS

Neal G. Anderson 7400 Griffin Lane, Jacksonville, Oregon

Andrew S. Bell Pfizer Global Research and Development, Sandwich, Kent, United Kingdom

Victor J. Cee Amgen, Inc., Thousand Oaks, California

Daniel P. Christen Transtech Pharma, High Point, North Carolina

David L. Gray Pfizer Global Research and Development, Ann Arbor, Michigan

Peter R. Guzzo Albany Molecular Research, Inc., Albany, New York

Arthur Harms Bausch and Lomb, Rochester, New York

Douglas S. Johnson Pfizer Global Research and Development, Ann Arbor, Michigan

Jie Jack Li Pfizer Global Research and Development, Ann Arbor, Michigan

Jin Li Pfizer Global Research and Development, Groton, Connecticut

Chris Limberakis Pfizer Global Research and Development, Ann Arbor, Michigan

John A. Lowe, III Pfizer Global Research and Development, Groton, Connecticut

Edward J. Olhava Millennium Pharmaceuticals, Cambridge, Massachusettes

Jeffrey A. Pfefferkorn Pfizer Global Research and Development, Ann Arbor, Michigan

Marta Piñeiro-Núñez Eli Lilly and Company, Indianapolis, Indiana

Stuart B. Rosenblum Schering-Plough Research Institute, Kenilworth, New Jersey

Larry Yet Albany Molecular Research, Inc., Albany, New York

Po-Wai Yuen Pfizer Global Research and Development, Ann Arbor, Michigan

1

THE ROLE OF MEDICINAL CHEMISTRY IN DRUG DISCOVERY

John A. Lowe, III

1.1 INTRODUCTION

This volume represents the efforts of the many chemists whose ability to master both synthetic and medicinal chemistry enabled them to discover a new drug. Medicinal chemistry, like synthetic chemistry, comprises both art and science. It requires a comprehensive mind to collect and synthesize mountains of data, chemical and biological. It requires the instinct to select the right direction to pursue, and the intellect to plan and execute the strategy that leads to the desired compound. Most of all, it requires a balance of creativity and perseverance in the face of overwhelming odds to reach the goal that very few achieve—a successfully marketed drug.

The tools of medicinal chemistry have changed dramatically over the past few decades, and continue to change today. Most medicinal chemists learn how to use these tools by trial and error once they enter the pharmaceutical industry, a process that can take many years. Medicinal chemists continue to redefine their role in the drug discovery process, as the industry struggles to find a successful paradigm to fulfill the high expectations for delivering new drugs. But it is clear that however this new paradigm works out, synthetic and medicinal chemistry will continue to play a crucial role. As the chapters in this volume make clear, drugs must be successfully synthesized as the first step in their discovery. Medicinal chemistry consists of designing and synthesizing new compounds, followed by evaluation of biological testing results and generation of a new hypothesis as the basis for further compound design and synthesis. This chapter will discuss the role of both synthetic and medicinal chemistry in the drug discovery process in preparation for the chapters that follow on the syntheses of marketed drugs.

The Art of Drug Synthesis. Edited by Douglas S. Johnson and Jie Jack Li

1.2 HURDLES IN THE DRUG DISCOVERY PROCESS

Although the tools of medicinal chemistry may have improved considerably (as discussed below), the hurdles to discovering a new drug have outpaced this improvement, accounting to a certain extent for the dearth of newly marketed drugs. Discussion of some of these hurdles, such as external pressures brought on by the public media and the stock market, lies outside the scope of this review. Instead, we will discuss those aspects of drug discovery under the control of the scientists involved.

One of the first challenges for the medicinal chemist assigned to a new project is to read the biology literature pertaining to its rationale. Interacting with biology colleagues and understanding the results from biological assays are critical to developing new hypotheses and program directions. Given the increasing complexity of current biological assays, more information is available, but incorporating it into chemistry planning requires more extensive biological understanding. This complexity applies to both the primary in vitro assay for the biological target thought to be linked to clinical efficacy, as well as selectivity assays for undesired off-target in vitro activities. Some of the same considerations apply to the increasingly sophisticated assays for other aspects of drug discovery, such as ADME (absorption, distribution, metabolism, and elimination) and safety, as summarized in Table 1.1.

The reader is referred to an excellent overview of the biology behind these assays, and their deployment in a typical drug discovery program (Lin et al., 2003). The tools for addressing each of these hurdles fall into two categories, in silico modeling and structure-based drug design, which are covered in Sections 1.3.1 and 1.3.2. Obviously, the final hurdle is in vivo efficacy and safety data, which generally determine a compound's suitability for advancement to clinical evaluation.

TABLE 1.1. Important Considerations for the Medicinal Chemists

In Vitro Target	In Vitro ADME[a]	Physical Properties	In Vivo	Safety
Primary assay	Microsomal stability (rat, human)	Rule-of-Five	Functional	Ames test
Whole cell assay	Hepatocyte stability (rat, human)	In silico ADME[a] (see Section 1.3.1)	Behavioral animal models (efficacy)	Micronucleus test
Functional assay	P450 substrate	Solubility	PK/PD[c]	HERG[d] IC_{50}
Selectivity assays	P450 inhibitor	Crystallinity (mp, stable polymorph)		P450 induction
	Permeability			Broad ligand screening
	Transporter efflux (e.g., P-gp[b])			Others (depending on project)
	Protein binding			

[a]Absorption, distribution, metabolism, and elimination; [b]P-glycoprotein; [c]Pharmacokinetics/pharmacodynamics; [d]Concentration for 50% inhibition of the function of the delayed rectifier K^+ channel encoded by the *human ether a-go-go related-gene* (HERG).

1.3 THE TOOLS OF MEDICINAL CHEMISTRY

1.3.1 In Silico Modeling

To overcome the many hurdles to discovering a new drug, medicinal chemists must focus on synthesizing compounds with drug-like properties. One of the first tools developed to help chemists design more drug-like molecules takes advantage of an area totally under the chemist's control—the physical properties of the compounds being designed. These are the rules developed by Chris Lipinski, sometimes referred to as the "Rule-of-Five" (Ro5), which describe the attributes drug-like molecules generally possess that chemists should try to emulate (Lipinski et al., 2001). The Ro5 states that drug-like molecules tend to exhibit four important properties, each related to the number 5 (molecular weight <500; cLogP, a measure of lipophilicity,<5; H-bond donors <5; and H-bond acceptors <10). The Ro5 can be applied all the way from library design in the earliest stages of drug discovery to the final fine-tuning process that leads to the compound selected for development. Correlating microsomal instability and/or absorption/efflux with Ro5 properties can also provide insight about the property most important for gaining improvement in these areas.

As is the case with any good model, the Ro5 is based on data, in this case from hundreds of marketed drugs. Using more specific data, models to address each of the hurdles in the drug discovery process have been developed (for comprehensive reviews, see Beresford et al., 2004; van de Waterbeemd and Gifford, 2003; Winkler, 2004). These include models of solubility (Cheng and Merz, 2003; Hou et al., 2004; Liu and So, 2001), absorption/permeability (Bergstroem, 2005; Stenberg et al., 2002), oral bioavailability (Stoner et al., 2004), brain penetration (Abbott, 2004; Clark, 2003) and P450 interaction (de Graaf et al., 2005). More recently, the solution of X-ray crystal structures of the P450 enzymes 3A4 (Tickle et al., 2005) and 2D6 (Rowland et al., 2006) should enable application of structure-based drug design (see below) to help minimize interactions with these metabolic enzymes. Models for safety issues, such as genotoxicity (Snyder et al., 2004) and HERG (*human ether a-go-go related-gene*) interaction (which can lead to cardiovascular side effects due to QT prolongation) (Aronov, 2005; Vaz and Rampe, 2005) are also being developed. Although this profusion of in silico models offers considerable potential for overcoming hurdles in the drug discovery process, the models are only as good as the data used to build them, and often the best models are those built for a single project using data from only the compounds prepared for that specific project.

The models described above can be used, alone or in combination with structure-based drug design (see Section 1.3.2), to screen real or virtual libraries of compounds as an integral part of the design process. These improvements in library design, coupled with more efficient library synthesis and screening, provide value in both time and cost savings. The move towards using this library technology has been accelerated by the availability of a new resource for library generation: outsourcing (Goodnow, 2001). Contract research organizations (CROs) in the United States or offshore provide numerous synthetic services such as synthesis of literature standards, templates and monomers for library preparation, and synthesis of libraries (D'Ambra, 2003). These capabilities can relieve in-house medicinal chemists of much of the routine synthetic chemistry so they can focus on design and synthesis to enable new structure-activity relationships (SAR) directions. For an overview of the process as it fits together for the successful discovery of new drugs, see Lombardino and Lowe, 2004.

1.3.2 Structure-Based Drug Design (SBDD)

Progress in SBDD has been steady over the past two decades such that it has become a generally accepted strategy in medicinal chemistry, transforming the way medicinal chemists decide how to pursue their series' SAR. Although obtaining X-ray crystallographic data for SBDD was achieved early on, it has taken many years to learn how to interpret, and not over-interpret, this data. Structural information on the protein target provided by X-ray crystallography offers the greatest structural resolution for docking proposed ligands, but other spectroscopic techniques, such as nuclear magnetic resonance (NMR), have demonstrated their utility as well. X-ray crystallography, however, is generally restricted to analysing soluble proteins such as enzymes. Also required is a ready source of large quantities of the target protein for crystallization, as is often the case for proteins obtained from microorganisms grown in culture.

Bacterial proteins are an ideal starting point for SBDD, as in the case of the β-ketoacyl carrier protein synthase III (FabH), the target for a recent SBDD-based approach (Nie et al., 2005). FabH catalyzes the initiation of fatty acid biosynthesis, and a combination of X-ray data along with structures of substrates and known inhibitors led to selection of a screening library to provide a starting point for one recent study. Following screening, co-crystallization of selected inhibitors then guided the addition of functionality to take advantage of interactions with the enzyme visualized by X-ray and docking studies. A 50-fold improvement in enzyme inhibitory potency was realized in going from structure **1** to **2**, accounted for by amino acid side-chain movements revealed by X-ray co-crystal structures of both compounds with the enzyme. Although much remains to be learned so that these side-chain movements can be predicted and exploited for new compound design, the study nonetheless provides a successful example of the implementation of SBDD in drug design.

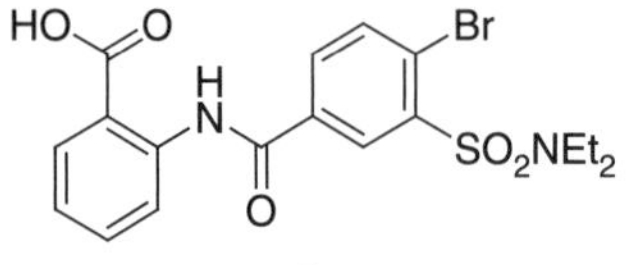

1 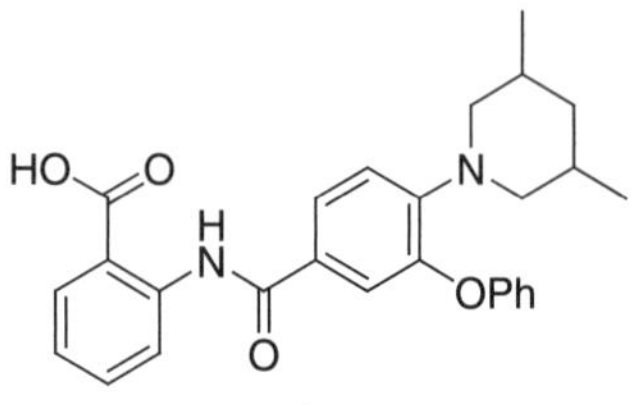 **2**

Although human proteins are more challenging to obtain in sufficient quantity for crystallization, modeling based on X-ray crystal structures has been successfully applied to many human targets. Probably the best-known efforts have been in the kinase area in search of anticancer drugs, which has been reviewed recently (Ghosh et al., 2001). For example, X-ray crystallographic data revealed important aspects of the binding of the anticancer drug Gleevec (**3**) to its target, the Bcr-Abl kinase, including the role of the pendant piperazine group, added originally to improve solubility, and the requirement for binding to an inactive conformation of the enzyme (Schindler et al., 2000). Combined with studies of the mutations responsible for Gleevec-resistant variants of Bcr-Abl, these studies enabled design of a new compound, BMS-354825 (**4**), active against most of these resistant mutants (Shah et al., 2004). More recently, non-ATP binding site inhibitors have been discovered and modeled by SBDD. For example, SBDD helped to characterize a new class of p38 kinase inhibitors that bind to a previously unobserved conformation of the enzyme that is incompatible with ATP binding (Pargellis et al., 2002). Insights from

SBDD then guided design of a picomolar p38 kinase inhibitor based on binding to this site, BIRB 796 (**5**).

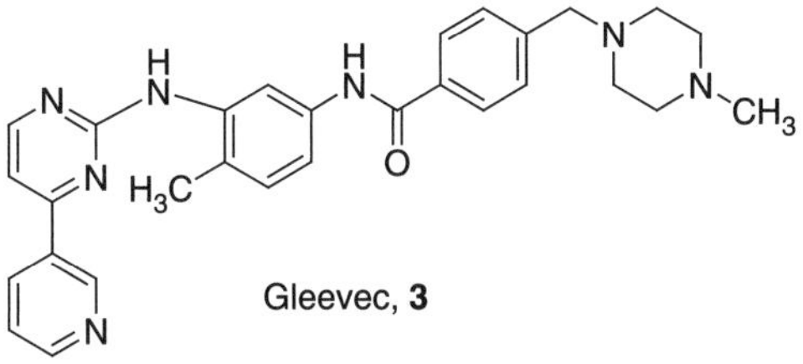

Gleevec, **3**

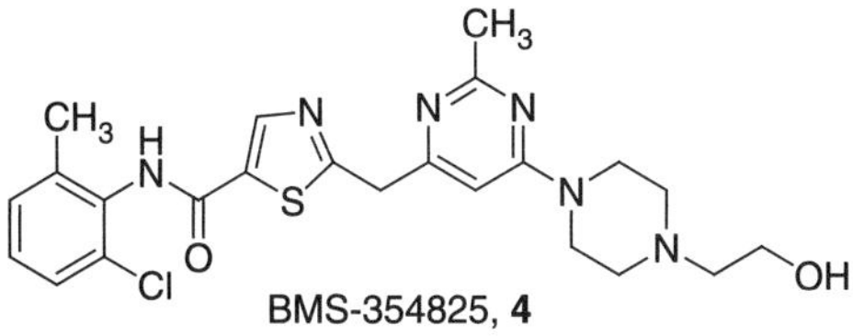

BMS-354825, **4**

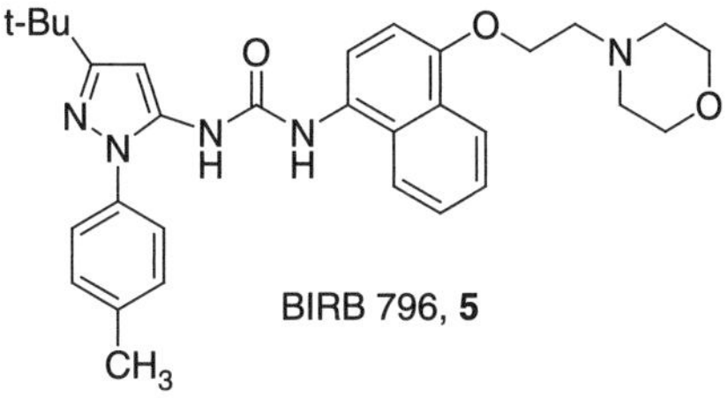

BIRB 796, **5**

SBDD approaches to other soluble proteins have produced inhibitors of the tissue factor VIIa complex (Parlow et al., 2003) and cathepsin G (Greco et al., 2002). In the case of factor VIIa inhibitors, X-ray data provided information for both designing a new scaffold for inhibitors and for simultaneously improving binding affinity and selectivity over thrombin. Compound **6** from this work was advanced to clinical trials based on its potency and selectivity for factor VIIa inhibition. The cathepsin G inhibitor program revealed a novel binding mode for an alpha-keto phosphonate to the enzyme's oxyanion hole and active site lysine, as well as an opportunity to extend groups into a vacant binding site to improve potency. The result was a nearly 100-fold increase in inhibition following an SAR study of this direction using the amide group in compound **7**.

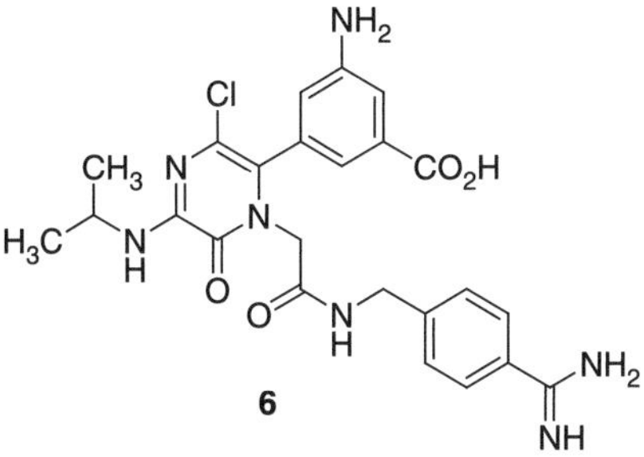

6

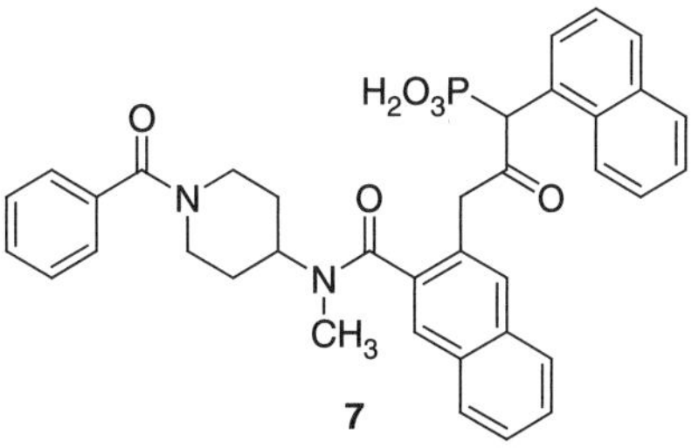

7

Another spectroscopic technique that has been widely applied to drug design is nuclear magnetic resonance (NMR) spectroscopy (Homans, 2004). Both X-ray crystallography and NMR can be used to take advantage of the opportunity to screen fragments, small molecules with minimal enzyme affinity, but which can be linked together with structural information to form potent inhibitors (Erlanson et al., 2004). For example, a recent approach to caspase inhibitors generated its lead structure by tethering an aspartyl moeity to a salicylic acid group; an X-ray co-crystal structure of the most potent compound **8** was found to mimic most of the interactions of the known peptidic caspase inhibitors

(Choong et al., 2002). Another example explored replacement of the phosphate group found in most Src SH2 domain inhibitors with various heteroatom-containing groups by soaking fragments into a large crystal and obtaining X-ray data, leading to the 5 nM malonate-based inhibitor **9** (Lesuisse et al., 2002).

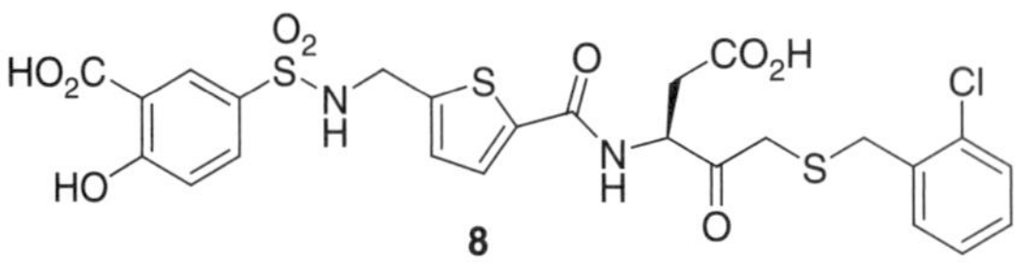

8

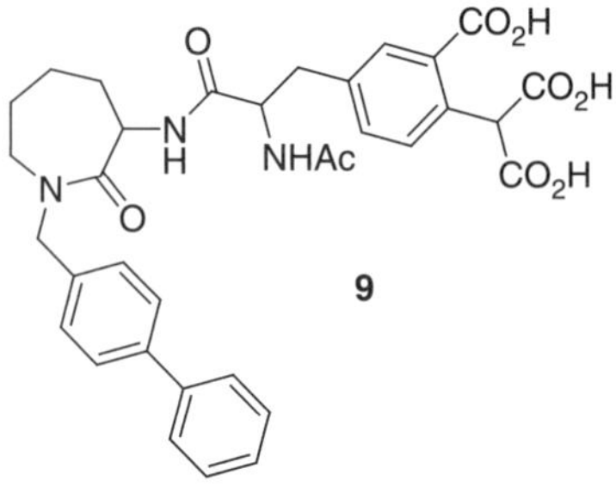

9

For proteins that are not water soluble, such as membrane proteins, techniques that depend on crystallization are very challenging. Homology modeling is an alternative that can be applied to transmembrane proteins such as the G-protein-coupled receptors (GPCRs), which are the target of many marketed drugs. Based on X-ray data for a prototype member of this family of proteins, bovine rhodopsin, a number of homology models for therapeutically relevant GPCRs have been built. In the case of the chemokine GPCR CCR5, a target for AIDS drugs, a homology model afforded an appreciation of the role of aromatic interactions and H-bonds involved in binding antagonists (Xu et al., 2004). A three-dimensional QSAR model was next developed based on a library of potent antagonists, and then combined with the homology model to confirm important interactions and indicate directions for new compound design, resulting in compound **10**, a subnanomolar CCR5 antagonist. A more sophisticated approach based on docking of virtual compounds to a homology model for the neurokinin NK-1 receptor for the neurotransmitter peptide substance P has revealed structurally novel antagonists (Evers and Klebe, 2004). The most potent of these, ASN-1377642 (**11**), overlaps nicely with CP-96,345, the literature NK-1 receptor antagonist on which the pharmacophore used for virtual screening was based. Similar combinations of SBDD-based technology are providing insights for new compound design in numerous areas of medicinal chemistry.

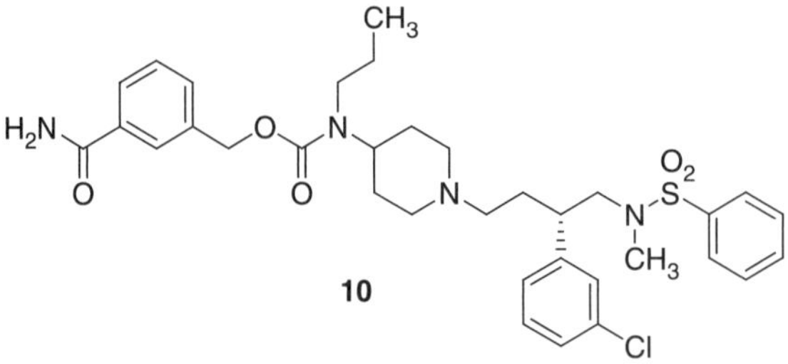

10

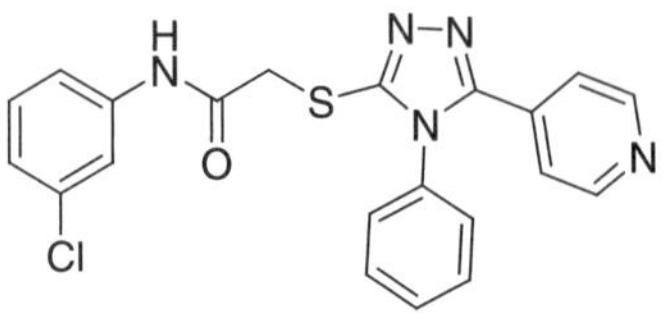

ASN-1377642, **11**

1.4 THE ROLE OF SYNTHETIC CHEMISTRY IN DRUG DISCOVERY

Some may ask why anything needs to be said about synthetic chemistry as a tool for drug discovery; after all, it is common to hear that "we can make anything." On the other hand, we can only carry out biological evaluation of compounds that have been synthesized.

Once the evaluation of biological activity and physical properties has been used to design new targets, a suitable synthetic route must be developed. However, considerations of what can be readily prepared factor into design much earlier. Chemists typically recognize familiar structural features for which they know a feasible synthetic route as they analyze data and properties. Design is guided by what can be readily made, especially what can be prepared as a library of compounds, so that work can begin immediately toward initiating the next round of biological testing.

Although there will always be limitations to what can be synthesized based on our imperfect knowledge, recent developments in two areas have facilitated the chemist's job: analysis/purification and synthetic methodology. In the first area, routine high-field NMR instruments allow 1H-NMR and 13C-NMR characterization of small amounts (<10 mg) of organic compounds. Liquid-chromatography/mass spectroscopy (LCMS) and other rapid analytical techniques, combined with medium- and high-pressure chromatography, allow for ready separation of reaction mixtures. New technologies such as reactor chips and miniaturization, supercritical fluids and ionic fluid reaction solvents, and chiral separation techniques will continue to improve synthetic capabilities.

In the second area, two recent advances have transformed synthetic methodology: transition-metal catalyzed cross-coupling reactions (Nicolaou et al., 2005) and olefin-metathesis technology (Grubbs, 2004). The formation of carbon–carbon bonds is probably the most fundamental reaction in synthetic chemistry. For the first several decades of the twentieth century, this reaction depended primarily on displacement of electrophilic leaving groups by enolate anions (or enamines) or addition of organometallic (e.g., Grignard) reagents. The advent of palladium-catalyzed coupling of more stable derivatives, such as olefins and acetylenes, boronic acids/esters, and tin or zinc compounds changed this simple picture. At the same time, the development of air-stable catalysts for producing complex carbon frameworks by metathesis of olefins expanded the chemist's repertoire. These methods allow much greater flexibility and tolerance for sensitive functional groups, enabling construction of more complicated, highly functionalized carbon frameworks.

Assembling this methodology, along with that developed over the previous century, into library-enabled synthesis allows the preparation of the large numbers of compounds favored for today's search for lead compounds using high-throughput screening (HTS) and in lead compound follow-up. Combinatorial chemistry was initially facilitated by developments in robotic handling technology and, for solid-phase synthesis, by Merrifield peptide synthesis. Both solution-phase (Selway and Terret, 1996) and solid-phase (Ley and Baxendale, 2002) parallel syntheses allow generation of large chemical libraries. The emphasis on these new technologies, combined with the cross-coupling and olefin metathesis synthetic methodologies, facilitates the synthesis of new classes of compounds with complex carbon frameworks. Their emergence as lead series and the ensuing follow-up are largely the result of their preponderance in the collection of compounds screened. In other words, it can be argued that synthetic methodology creates the chemical space that is available for screening and hence influences in a very profound way the medicines available to mankind. As the syntheses in the succeeding chapters make clear, synthetic chemistry plays a significant role alongside medicinal chemistry in the drug discovery process.

REFERENCES

Abbott, N. J. (2004). *Drug Discovery Today: Technologies*, 1(4): 407–416.

Aronov, A. M. (2005). *Drug Discovery Today*, 10(2): 149–155.

Beresford, A. P., Segall, M., and Tarbit, M. L. H. (2004). *Curr. Opin. Drug Discov. Dev.*, 7(1): 36–42.

Bergstroem, C. A. S. (2005). *Basic Clin. Pharmacol. Toxicol.*, 96(3): 156–161.

Cheng, A. and Merz, K. M., Jr. (2003). *J. Med. Chem.*, 46(17): 3572–3580.

Choong, I. C., Lew, W., Lee, D., Pham, P., Burdett, M. T., Lam, J. W., Wiesmann, C., Luong, T. N., Fahr, B., DeLano, W. L., McDowell, R. S., Allen, D. A., Erlanson, D. A., Gordon, E. M., and O'Brien, T. (2002). *J. Med. Chem.*, 45(23): 5005–5022.

Clark, D. E. (2003). *Drug Discovery Today*, 8(20): 927–933.

D'Ambra, T. E. (2003). *Abstracts, 36th Middle Atlantic Regional Meeting of the American Chemical Society*, Princeton, NJ, United States, June 8–11, 16.

de Graaf, C., Vermeulen, N. P. E., and Feenstra, K. A. (2005). *J. Med. Chem.*, 48(8): 2725–2755.

Erlanson, D. A., McDowell, R. S., and O'Brien, T. (2004). *J. Med. Chem.*, 47(14): 3463–3482.

Evers, A. and Klebe, G. (2004). *J. Med. Chem.*, 47(22): 5381–5392.

Ghosh, S., Liu, X.-P., Zheng, Y., and Uckun, F. M. (2001). *Curr. Cancer Drug Targets*, 1(2): 129–140.

Goodnow, R. A., Jr. (2001). *J. Cellular Biochem.*, (Suppl. 37): 13–21.

Greco, M. N., Hawkins, M. J., Powell, E. T., Almond, H. R., Jr., Corcoran, T. W., de Garavilla, L., Kauffman, J. A., Recacha, R., Chattopadhyay, D., Andrade-Gordon, P., and Maryanoff, B. E. (2002). *J. Am. Chem. Soc.*, 124(15): 3810–3811.

Grubbs, R. H. (2004). *Tetrahedron*, 60(34): 7117–7140.

Homans, S. W. (2004). *Angew. Chem., Int. Ed.*, 43(3): 290–300.

Hou, T. J., Xia, K., Zhang, W., and Xu, X. J. (2004). *J. Chem. Inf. Comput. Sci.*, 44(1): 266–275.

Lesuisse, D., Lange, G., Deprez, P., Benard, D., Schoot, B., Delettre, G., Marquette, J.-P., Broto, P., Jean-Baptiste, V., Bichet, P., Sarubbi, E., and Mandine, E. (2002). *J. Med. Chem.*, 45(12): 2379–2387.

Ley, S. V. and Baxendale, I. R. (2002). *Nat. Rev. Drug Discov.*, 1(8): 573–586.

Lin, J., Sahakian, D. C., de Morais, S. M. F., Xu, J. J., Polzer, R. J., and Winter, S. M. (2003). *Curr. Top. Med. Chem.*, 3(10): 1125–54.

Lipinski, C. A., Lombardo, F., Dominy, B. W., and Feeney, P. J. (2001). *Adv. Drug Deliv. Rev.*, 46(1–3): 3–26.

Liu, R. and So, S.-S. (2001). *J. Chem. Inf. Comput. Sci.*, 41(6): 1633–1639.

Lombardino, J. G. and Lowe, J. A. (2004). *Nat. Rev. Drug Discov.*, 3(10): 853–862.

Nicolaou, K. C., Bulger, P. G., and Sarlah, D. (2005). *Angew. Chem., Int. Ed.*, 44(29): 4442–4489.

Nie, Z., Perretta, C., Lu, J., Su, Y., Margosiak, S., Gajiwala, K. S., Cortez, J., Nikulin, V., Yager, K. M., Appelt, K., and Chu, S. (2005). *J. Med. Chem.*, 48(5): 1596–1609.

Pargellis, C., Tong, L., Churchill, L., Cirillo, P. F., Gilmore, T., Graham, A. G., Grob, P. M., Hickey, E. R., Moss, N., Pav, S., and Regan, J. (2002). *Nat. Struct. Bio.*, 9(4): 268–272.

Parlow, J. J., Case, B. L., Dice, T. A., Fenton, R. L., Hayes, M. J., Jones, D. E., Neumann, W. L., Wood, R. S., Lachance, R. M., Girard, T. J., Nicholson, N. S., Clare, M., Stegeman, R. A., Stevens, A. M., Stallings, W. C., Kurumbail, R. G., and South, M. S. (2003). *J. Med. Chem.*, 46(19): 4050–4062.

Rowland, P., Blaney, F. E., Smyth, M. G., Jones, J. J., Leydon, V. R., Oxbrow, A. K., Lewis, C. J., Tennant, M. G., Modi, S., Eggleston, D. S., Chenery, R. J., and Bridges, A. M. (2006). *J. Biol. Chem.*, 281: 7614–7622.

Schindler, T., Bornmann, W., Pellicena, P., Miller, W. T., Clarkson, B., and Kuriyan, J. (2000). *Science*, 289: 1938–1942.

Selway, C. N. and Terrett, N. K. (1996). *Bioorg. Med. Chem.*, 4(5): 645–654.

Shah, N. P., Tran, C., Lee, F. Y., Chen, P., Norris, D., and Sawyers, C. L. (2004). *Science*, 305: 399–402.

Snyder, R. D., Pearl, G. S., Mandakas, G., Choy, W. N., Goodsaid, F., and Rosenblum, I. Y. (2004). *Environmental and Molecular Mutagenesis*, 43(3): 143–158.

Stenberg, P., Bergstroem, C. A. S., Luthman, K., and Artursson, P. (2002). *Clin. Pharmacokinet.*, 41(11): 877–899.

Stoner, C. L., Cleton, A., Johnson, K., Oh, D.-M., Hallak, H., Brodfuehrer, J., Surendran, N., and Han, H.-K. (2004). *Int. J. Pharmaceutics*, 269(1): 241–249.

Tickle, I. J., Vonrhein, C., Williams, P. A., Kirton, S. B., and Jhoti, H. (2005). (Astex Technology Ltd., UK), US patent 2005159901.

van de Waterbeemd, H. and Gifford, E. (2003). *Nat. Rev. Drug Discovery*, 2(3): 192–204.

Vaz, R. J., Li, Y., and Rampe, D. (2005). *Prog. Med. Chem.*, 43: 1–18.

Winkler, D. A. (2004). *Drugs Fut.*, 29(10): 1043–1057.

Xu, Y., Liu, H., Niu, C., Luo, C., Luo, X., Shen, J., Chen, K., and Jiang, H. (2004). *Bioorg. Med. Chem.*, 12(23): 6193–6208.

2

PROCESS RESEARCH: HOW MUCH? HOW SOON?

Neal G. Anderson

2.1 INTRODUCTION

When one treats 1,2,3-trichloropropane with alkali and a little water the reaction is violent; there is a tendency to deposit the reaction product, the raw materials and the apparatus on the ceiling and the attending chemist. I solved this by setting up duplicate 12-liter flasks, each equipped with double reflux condensers and surrounding each with a half dozen large tubs. In practice, when the reaction took off I would flee through the door or window and battle the eruption with water from a garden hose. The contents flying from the flasks were deflected by the ceiling and collected under water in the tubs. I used towels to wring out the contents that separated, shipping the lower layer to [the client]. They complained of solids suspended in the liquid, but accepted the product and ordered more. I increased the number of flasks to four, doubled the number of wash tubs, and completed the new order.

They ordered a 55 gallon drum [of the product]. At best, with myself as chemist and supervisor, I could make a gallon a day, arriving home with skin and lungs saturated with 2,3-dichloropropene. I needed help. An advertisement in the local newspaper resulted in an interview with a former producer of illicit spirits named Preacher who had just done penance at the local penitentiary. He listened carefully and approved of my method of production, which he said might be improved with copper coils. Immediately he began to enlarge our production room by removing a wall, putting in an extra table, and increasing the number of washtubs and reaction set-ups. It was amazing to see Preacher in action (I gave him encouragement through the window); he would walk up the aisles from set-up to set-up putting in first the caustic then the water, then fastening the rubber stoppers and condenser, then using the hose. At this stage the room was a swirling mass of steam and 2,3-dichloropropene. We made a vast amount of material and shipped the complete order to [the client]—on schedule.

(Max Gergel, 1979)

The Art of Drug Synthesis. Edited by Douglas S. Johnson and Jie Jack Li

Chemical process research and development has greatly evolved over the past six decades, and a number of resources are available (Anderson, 2000; Blaser and Schmidt, 2004; Cabri and Di Fabio, 2000; Collins et al., 1997; Gadamasetti, 1999; Lednicer, 1998; McConville, 2002; Rao, 2004; Repic, 1998; Weissermel and Arpe, 1997). In the above account of scale-up in the early 1950s, as described by the head of Columbia Organics, safety considerations, in-process controls, purification, and analyses were essentially nonexistent. Today we are concerned not only for containing the product in the process equipment, but also for keeping contaminants out of the batch. Today, such an operation would be conducted only after safety hazard analysis, selection of suitable reactors and protective personnel equipment (PPE), successful small-scale runs in the laboratory (use-tests), development of critical in-process controls, and thorough analyses of the product from the small runs. Then the process would be detailed in a log sheet or batch record, which would be approved by management. After completing the large-scale run, the product would be analyzed and its quality documented. Despite the changes that have evolved over the decades, it is important to note that both earlier and current processes have a key feature in common: delivery must be on time.

In the continuum that is drug development, timeliness is crucial. Delaying the introduction of a drug by six months may reduce the lifetime profits by 50% (Ritter, 2002). As a drug candidate moves closer to launch, more material is required, and more resources (expensive starting materials, attention of personnel, and so on) must be invested (Fig. 2.1). Timely process research and development (R&D) can avoid costly surprises that delay drug introduction. Because fewer than 10% of all drug candidates progress from pilot plant scale-up to successful launch (Mullin, 2006), people are justifiably cautious about investing time and money too early into process R&D.

Effective chemical process R&D speeds a drug to market. In the discovery laboratory, paying attention to the practices of process research is likely to improve yields of laboratory reactions, reproduce small-scale runs more easily, and scale up to 100 + g runs more efficiently. Observations may lead to better processes in later development, for example, by minimizing byproducts, easing work-ups and purification, and by detecting polymorphs.

Scale-up from grams to 100 g and more may lead to unexpected problems. Safe operations are essential to minimize risk during scale-up: with scale-up there is always increased liability from accidents, including injury to personnel, loss of equipment, delay of key deliveries, damage to a company's reputation, and more. Some companies require that safety hazard assessments be completed before any process is run in a pilot plant; others require safety hazard assessments before a process is scaled up to greater than a given threshold amount. Testing for such assessments may be conducted on

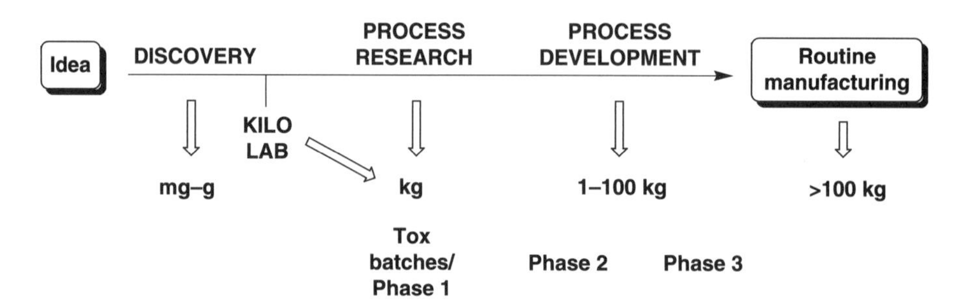

Figure 2.1. Batch sizes for compounds during drug development.

milligram amounts in the earliest phases, for example, differential scanning calorimetry studies to anticipate exothermic processes (Rowe, 2002). Later stages of testing may require multigram amounts of materials, which can only be prepared after some delay of time. Companies contacting a contract research organization (CRO) or contract manufacturing organization (CMO) should anticipate that these organizations will conduct safety hazard assessments before producing large batches of materials.

Stationary equipment used in pilot plants has limitations that often preclude the direct scale-up of discovery processes. With immobile vessels it becomes necessary to design processes so that mixtures can be removed through the bottom valve or by suction through a tube. No longer can a viscous oil or heavy slurry be removed by scraping the contents from the vessel using a spatula, and transferability becomes a major consideration. Peering into such a jacketed vessel may be possible only through a small glass porthole, which is often obscured by vapors or suspended solids above the reaction. Although agitators do not touch the bottom of a reactor (anchor agitators almost touch the surface of their reactor, but these agitators are rarely used in the pharmaceutical and fine chemicals industries), agitation is usually more effective than that observed in the laboratory with magnetic stirbars. Stripping off solvents to provide a residue product is rarely done on scale; clearly one must consider the minimum agitation volume for efficient, effective stripping. Often such considerations are not appreciated until one is responsible for the scale-up of a process in a pilot plant.

Large-scale operations, for example, 50 L or more, require extended times for routine operations that are normally carried out on a gram-scale without a second thought. This is primarily due to the slower rate of heat transfer in the cylindrical and spherical vessels used routinely for scale-up: although the volume increases proportional to the radius cubed (for a sphere), the surface area increases proportional to the radius squared. As heat control in these vessels is usually mediated by circulating liquids through the external jackets of these vessels, the rate of heat transfer (as a function of surface area) increases more slowly than the volume. Both mathematical calculations and actual plant experience show that as the volume increases 10-fold, the time to conduct an operation will at least double (Anderson, 2004). Dose-controlled additions are used to moderate exotherms in slow reactions, those requiring hours. Typical addition times on scale are 20 minutes to several hours, and extended additions are easily carried out and controlled using pumps and various feedback mechanisms. Similarly, the times necessary to strip off a solvent or to cool a crystalline slurry will also be extended. The prepared chemist will anticipate any processing difficulties that would arise on scale by subjecting the process to extended operations in the laboratory first.

Two examples that demonstrate the impact of extended additions on scale are highlighted in Scheme 2.1. The Swern oxidation was developed in the laboratory at $-15°C$ in anticipation that cooling in pilot-plant batches would be difficult to achieve at colder temperatures. When the process was conducted at $-15°C$ in the pilot plant, the yield of **2** dropped from the expected 55% to only 31%. The reduced yield was attributed to decomposition of the activated intermediates during the extended addition times (two hours for each addition of oxalyl chloride and triethylamine); by cooling to $-40°C$ in the pilot plant, these components were added more rapidly, and an acceptable yield of 51% was reached (Horgan et al., 1999; successful scale-up through semi-continuous processing gave the same yield, but with improved processing: McConnell, 2003). In the chlorination of the hydroxypyrimidine **3**, $POCl_3$ was added as the limiting reagent. When the addition was carried out over 2–3 minutes, the amount of dimeric impurity **5** formed was about 3%; when the addition was extended in the lab to 30 minutes, a realistic

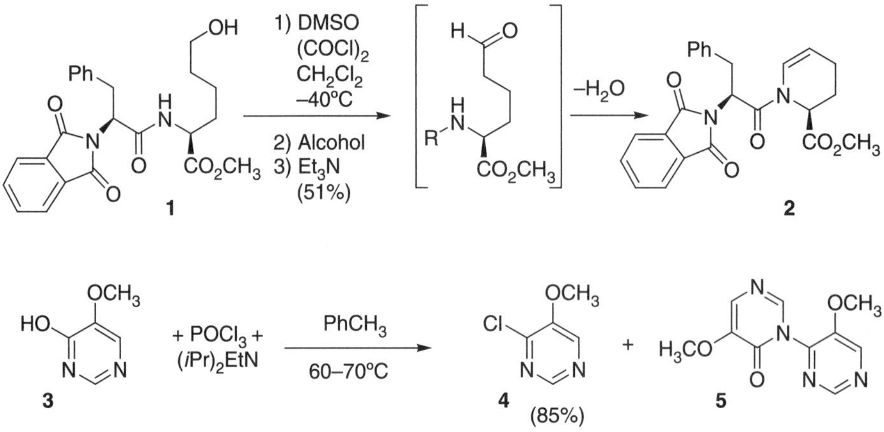

Scheme 2.1. Processes affected by times of additions of key reagents.

amount of time for pilot-plant operations, the dimer was present at about 13%. By modifying the addition protocol, only about 4% of the dimer was found in pilot-plant runs, resulting in 85–87% yields of the chloropyrimidine **4** (Anderson et al., 1997). Mechanistic understandings such as these may improve the chances of successful scale-up.

In the course of laboratory experiments to develop a rugged process for the *O*-acetoxy magnesium carboxylate **10**, investigators noticed that increased ageing of the reaction mixtures before adding Ac_2O led to reduced yields of **10** and increased yields of the nonproductive lactone **11** (Scheme 2.2). Fourier transform infrared (FTIR) investigations and other experiments showed that the $[CO_2]$ in solution (initially $\sim$0.25 M) similarly decreased with time, and the intermediate magnesium carbonate **9** was implicated. KO*t*Bu was added to decompose **9** (effectively scavenging the CO_2), leading to an 85% yield of **10** (Engelhardt et al., 2006). If these researchers had not paid attention to the decreased yields associated with increased ageing of reaction streams and hence developed the work-up using KO*t*Bu, low yields might have resulted upon scale-up.

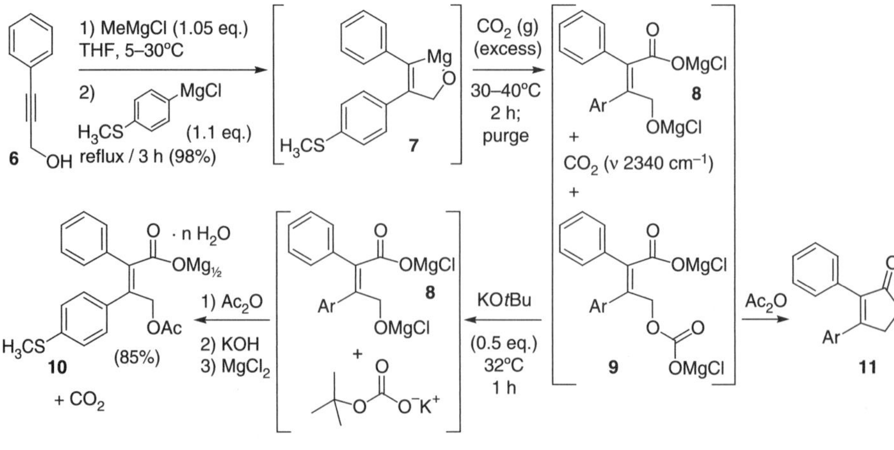

Scheme 2.2. Rugged processing for a Grignard-carboxylation-acetylation sequence.

Suffice it to say that *almost* all operations on scale are possible if money and time are not issues; of course, money rarely flows freely, and time is one of the major issues for developing drugs. Operations are almost always changed in progressing from the discovery route to scale-up. The best approach for rapid, successful scale-up is to scale down operations to the lab, and then develop processes that can be scaled up by mimicking conditions that will subsequently be encountered on scale.

2.2 CONSIDERATIONS FOR SUCCESSFUL SCALE-UP TO TOX BATCHES AND PHASE 1 MATERIAL

Material for toxicology studies and Phase 1 clinical investigations is often prepared from the same batch. One benefit of preparing only one batch is that the labor charges are less than they would be for two batches, and in early phases of drug development the cost of labor is much greater than the cost of raw materials. A second benefit is that the impurity profile of material first going into man will be that of the material used for the tox batches; this is significant, because the tox studies qualify the impurities, and subsequent active pharmaceutical ingredient (API) batches for human use must have no new impurities, only impurities with levels no greater than those found in the tox batches. A disadvantage of preparing a large amount of material for both tox and Phase 1 is that the excess material will be wasted if the drug candidate fails during the tox studies.

For tox and Phase 1 batches it is essential to set reasonable goals for API purity, and use unit operations that would likely be employed on scale. A purity of 98% is reasonable; the FDA will readily permit companies to upgrade the routine quality of their API specifications after anticipated optimization, but a request to downgrade the drug substance purity specification will likely not be encouraged. As a worst-case scenario, suppose that a purity of 99.8% could not be met because scrutiny of a new HPLC method showed that a highly purified reference standard was only 99.0% pure. Under such circumstances additional steps, such as treatment with activated carbon or recrystallization, may be necessary, increasing the overall cost of goods (COG). In order to prepare material of ~98% purity for the tox batches, it is necessary to understand and control the isolation of the drug candidate. Once optimal isolation conditions have been identified, the batch may be isolated under slightly less than optimal conditions, for example, by cooling the crystallization slurry rapidly or by applying smaller volumes of washes to the wet cake on the filter. Another approach is to crystallize the product in the presence of additional typical process impurities; for instance, the product may be crystallized in the presence of impurities added as a portion of the mother liquor from a prior batch. Other impurities that need to be controlled in tox batches and subsequent batches include metals, such as palladium, Class 1 and Class 2 residual solvents (www.fda.gov), and any by-products known to be toxic. [Note that, in the passive approach to removing palladium residues, a step using Pd is positioned early in a route and the amount of Pd in process streams is reduced by attrition in the work-ups of successive steps. A number of approaches are available to actively remove Pd, and these have been recently reviewed. (Garrett and Prasad, 2004; Thayer, 2005; Welch et al., 2005)].

Ideally all subsequent batches will be prepared by the route and process used for tox and/or Phase 1 batches, so that on-scale impurities and impurity profiles will meet the guidelines above. Of course it is difficult to predict the final optimized process for a drug candidate. The best approach to control impurities is to determine the optimal starting materials, reagents, process, and final form (salt, polymorph) early ("freeze" the final step to the API), and vary only routes to the final intermediates. If the impurities and impurity profiles change in batches prepared after the qualifying tox batch has been made, and an acceptable procedure cannot be devised to remove or reduce the offending impurities, a bridging tox study will be necessary.

Tox/Phase 1 batches are often prepared in the kilo lab, and most operations that are used in the discovery lab can be used there. An expedient route (often the only route at that time) is employed, and expensive reagents, low yields, and laboratory manipulations such as trituration and drying over Na_2SO_4 are often tolerated to reach the timelines. Often, very little is known about acceptable operating ranges, with great effort made to maintain a narrow parameter range. (Anyone who has tried to carry out an exothermic addition at $-70°C \pm 2°C$ knows how this can be a grueling experience.) As insurance against unforeseen difficulties in preparing tox/Phase 1 batches, people often plan to prepare an extra 10–20% of material for contingencies. Often, little is invested in process research at the time that tox/Phase 1 batches are being prepared, but the overriding goal is to provide acceptable material by the deadline.

Preparative chromatography should be considered as enabling technology appropriate for tox/Phase 1 batches. Extensive crystallization studies, for example, to purge an impurity difficult to remove by crystallization, or to prepare an enantiomerically pure intermediate by classical resolution, may not be justified for preparing material in early drug development (Welch et al., 2004).

2.3 CONSIDERATIONS FOR PHASE 2 MATERIAL AND BEYOND

Desirable qualities for processes to prepare drug candidates and API include on-time delivery with expected yield and quality. In-process controls (IPC) are developed to ensure quality and productivity, and these are essential for a new drug application (NDA). Detailed analytical methods and specifications are developed for the drug candidate and intermediates. Rugged operations are developed, simple and forgiving, with wide operating ranges. Scalable processes are developed, so that technology transfer to larger equipment can proceed efficiently. Some key considerations for rapid scale-up are found in Table 2.1.

Some process conditions may not be possible in some scale-up equipment, and often processes are developed to accommodate existing equipment. For instance, a route may be redesigned to avoid a hydrogenation if a vessel that can contain hydrogen under pressure is not available. Most general-purpose scale-up equipment can accommodate temperatures ranging from about $-50°C$ to $+140°C$. Processes that run colder than $-50°C$ are generally conducted in expensive reactors made of Hastelloy cooled through heat transfer fluids circulating in external jackets, and large-scale reactions have been run at about $-100°C$ by pumping liquid nitrogen into the reaction mixture. Safety is a primary consideration for processes run above 140°C (the upper limit for heating by steam circulating through external jackets), and continuous operations may be the best choice (Anderson, 2001). In other cases equipment may be purchased for a process as an investment, such as a pH meter and automatic titrator used to maintain the pH of a process stream.

2.3.1 Reagent Selection

Reagents are chosen to minimize cost, minimize waste, and to maintain safe operations, among other considerations. Less hazardous reagents may be chosen in order to minimize

TABLE 2.1. Scale-Up Criteria from the Viewpoint of Out-Sourcing Chemical Firms

Criteria	Considerations
1. Reagents	Cost/purity/availability Specialized equipment needed Toxicity: restricting exposure requires additional PPE and monitoring Chemical hazard: liability worse with scale-up
2. Solvent	Safety Ease of work-up Ease of removing solvent from final product
3. Work-up and isolation	Minimal reliance on chromatography No concentrations to dryness Minimal number of repeated steps, for example, extractions Facile solvent displacements; that is, chase lower bp solvent with a higher bp solvent Controlled crystallization preferred over precipitation Crystallizations used to upgrade key intermediates Minimal dilution, to minimize V_{max} and increase productivity
4. Reaction ruggedness	Exotherms and gas evolution: liability worse with scale-up Cryogenic or very high temperatures require specialized equipment Tight control of reaction conditions, for example, temperature or pH, requires attention Excluding moisture or oxygen may require additional considerations Rapid additions and short reaction times require nonstandard equipment
5. Yield	Low yields require that more intermediates be prepared for multistep processes Low yields indicate areas for improvements
6. Route	Convergent routes preferred, in principle If chiral, location of resolution step or integrity of induced stereocenters Minimal number of steps generally preferred
7. Analytical	Solids have melting points described Final form well defined (salt, polymorph) Spectra and chromatography information available, with detailed methods

time spent wearing PPE that hinders movement. Substitutes may be sought for air-sensitive reagents that pose handling constraints (Van Arnum et al., 2004). Less expensive reagents may be employed, such as substituting mixed anhydride couplings with pivaloyl chloride instead of using HBTU for forming amides.

A reaction commonly developed in the discovery lab is removing a Boc group from an amine using trifluoroacetic acid (TFA). Because TFA is highly corrosive, difficult to contain and recover (bp 72°C), and poses incineration problems due to the generation of HF, alternative conditions are usually explored for scale-up. Strong acids used instead of TFA include HCl, H_2SO_4, methanesulfonic acid, and toluenesulfonic acid

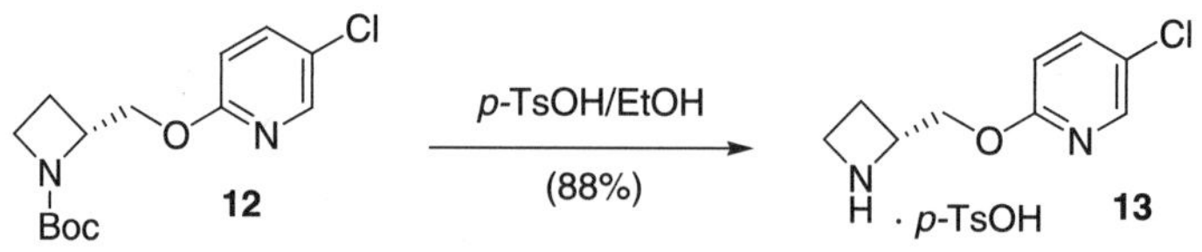

Scheme 2.3. Removing a Boc group with p-TsOH.

(p-TsOH). In Scheme 2.3, the tosylate salt **13** crystallized directly from the deprotection, and proved to be more stable than either the HCl or TFA salt (Lynch et al., 1998).

2.3.2 Solvent Selection

Once the reagents are selected, compatible solvents are chosen. Some solvents avoided for scale-up are shown in Table 2.2. 1,2-Dichloroethane, a Class 1 solvent with toxicity similar to benzene, should be avoided. Solvents more commonly used are shown in Table 2.3. Characteristics considered include toxicity, polarity, boiling point, freezing point, miscibility with water, and ability to remove water (and other compounds) by

TABLE 2.2. Solvents Avoided on Scale

Solvent	Undesirable Characteristic	Alternative Solvent
Et_2O	Flammable	MTBE
$(i\text{-}Pr)_2O$	Peroxides	MTBE
HMPA	Toxicity	NMP
Pentane	Flammable	Heptane
Hexane	Electrostatic discharge; neurological toxicity	Heptane
$CHCl_3$	Mutagenicity, toxicity ($COCl_2$), environmental	CH_2Cl_2; $PhCH_3$ with CH_3CN, n-BuOH or DMF
CCl_4, $ClCH_2CH_2Cl$	Mutagenicity, environmental	CH_2Cl_2
PhH	Carcinogen	$PhCH_3$
Dioxane	Carcinogen	THF
Acetonitrile	Animal teratogen	i-PrOH
Ethylene glycol	Toxicity	1,2-Propanediol
$HOCH_2CH_2OR$ (R = Me, Et)	Teratogen	1,2-Propanediol
$MeOCH_2CH_2OMe$ (DME)	Teratogen	THF, 2-Me-THF

TABLE 2.3. Solvents Commonly Used on Scale

Water	DMSO	MIBK	MTBE
MeOH	DMF	DME	$PhCH_3$
1,2-Propanediol	t-BuOH	EtOAc	Et_3N
EtOH	NMP	THF	Xylenes
AcOH	Acetone	i-PrOAc	Heptane
n-BuOH	t-AmOH	PhCl	Cyclohexane
i-PrOH	CH_2Cl_2	2-Me-THF	Methylcyclohexane
Acetonitrile	Pyridine	i-BuOAc	

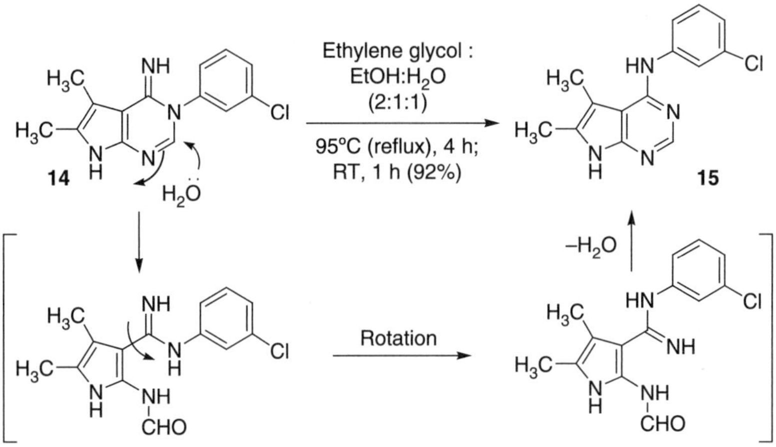

Ethylene glycol : EtOH:H_2O (2:1:1)
95°C (reflux), 4 h; RT, 1 h (92%)
Rotation
$-H_2O$

Scheme 2.4. Process for preparing a pyrimidine by Dimroth "rearrangement."

azeotroping. Not all CROs and CMOs will use CH_2Cl_2 due to safety considerations; sometimes either CH_2Cl_2 must be used, or the chemistry must be changed. Acetonitrile and DME are found in each table, as some firms will use these solvents on scale.

From a practical standpoint, the best solvent is the one that permits the direct isolation of the product, without any extractive work-up (Chen and Singh, 2001). A good example is the preparation of the pyrimidine in Scheme 2.4 (a tyrosine kinase inhibitor) by a Dimroth "rearrangement" (Fischer and Misun, 2001). Some water in the mixture is necessary for the hydration–dehydration sequence, and the product crystallizes upon cooling. The combination of water, ethylene glycol, and EtOH was chosen to afford a rapid reaction through a conveniently high temperature and to allow good crystallization of the product.

2.3.3 Unit Operations

Rugged process conditions are sought to define safe operating ranges and improve process reliability. When time is available, abuse tests are run to define the ranges of acceptable processing. Unnecessary operations are eliminated to decrease processing cycle times and produce more material on scale. One example is eliminating the operation of drying rich organic extracts over Na_2SO_4; this step is often redundant, because H_2O is removed by azeotropic distillation with solvents commonly used for extractions.

As shown in Scheme 2.5, optimizing an addition sequence can lead to very different products, and have a great effect on the ease of purifying a compound (Zook et al., 2000). The mesylate **17**, a key intermediate in one route to nelfinavir mesylate, was prepared from **16**. When CH_3SO_2Cl was added to a mixture of **16** and Et_3N, routine conditions for

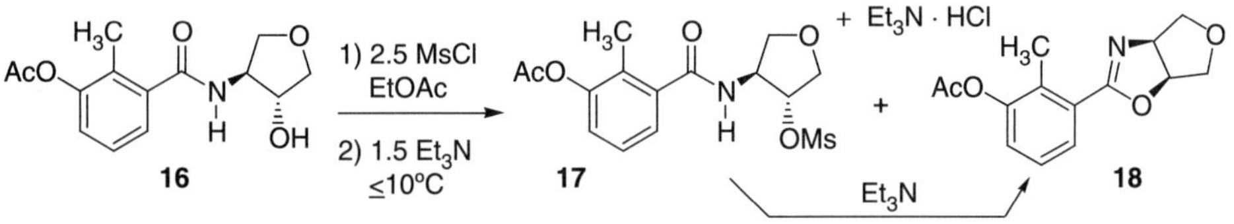

1) 2.5 MsCl EtOAc
2) 1.5 Et_3N ≤10°C
+ $Et_3N \cdot HCl$
Et_3N

Scheme 2.5. Adding the base last as a key operation of a mesylation.

preparing a mesylate, the sole product was the oxazoline **18**. By simply adding the Et_3N gradually after adding the CH_3SO_2Cl, thus minimizing the amount of free base present in the reaction mixture, the ratio of mesylate to oxazoline was 95 : 5 (see Rao, 2004, Chapter 5, for additional examples on the power of altering additions). Note that a recent paper by Li et al., (2006) showed that a parallel addition of starting material and base provided a superior yield in a semi-batch process.

Understanding and controlling operations are necessary for developing a rugged process. In the oxidation of the diarylacetylene **19** to the diketone **20**, a nice example of an oxidation that does not use inorganic oxidants or transition metal catalysts, the reaction is heated at 105–110°C to remove the byproduct dimethylsulfide by co-distillation with solvents (Scheme 2.6). Removing this byproduct is necessary to complete the oxidation. Even though this process is heated significantly above the boiling point of dimethylsulfide, this byproduct is not effectively removed from the reaction mixtures in the laboratory unless distillation occurs (Wan et al., 2006). Other approaches may be useful on scale to remove dimethylsulfide, such as adding a surfactant to decrease the surface tension of the mixture, sparging with nitrogen, charging the reactor to about one-third of the total volume to increase headspace, or optimizing reactor geometry and agitation. Key information from laboratory runs, such as the need to remove dimethylsulfide and the method used, should be included in any technology transfer package.

Most operations scaled up in the pharmaceutical industry use semibatch (and batch) processing in the general-purpose equipment. Such operations allow for fine control of slow unit operations, for example, reactions needing hours to complete, fermentation, and crystallization, and such fine control may be necessary to ensure high quality and productivity.

Batch operations cannot be used to readily scale up three types of processes: those that involve rapid heat transfer, efficient micromixing (for heterogeneous processes or rapid reactions, both heterogeneous and homogeneous), or contact with localized energy sources or catalysts (photolyses, microwave reactions, sonolyses, thermolyses, and others). Laboratory investigations can identify potential processing difficulties. The warning signs are evidenced by changes in product yield or quality with changes in laboratory mixing speed, addition rate or position of feed stream, or with scale-up to a vessel of different geometry. (If the yield or quality of key materials drops through premature scale-up, key deliveries may be delayed.) Continuous processing is often the only way to effectively scale up such processes, and inexpensive laboratory equipment can be used (Anderson, 2001). Commercially available microreactors have been used for continuous processing, and also offer enhanced selectivity through improved mixing and heat transfer (Ehrfeld et al., 2000). Many continuous reactors are cGMP-friendly: inexpensive static

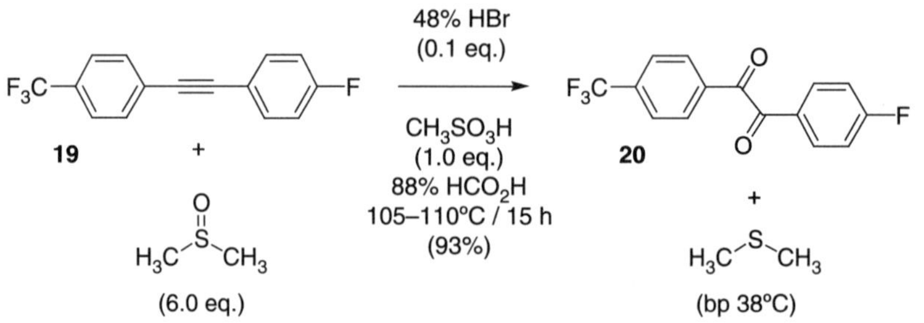

F_3C–Ar–C≡C–Ar–F **19** + H_3C–S(=O)–CH_3 (6.0 eq.) → [48% HBr (0.1 eq.); CH_3SO_3H (1.0 eq.); 88% HCO_2H; 105–110°C / 15 h; (93%)] → F_3C–Ar–CO–CO–Ar–F **20** + H_3C–S–CH_3 (bp 38°C)

Scheme 2.6. DMSO as bulk oxidant in the HBr-catalyzed oxidation of a diaryl acetylene.

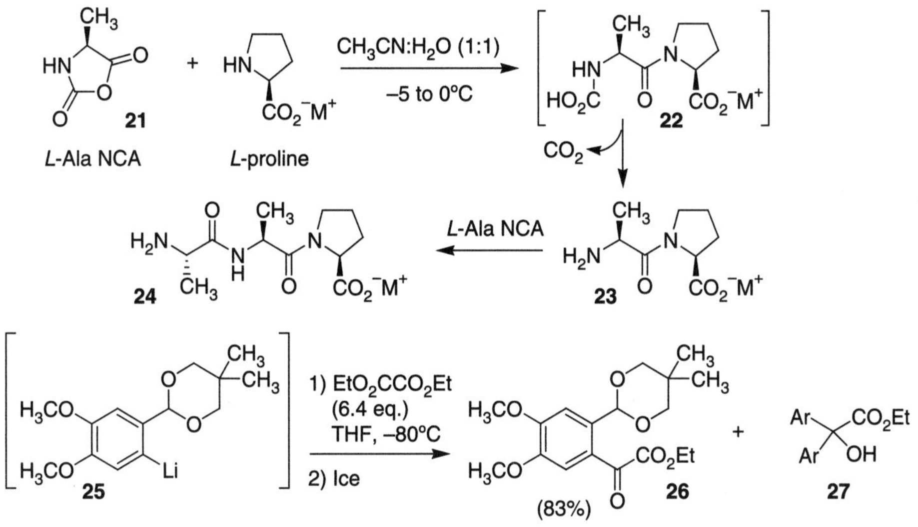

Scheme 2.7. Successful scale-up through continuous processing.

mixers can be purchased for each campaign. Continuous processing fits well with the FDA's process analytical technology (PAT) initiative. A benefit of continuous processes is that large amounts of material can be produced by simply "numbering up," that is, running operations for a longer time or in parallel. Continuous operations may also provide a safe way to conduct scale-up, in particular when high temperatures are involved.

Two examples that demonstrate the utility of continuous processing when dealing with reactive intermediates or products are shown in Scheme 2.7. In the first case, a route to enalapril was developed using the *N*-carbonic anhydride (NCA) of *L*-alanine (**21**). Forming the NCA both protects the amine and activates the carbonyl to attack, saving two steps relative to the classical approach of protecting the amine of alanine with a robust group such as carbobenzoxy or Boc, activating the carbonyl group and coupling, and deprotecting the amine of the product. When a solution of the NCA was added to a salt of *L*-proline in less than 5 s, a 90% yield of the dipeptide **23** was produced; however, when the addition was extended to 3 min, the yield dropped to 65%. With a 50-fold scale-up, the yield of dipeptide dropped. The main impurity was the tripeptide **24**, formed by reaction of the product with the NCA; physically removing the desired product from the starting material minimized this side reaction. Successful scale-up conditions were developed by combining the starting materials in an in-line static mixer (1 s residence time) and rapidly quenching the product. There was no change in selectivity with scale-up using the static mixer approach (Blacklock et al., 1988; Paul, 1988). In the second example the product (keto ester **26**) was more reactive than one of the starting materials (diethyl oxalate). When the condensation of **25** was conducted in a semibatch mode with an excess of diethyl oxalate, keto ester **26** was isolated in only 23% yield, with the major product being the tertiary alcohol **27** arising from reaction of the product with **25**. By conducting the reaction in a continuous mode (a 1 mL syringe barrel was adapted and used as a static mixer) and immediately quenching the reaction stream, the keto ester **26** was produced in 83% yield (Hollwedel and Koßmehl, 1998). When semibatch operations were employed in each of these examples, scale-up gave poor results, and choices were to made to employ continuous operations rather than change the conditions, reagents, or route. In some cases continuous operations may be the quickest way to prepare large amounts of material.

2.3.4 Developing Simple, Effective, Efficient Work-Ups and Isolations

Work-up becomes a major consideration in designing processes for preparation of all phases of drug development after Phase 1, as "60 to 80% of both capital expenditures and operating costs go to separations." (Eckert, 2000) *Work-up conditions can limit the selection of reagents and routes.* Simple work-ups with a minimal number of transfers decrease the number of opportunities for physical losses and contamination.

The criteria for an efficient work-up are understandably different for discovery processes and cost-effective large-scale processes. The procedure in Scheme 2.8 provides an efficient, practical route to unsymmetrical ureas (e.g., **30**). Volatilizing the acetone byproduct drives the reaction to completion and minimizes the formation of the symmetrical urea byproducts (Gallou et al., 2005). Simply evaporating the free *N*-methyl pyrrolidine and solvent afforded the products, with purities acceptable for assembling libraries of compounds.

Although the work-up above is convenient even for kilo lab operations, other considerations prevail for pilot-plant and manufacturing equipment. For instance, concentrating to a residue product can be time-consuming, with the risk that the product will decompose during a lengthy operation. The condensate would likely contain a mixture of *N*-methyl pyrrolidine and THF, and mixed condensates require additional purification and/or analyses before they can be recovered and reused in subsequent manufacturing batches. When a reaction product is nicely crystalline, adding an antisolvent may crystallize the product directly, and this is often preferred for a pilot-plant campaign and manufacturing (Chen and Singh, 2001). Under direct isolation conditions the advantages of decreased processing time outweigh the disadvantages of having mixed solvent streams (Anderson, 2004).

Purification by preparative chromatography is rarely considered for Phase 2 material and further scale-up. The cost of preparative chromatography may range from \$1200 to \$5000/kg (one step, 50 kg/batch, 20 batches/yr), usually relegating chromatographic purification for preparing only small amounts of high-value API. (In calculating the cost of preparative chromatography, variables considered include loading of crude material onto the stationary phase, cost of stationary phase and solvents, the percentage of solvents that can be recovered and recycled, labor costs, waste disposal, and the cost and amortization of chromatography equipment and equipment used to remove solvents

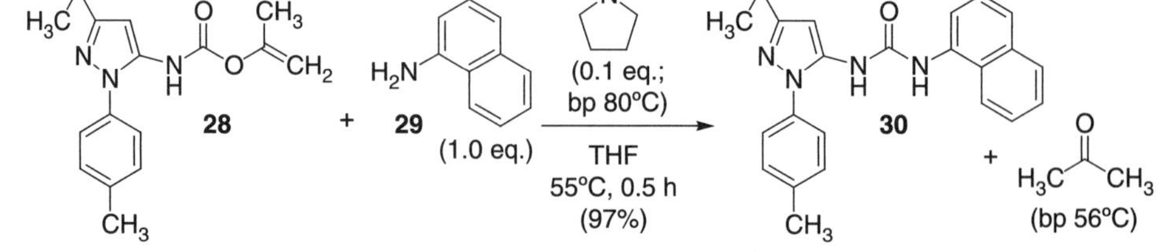

Scheme 2.8. Formation of an unsymmetrical urea.

from the eluted fractions.) A simple alternative is to slurry an absorbent in a solution of an impure mixture, and filter off the absorbent with the key impurities bound to it. Another alternative is a hybrid operation in which the impurities are removed as the solution passes through a bed of an absorbent contained in a filter ("plug chromatography"). The future will see judicious use of adsorbents to remove organic and inorganic impurities to provide material for both short-term and long-term needs (Welch et al., 2002, 2005).

Direct isolation ("one-pot") processes should be considered for materials prepared for Phase 2 and later development. Examples of this were shown in Schemes 2.3 and 2.4 for the tosylate salt **13** and the product **15** from the Dimroth "rearrangement." Considerable processing time may be saved, resulting in lower COG from reduced labor costs. The attendant savings on solvent costs and waste disposal can also reduce COG. Some research time may be necessary to develop these processes, in particular fine-tuning the processes to purge impurities. In early phases of drug development the fastest scale-up may be through conventional extractive work-up, concentration, and crystallization.

2.3.5 The Importance of Physical States

The physical forms of intermediates and APIs can significantly affect (1) the ease of unit operations such as filtration and drying; (2) the ability to upgrade material, usually by crystallization; (3) the stability of process streams and isolated compounds; (4) the ease and reliability of formulating the API; and (5) the bioavailability of the API in the drug product. Protecting groups are often chosen to afford crystalline intermediates (Sledeski et al., 2000). Foams and viscous liquids are avoided, as it is difficult to prepare and transfer these materials on scale. Compounds melting below 50°C will be difficult to crystallize.

The physical form of solids is especially important for routine processing of isolated compounds. Crystalline needles are often encountered, and under the pressures of isolation these relatively fragile forms can compact, extending filtration times and decreasing the ability to remove impurities by washing the wet cake. Crystal engineering can become a major development effort, involving temperature cycling and even sonication (Kim et al., 2005). Thermochemical behavior (mp or DSC) should be generated for all isolated solids during early process research.

Polymorphs and pseudopolymorphs (solvates) may be discovered for any compound, not just APIs. For instance, more than 100 solvates have been identified for sulfathiazole (Bingham et al., 2001). New forms of a drug candidate can present opportunities for expanding intellectual property, and may also provide definitive proof of structure by single-crystal X-ray analysis. The detection of undesired polymorphs or pseudopolymorphs is key to avoid interrupted sales of drug product. Ritonavir (**31**) was aggressively developed by Abbott and approved by the FDA in 1996 (Scheme 2.9). About two years later a new polymorph (Form II) was discovered in the drug product, and crystallization to give Form I could not be controlled in any manufacturing plant. The undesired Form II may have been associated with the urethane **32**, an impurity that was present in the

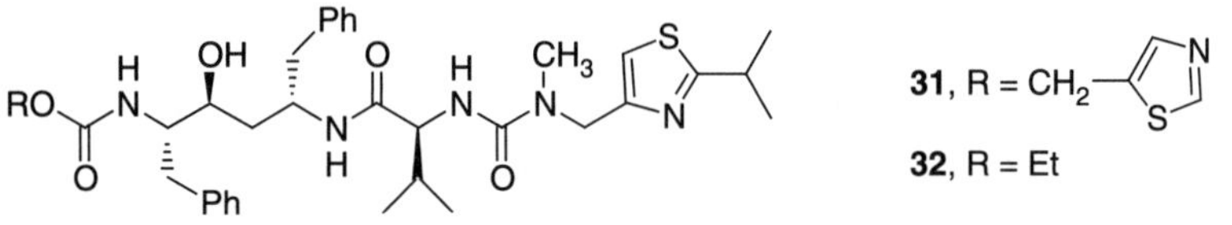

31, R = CH_2–(thiazolyl)

32, R = Et

Scheme 2.9.

optimized route to ritonavir. The drug product was reformulated to accommodate Forms I and II, and, fortunately, no market hiatus occurred (annual sales were then about $250 million) (Chembukar et al., 2000).

Selection of the desired (usually most stable) polymorph and design of a rugged process to prepare the desired form will reduce time to market. There are many negative ramifications of new API polymorphs: Batches out of compliance with NDA may lead to interrupted sales; altered bioavailability may make the drug product less efficacious or even more toxic; and a generic competitor may develop a formulation of a new polymorph (Remenar et al., 2003). In 2003, the cost of polymorph screening was $25,000–$100,000 (Rouhi, 2003). Investing early in selecting the optimal form is wise.

2.3.6 Route Design and Process Optimization to Minimize COG

Thorough process R&D can dramatically decrease the cost of clinical supplies and API. Crystallization-induced dynamic resolution (CIDR) is a powerful tool that can produce the desired enantiomer theoretically in 100% yield, which dramatically reduces the COG of an enantiomer as compared to the 50% maximum yield obtained through classical resolution of a racemic mixture with a chiral resolving agent. This CIDR is just one type of crystallization-induced asymmetric transformation (CIAT), which can increase the efficiency of processes to form diastereomers, olefins, and other compounds. The processes of CIAT and CIDR have even led to the crystallization and isolation of what was a minor reaction product in solution (Pye et al., 2002). Developing CIAT and CIDR processes can markedly decrease the COG of APIs (Anderson, 2005; Brands and Davies, 2006).

Impressive examples of CIAT processes are shown in Schemes 2.10 and 2.11. In an early synthesis of tadalafil, a Pictet–Spengler reaction was used to prepare the *cis*-ß-carboline intermediate **35** in 42% yield (TFA, CH_2Cl_2, 4°C/5 days; chromatography) or 58% yield (AcOH, H_2O, 50°C/4 days; crystallization) (Daugan, 1999; Orme et al., 2005). A much improved route utilizing CIAT is shown in Scheme 2.10 (Orme et al., 2005). A dynamic equilibrium was set up between **35** and **36** and the undesired diastereomer **36** stayed dissolved in the reaction medium while the product **35** crystallized out, driving the process to completion and raising the yield to 92%.

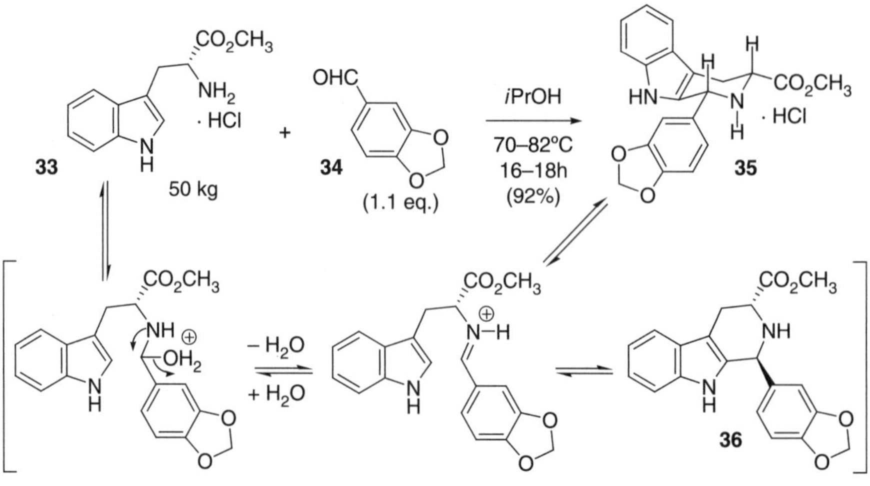

33 50 kg + **34** (1.1 eq.) → *i*PrOH, 70–82°C, 16–18h (92%) → **35** · HCl; – H_2O / + H_2O; **36**

Scheme 2.10. Preparation of a tadalafil intermediate through CIAT processing.

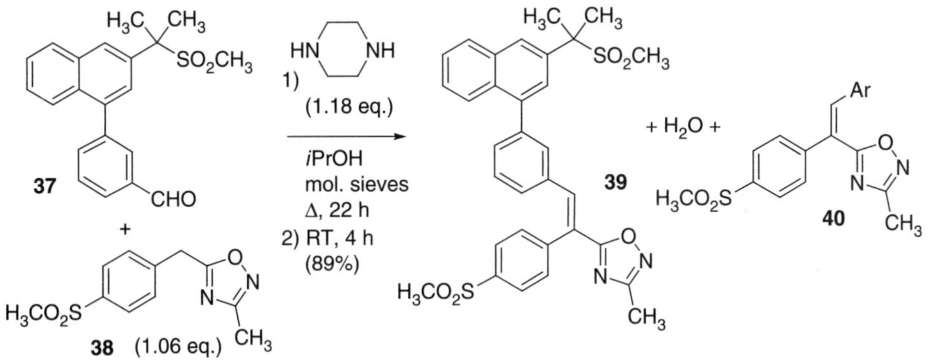

Scheme 2.11. CIAT processing used for a Knoevenagel reaction.

A recent CIAT example from Merck employing the venerable Knoevenagel reaction is highlighted in Scheme 2.11. The condensation is reversible, and must be driven by removing the byproduct H_2O by refluxing and passing the condensate through molecular sieves. The product **39** crystallized over 22 h and the undesired isomer **40** remained in the supernatant (Conlon et al., 2006). Usually, homogeneous reaction conditions are designed for rapid scale-up. As this reaction was a suspension throughout, process conditions were probably developed for optimal equilibration and crystallization of the product.

The CIAT and CIDR processes should be considered when appropriate before Phase 2 in order to decrease the cost of labor and raw materials for large amounts of materials. If a CIAT or CIDR process is identified earlier, it may be wise to devote time and resources to develop and exploit it. The potential for CIAT and CIDR processes may be uncovered by simply extending and carefully monitoring processes in the laboratory.

Telescoping, selecting reagents, and reaction optimization can afford significant decreases in the COG. The process route initially used to prepare 200 kg of the API intermediate acid **44** involved seven steps, four solvents, bromination, diazotization, and reaction with CO at 120°C, and produced the drug candidate in 20% yield. The optimized route shown in Scheme 2.12 was operationally simpler (using the same vessel for every

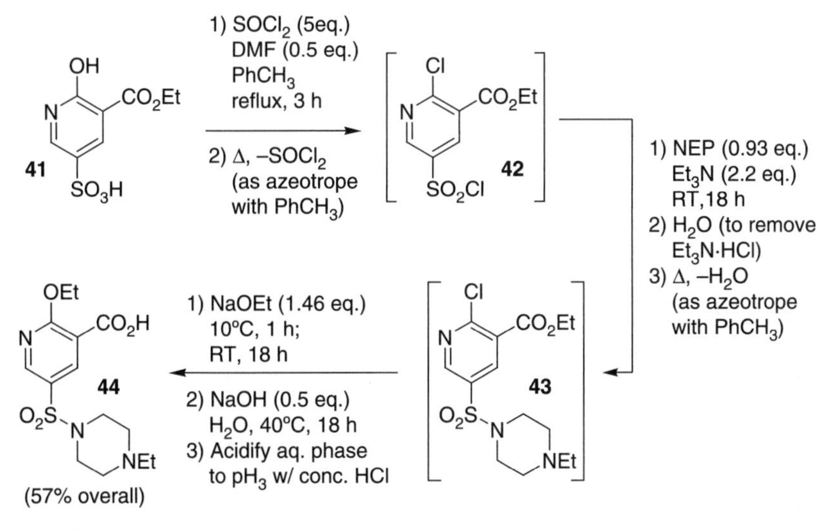

Scheme 2.12. An optimized process for an intermediate of a PDE5 inhibitor.

step except for crystallizing the product); used only one organic solvent; eliminated the hazardous chemistry above; and *reduced the COG of the drug candidate to 25% of the discovery route* (Dale et al., 2002). This is a powerful example of cost-effective process R&D, judiciously employing telescoped processing. Telescoping is a viable option when processing is understood and controlled; otherwise, impurities complicate work-ups and isolations downstream. Other considerations to develop the best commercial route can be found in an excellent review from AstraZeneca, GlaxoSmithKline, and Pfizer (Butters et al., 2006).

2.4 SUMMARY

Art is elegance in simplicity for scale-up operations. Many of the examples shown have resulted in simplified processes through detailed investigations and thorough process development. Simplified processes that control the key parameters are the most reliable for scale-up operations.

As drug development is an expensive process, people are always trying to balance the benefits of process research against the time and resources needed for that research. During the development of a drug candidate, the criteria for a successful route change. To prepare material early in a drug development campaign, the best route will be the most expedient one, but for a marketed API the best route is the one that is most economical for large-scale operations. The development of each drug candidate presents unique challenges and priorities, and timely process R&D even for preparing Phase 1 and tox batches can avoid costly surprises that delay drug development. Some difficulties may be anticipated and investigated in the laboratory, such as the effects of extended processing; other difficulties may be encountered only upon scale-up to multikilo batches. By investing time and money early in process research to develop a drug candidate, development problems can be avoided and material can be prepared in a timely manner.

REFERENCES

Anderson, N. G. (2000). *Practical Process Research & Development*, Academic Press, San Diego.

Anderson, N. G. (2001). *Org. Process Res. Dev.*, 5: 613.

Anderson, N. G. (2004). *Org. Process Res. Dev.*, 8: 260.

Anderson, N. G. (2005). *Org. Process Res. Dev.*, 9: 800.

Anderson, N. G., Ary, T. D., Berg, J. L., Bernot, P. J., Chan, Y. Y., Chen, C.-K., Davies, M. L., DiMarco, J. D., Dennis, R. D., Deshpande, R. P., Do, H. D., Droghini, R., Early, W. A., Gougoutas, J. Z., Grosso, J. A., Harris, J. C., Haas, O. W., Jass, P. A., Kim, D. H., Kodersha, G. A., Kotnis, A. S., LaJeunesse, J., Lust, D. A., Madding, G. D., Modi, S. P., Moniot, J. L., Nguyen, A., Palaniswamy, V., Phillipson, D. W., Simpson, J. H., Thoraval, D., Thurston, D. A., Tse, K., Polomski, R. E., Wedding, D. L., and Winter, W. J. (1997). *Org. Process Res. Dev.*, 1: 300.

Bingham, A. L., Hughes, D. S., Hursthouse, M. B., Lancaster, R. W., Tavener, S., and Threlfall, T. L. (2001). *Chem. Commun.*, 603.

Blacklock, T. J., Shuman, R. F., Butcher, J. W., Shearin, W. E. Jr., Budavari, J., and Grenda, V. J. (1988). *J. Org. Chem.*, 53: 836.

Blaser, H. U. and Schmidt, E., (Eds) (2004). *Asymmetric Catalysis on Industrial Scale*, Wiley-VCH, Weinheim.

Brands, K. M. J. and Davies, A. J. (2006). *Chem. Rev.*, 106: 2711.

Butters, M., Catterick, D., Craig, A., Curzons, A., Dale, D., Gillmore, A., Green, S. P., Marziano, I., Sherlock, J.-P., and White, W. (2006). *Chem. Rev.*, 106: 3002.

Cabri, W. and Di Fabio, R. (2000). *From Bench to Market: The Evolution of Chemical Synthesis*, Oxford University Press, New York.

Chembukar, S. R., Bauer, J., Deming, K., Spiwek, H., Patel, K., Morris, J., Henry, R., Spanton, S., Dziki, W., Porter, W., Quick, J., Bauer, P., Donaubauer, J., Narayanan, B. A., Soldani, M., Riley, D., and McFarland, K. (2000). *Org. Process Res. Dev.*, 4: 413.

Chen, C.-K., and Singh, A. K. (2001). *Org. Process Res. Dev.*, 5: 509.

Collins, A. N., Sheldrake, G. N., and Crosby, J., (Eds) (1997). *Chirality in Industry II*, Wiley, New York.

Conlon, D. A., Drahus-Paone, A., Ho, G.-J., Pipik, B., Helmy, R., McNamara, J. M., Shi, Y.-J., Williams, J. M., Macdonald, D., Deschênes, D., Gallant, M., Mastracchio, A., Roy, B., and Scheigetz, J (2006). *Org. Process Res. Dev.*, 10: 36.

Dale, D. J., Draper, J., Dunn, P. J., Hughes, M. L., Hussain, F., Levett, P. C., Ward, G. B., and Wood, A. S. (2002). *Org. Process Res. Dev.*, 6: 767.

Daugan, A. C.-M. (1999). U. S. Patent 5,859,006 (ICOS Corp.).

Eckert, C. A., in Zurer, P. (2000). *Chem. Eng. News*, 79(1): 26.

Ehrfeld, W., Hessel, V., and Löwe, H. (2000). *Microreactors: New Technology for Modern Chemistry*, Wiley, New York.

Engelhardt, F. C., Shi, Y.-J., Cowden, C. J., Conlon, D. A., Pipik, B., Zhou, G., McNamara, J. M., and Dolling, U.-H. (2006). *J. Org. Chem.*, 71: 480.

FDA, www.fda.gov/cder/guidance/1907fnl.pdf.

Fischer, R. W., and Misun, M. (2001). *Org. Process Res. Dev.*, 5: 581.

Gadamasetti, K. G., (Ed.) (1999). *Process Chemistry in the Pharmaceutical Industry*, Marcel Dekker, Inc., New York.

Gallou, I., Eriksson, M., Zeng, X., Senanayake, C., and Farina, V. (2005). *J. Org. Chem.*, 70: 6960.

Garrett, C., and Prasad, K. (2004). *Adv. Synth. Catal.*, 346: 889

Gergel, M. (1979). *Excuse Me Sir, Would You Like to Buy a Kilo of Isopropyl Bromide?*, Pierce Chemical Co., Rockford, IL, pp. 127–128.

Hollwedel, F., and Koßmehl, G. A. (1998). *Synthesis*, 1241.

Horgan, S. W., Burkhouse, D. W., Cregge, R. J., Freund, D. W., LeTourneau, M., Margolin, A., and Webster, M. E. (1999). *Org. Process Res. Dev.*, 3: 241.

Kim, S., Wei, C.-K., and Kiang, S. (2005). *Org. Process Res. Dev.*, 7: 997.

Lednicer, D. (1998). *Strategies for Organic Drug Synthesis and Design*, Wiley, New York.

Li, W., Wayne, G. S., Lallaman, J. E., Chang, S.-J., and Wittenberger, S. J. (2006). *J. Org. Chem.*, 71: 1725.

Lynch, J. K., Holladay, M. W., Ryther, K. B., Bai, H., Hsiao, C.-N., Morton, H. E., Dickman, D. A., Arnold, W., and King, S. A. (1998). *Tetrahedron: Asymmetry*, 9: 2791.

McConnell, J. R., in Rouhi, A. M. (2003). *Chem. Eng. News*, 81(28): 37.

McConville, F. (2002). *The Pilot Plant Real Book*, FXM Engineering & Design, Worcester MA.

Mullin, R. (2006). *Chem. Eng. News*, 84(4): 9.

Orme, M. W., Martinelli, M. J., Doecke, C. W., Pawlak, J. M., and Chelius, E. C. (2005). EP 1546149 (Lilly ICOS).

Orme, M. W., Sawyer, J. S., Schultze, L. M., Daugan, A. C.-M., and Gellibert, F. (2005). U. S. Patent 6,911,542 (Lilly ICOS).

Paul, E. L. (1988). *Chem. Eng. Sci.*, 43: 1773.

Pye, P. J., Rossen, K., Weissman, S. A., Maliakal, A., Reamer, R. A., Ball, R., Tsou, N. N., Volante, R. P., Reider, P. J. (2002). *Chem. Eur. J.* 8: 1372.

Rao, C. S. (2004). *The Chemistry of Process Development in Fine Chemical & Pharmaceutical Industry*, Asian Books Private Ltd., New Delhi.

Remenar, J. F., MacPhee, J. M., Larson, B. K., Tyagi, V. A., Ho, J. H., McIlroy, D. A., Hickey, M. B., Shaw, P., and Almarsson, Ö. (2003). *Org. Process Res. Dev.*, 7: 990.

Repic, O. (1998). *Principles of Process Research and Chemical Development in the Pharmaceutical Industry*, Wiley, New York.

Ritter, S. K. (2002). *Chem. Eng. News*, 80(47): 19.

Rouhi, A. M. (2003). *Chem. Eng. News*, 81(8): 32.

Rowe, S. M. (2002). *Org. Process Res. Dev.*, 6: 877.

Sledeski, A. W., Kubiak, G. G., O'Brien, M. K., Powers, M. R., Powner, T. H., and Truesdale, L. K. (2000). *J. Org. Chem.*, 65: 8114.

Thayer, A. (2005). *Chem. Eng. News*, 83(36): 55.

Van Arnum, S. D., Moffet, H., and Carpenter, B. K. (2004). *Org. Process Res. Dev.*, 8: 769.

Wan, Z., Jones, C. D., Mitchell, D., Pu, J. Y., and Zhang, T. Y. (2006). *J. Org. Chem.*, 71: 826.

Weissermel, K., and Arpe, H.-J. (1997). *Industrial Organic Chemistry*, 3rd ed., VCH, Weinheim.

Welch, C. J., Albaneze-Walker, J., Leonard, W. R., Biba, M., DaSilva, J., Henderson, D., Laing, B., Mathre, D. J., Spencer, S., Bu, X., and Wang, T. (2005). *Org. Process Res. Dev.*, 9: 198.

Welch, C. J., Fleitz, F., Antia, F., Yehl, P., Waters, R., Ikemoto, N., Armstrong, III, J. D., and Mathre, D. J. (2004). *Org. Process Res. Dev.*, 8: 186.

Welch, C. J., Shalmi, M., Biba, M., Chilenski, J. R., Szumigala, Jr., R. H., Dolling, U., Mathre, D. J., and Reider, P. J. (2002). *J. Sep. Sci.*, 25: 847.

Zook, S. E., Busse, J. K., and Borer, B. C. (2000). *Tetrahedron Lett.*, 41: 7017.

I

CANCER AND INFECTIOUS DISEASES

3

AROMATASE INHIBITORS FOR BREAST CANCER: EXEMESTANE (AROMASIN®), ANASTROZOLE (ARIMIDEX®), AND LETROZOLE (FEMARA®)

Jie Jack Li

1

USAN: Exemestane
Trade name: Aromasin®
Farmitalia Carlos Erba S.r.l./Pfizer
Launched: 1999
M.W. 296.40

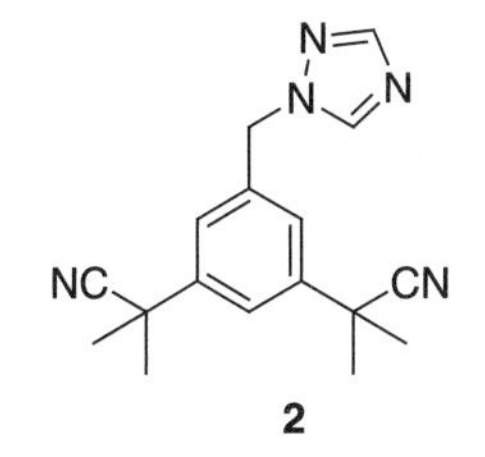

2

USAN: Anastrozole
Trade name: Arimidex®
Astra-Zeneca PLC
Launched: 1995
M.W. 293.37

3

USAN: Letrozole
Trade name: Femara®
Novartis AG
Launched: 1996
M.W. 285.30

The Art of Drug Synthesis. Edited by Douglas S. Johnson and Jie Jack Li

3.1 INTRODUCTION (Bazell, 1998; Buzdar et al., 2002; Crotty, 2001; Miller and Iqbal, 2001; Saji and Toi, 2002; Vasella and Slater, 2003)

Cancer is the uncontrolled growth of abnormal cells. It is as old as the existence of animals—cancers are found even in dinosaur bones. Approximately 110 types of cancer have been characterized. In particular, breast cancer (second only to lung cancer in terms of fatality rate) strikes one in eight women and there are approximately 200,000 annual incidents in the United States alone. Twenty-five percent of women with breast carcinoma will eventually die from their disease. The current arsenal of treatment for breast cancer includes

1. Surgery, that is, mastectomy or lumpectomy;
2. Radiation;
3. Chemotherapy;
4. Hormone treatment.

An aggressive chemotherapy regimen typically includes Cytoxan® (**4**), 5-FU (**5**, fluorouracil), and methotrexate (**6**). Cytoxan® (**4**), whose generic name is cyclophosphamide, is a nitrogen mustard and an alkylating agent discovered in the 1940s. These alkylating agents kill both resting and multiplying cancer cells. Specifically, they work by interfering with the chemical growth processes of cancer cells, thus preventing DNA from uncoiling, thereby blocking DNA replication and cell division. Unfortunately, they also destroy healthy cells indiscriminately. The use of such drugs has often been compared to a carpet bombing strategy. By the same mechanism, a common side effect to such cancer treatment is hair loss—alkylating agents also kill hair follicle cells along with active dividing cancer cells. These nitrogen mustard drugs are toxic; therefore they should be injected cleanly into a vein. On the other hand, 5-FU (**5**) and methotrexate (**6**), a folic acid analog, are antimetabolites. They prevent cancer cells from metabolizing nutrients and other essential substances, thus blocking processes within the cell that lead to cell division.

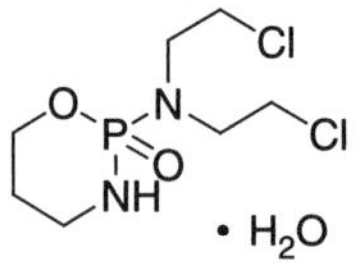

Cl Cl O P N O NH • H_2O

4, Cyclophosphamide (Cytoxan®)

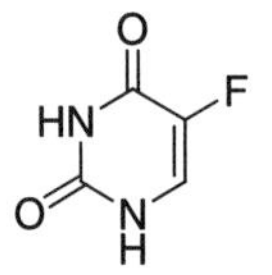

O F HN O N H

5, Fluorouracil (5-FU)

N N NH_2 N N N NH_2 H N HO_2C HO_2C O

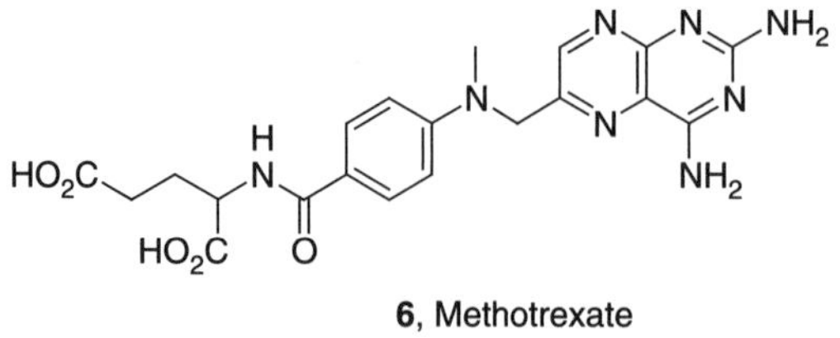

6, Methotrexate

The female hormone (estrogen) can fuel the growth of breast cancer cells, and pregnancy should be avoided for breast cancer patients. Tamoxifen (**7**, Nolvadex®) was initially developed by Imperial Chemicals Industries (ICI) as a birth-control pill. Although it was ineffective as a contraceptive in animal models, tamoxifen is now the most frequently prescribed anticancer drug in the world. In 1977, the FDA approved tamoxifen (**7**) for the treatment of metastatic breast cancer. It is a partial antagonist and partial agonist of the estrogen receptor (ER). It blocks the generation of estrogen in some parts of the body, but acts like estrogen in some other parts of the body. More specifically, it is a SERM (i.e., a Selective Estrogen Receptor Modulator). Tamoxifen is very well tolerated, but recently has been shown to sometimes lead to blood clots and endometrial cancer. As estrogen is a key trigger in two-thirds of all breast cancers, after surgery, radiation, and chemotherapy, the estrogen-dependent breast cancer patient is often hormonally treated with tamoxifen for 5 years. In 1998, the FDA also approved tamoxifen for prophylactic use in women at high risk of developing breast cancer (those who have ER-positive tumors)—patients taking tamoxifen are 45% less likely to get breast cancer recurrence. Another newer SERM, raloxifene (**8**, Evista®) renders a 58% reduction of breast cancer.

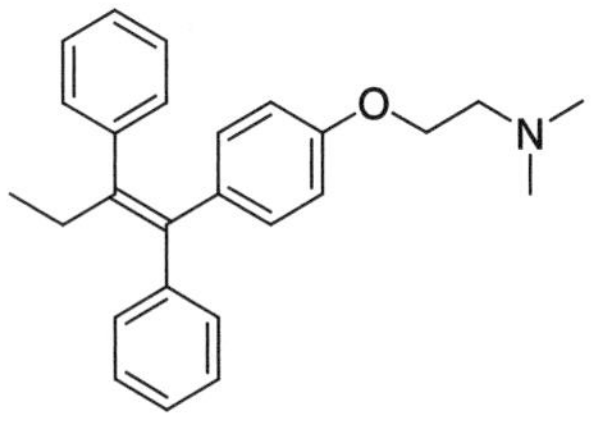

7, Tamoxifen (Nolvadex®)

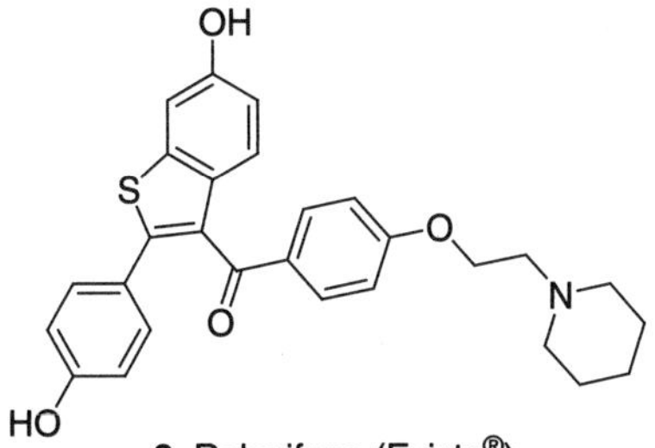

8, Raloxifene (Evista®)

Despite the significant benefit that tamoxifen has bestowed on breast cancer patients, the third-generation aromatase inhibitors are rapidly replacing tamoxifen as the first-line treatment for breast cancers. In this chapter, focus will be given to three representative small-molecule aromatase inhibitors for breast cancer: exemestane (**1**, Aromasin®), anastrozole (**2**, Arimidex®), and letrozole (**3**, Femara®).

Aromatase catalyzes the conversion of C_{19} steroids (androgens) into C_{18} steroids (estrogens) containing a phenolic A ring (Scheme 3.1). It is an enzyme complex comprised of two proteins. One is nicotinamide adenine dinucleotide phosphate

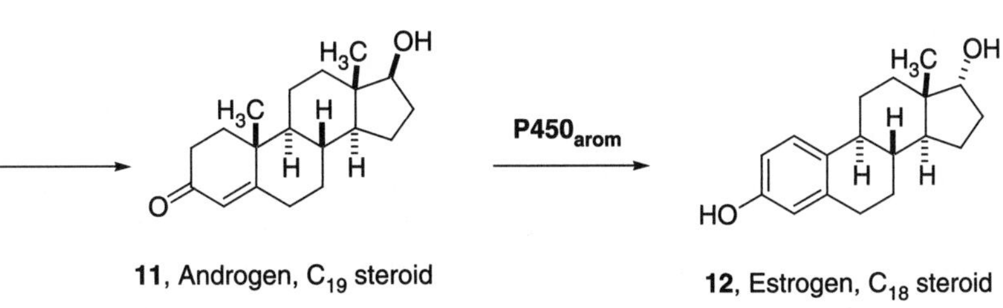

9, Cholesterol, C_{27} steroid

10, Progestogen, C_{21} steroid

P450$_{arom}$

11, Androgen, C_{19} steroid

12, Estrogen, C_{18} steroid

Scheme 3.1. Oxidative degradation of cholesterol to estrogens.

(NADPH)-cytochrome P450 reductase. The other is $P450_{arom}$, a hemoprotein of 530 amino acids with a molecular weight of ca. 55 kDa.

Aromatase inhibitors may be classified into two types. Type I aromatase inhibitors bind to the aromatase enzyme irreversibly, so they are called inactivators. In some cases they are dubbed mechanism-based or "suicide" inhibitors when they are metabolized by the enzyme into reactive intermediates that bind covalently to the active site. Type I aromatase inhibitors are usually steroidal in structure as represented by exemestane (**1**), formestane (**13**), and atamestane (**14**). Formestane (**13**) was launched by Ciba-Geigy in 1992. As formestane (**13**) is readily and extensively metabolized when administered orally, it is used as a depot formulation for injection.

Exemestane (**1**) is a type I aromatase inhibitor. As it irreversibly binds to the aromatase enzyme, it causes permanent inactivation of the enzyme even after the drug is cleared from the circulation system. The drug is highly selective, showing no in vitro effects on 20,22-desmolase and testosterone 5α-reductase. Due to the steroidal structure of exemestane (**1**), it possesses androgenic (male hormonal) properties. Its use has been associated with steroidal-like effects such as weight gain and acne. Also, because exemestane (**1**) acts as a weak male hormone, it is speculated that it may even promote bone density in addition to helping to control breast cancer.

Type I aromatase inhibitors **13** and **14** Type II aromatase inhibitor **15**

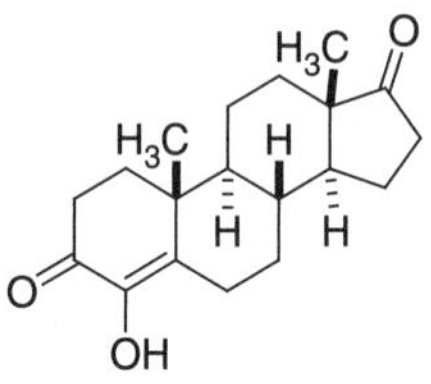

13, Formestane 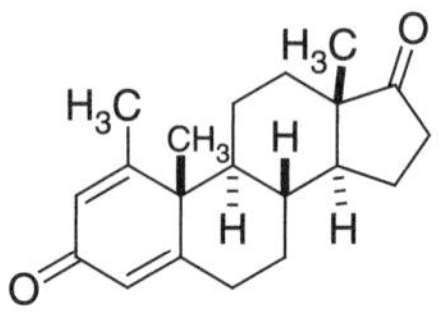 **14**, Atamestane 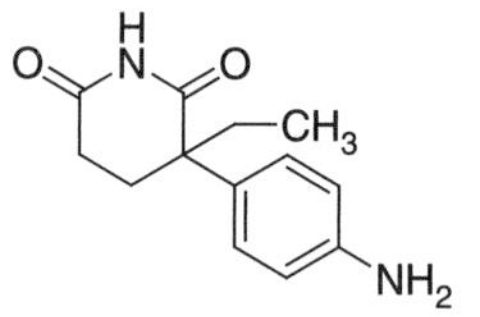 **15**, Aminoglutethimide

Type II aromatase inhibitors bind to the enzyme reversibly. The first marketed aromatase inhibitor for the treatment of breast cancer was aminoglutethimide (**15**), also by Ciba-Geigy in 1981. Aminoglutethimide is not very selective, binding to several steroidal hydroxylases that have the CYP prosthetic group. Thankfully, continued SAR development led to the latest type II aromatase inhibitors such as anastrozole (**2**) and letrozole (**3**), with exceptional specificity for the aromatase P450 enzyme. Therefore, there are fewer selectivity-related toxicities with the drugs. Although formestane (**13**) belongs to the second generation of aromatase inhibitors, anastrozole (**2**) and letrozole (**3**) are considered the third-generation aromatase inhibitors.

Exemestane (**1**), anastrozole (**2**), and letrozole (**3**) are given orally at once-daily doses of 25 mg, 1 mg, and 2.5 mg, respectively. Exemestane (**1**) has the shortest half-life ($T_{1/2}$) of all, 27 h, versus 41 h and 2–4 days for anastrozole (**2**) and letrozole (**3**), respectively. Exemestane (**1**) and anastrozole (**2**) attain a steady-state plasma level in 7 days, whereas it takes 60 days for letrozole (**3**) to reach steady-state plasma levels. Exemestane (**1**) is metabolized mostly by CYP3A4 and aldoketoreductases. Anastrozole (**2**) inhibits CYP1A1, CYP2C8/9, and CYP3A4 at relatively high concentrations, but it has no effect on CYP2A6 or CYP2D6. On the other hand, letrozole (**3**) strongly inhibits CYP2A6 and moderately inhibits CYP2C19, but it has low affinity for CYP3A4.

Therefore, there is potential for drug–drug interaction if patients are prescribed concomitant medication that interacts with these cytochrome P450 enzymes (Buzdar et al., 2002).

Overall, these three aromatase inhibitors do not seem to differ significantly with regard to their clinical efficacy. However, there are differences among them in terms of pharmacokinetics and their effects on plasma lipids, bone, and andrenosteroidogenesis.

3.2 SYNTHESIS OF EXEMESTANE (Buzzetti, 1989; Di Salle and Zaccheo, 1992; Longo and Lambardi, 1989)

Exemestane (**1**) was invented and synthesized by the Italian company Farmitalia Carlos Erba S.r.l. The synthesis of exemestane (**1**) is straightforward (Scheme 3.2). The Vilsmeier–Haack reagent (**16**) was prepared by refluxing paraformaldehyde and dimethylamine hydrochloride in isopentanol at 131°C while removing water from isopentanol using a Dean–Stark separator. After cooling the internal temperature to 10–15°C, commercially available boldenone (androsta-1,4-dien-17β-ol-3-one, **17**) was added and the reaction mixture was refluxed for an additional 15 h to give the 6-methylene derivative **18**, 6-methyleneandrosta-1,4-dien-17β-ol-3-one, in 31% yield. Subsequently, Jones oxidation of **18** in acetone at −10°C afforded exemestane (**1**) in 79% yield after recrystallization using a 65 : 35 mixture of ethanol and water.

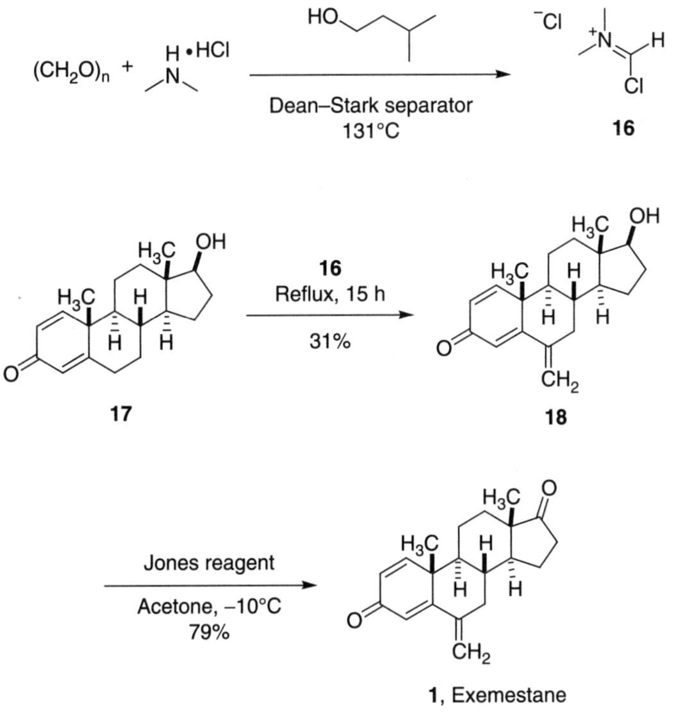

$(CH_2O)_n$ + H•HCl N → (HO isopentanol; Dean–Stark separator, 131°C) → ⁻Cl ⁺N=CH Cl **16**

17 → (**16**, Reflux, 15 h, 31%) → **18**

→ (Jones reagent, Acetone, −10°C, 79%) → **1**, Exemestane

Scheme 3.2.

3.3 SYNTHESIS OF ANASTROZOLE (Edwards and Large, 1990; Mokbel, 2003; Plourde, 1994; Plourde et al., 1995; Prous et al., 1995)

Anastrozole (**2**) and letrozole (**3**) are type II aromatase inhibitors. They competitively inhibit the conversion of androgens to estrogens. They are both potent and selective aromatase inhibitors. Anastrozole is the most extensively investigated third-generation aromatase inhibitor. It is very potent, with a daily dose of a mere 1 mg. Its safety profile is well developed; all doses evaluated up to 10 mg were well tolerated and no serious adverse events were attributed to it. No clinical or laboratory evidence of adrenal insufficiency was observed.

Anastrozole is well absorbed following oral administration. The maximum plasma concentration of the drug occurs at 2 h from dosing, that is, $C_{max} = 2$ h. It is cleared slowly, with a half-life of 30–50 h. The drug is metabolized extensively and only 10% of the drug is excreted in urine as the unchanged drug.

The synthesis of anastrozole (Scheme 3.3) began with an S_N2 displacement of commercially available 3,5-*bis*(bromomethyl)toluene (**19**) using potassium nitrile and a phase-transfer catalyst, tetrabutylammonium bromide (Edwards and Large, 1990). The resulting *bis*-nitrile **20** in DMF was then deprotonated with sodium hydride in the presence of excess methyl iodide to give the *bis*-dimethylated product **21**. Subsequently, a Wohl–Ziegler reaction on **21** was carried out using *N*-bromosuccinamide (NBS), and a catalytic amount of benzoyl peroxide (BPO) as the radical initiator. Finally, an S_N2 displacement of benzyl bromide **22** with sodium triazole in DMF afforded anastrozole (**2**) as a white solid.

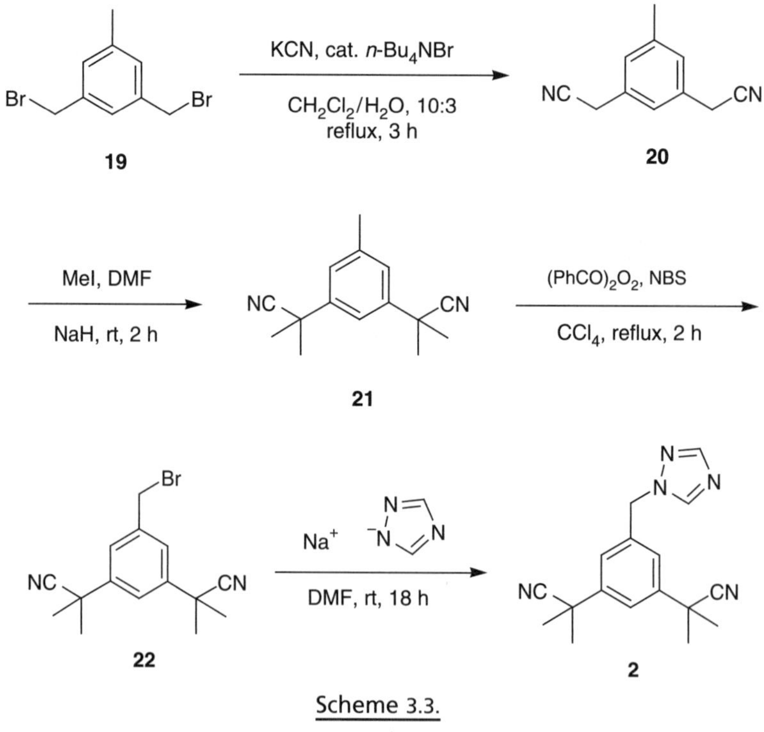

Br Br **19** — KCN, cat. n-Bu_4NBr / CH_2Cl_2/H_2O, 10:3, reflux, 3 h → NC CN **20**

MeI, DMF / NaH, rt, 2 h → NC CN **21** — $(PhCO)_2O_2$, NBS / CCl_4, reflux, 2 h →

Br NC CN **22** — Na^+ ^-N (triazole) / DMF, rt, 18 h → NC CN **2**

Scheme 3.3.

3.4 SYNTHESIS OF LETROZOLE (Bhatnagar, 1990; Bowman et al., 1990; Demers, 1994; Haynes et al., 2003; Lamb and Adkins, 1998; Prous et al., 1994; Sioufi et al., 1997)

Like anastrozole (**2**), letrozole (**3**) is also a type II aromatase inhibitor. In vitro, letrozole is more potent than many aromatase inhibitors. It is 170-fold more potent than aminoglutethimide (**15**), 19-fold more than anastrozole (**2**), and 6-fold more than formestane (**13**) against human aromatase. Letrozole also acts very selectively against aromatase: long-term administration did not affect basal levels of 17α-hydroxyprogestrone or aldosterone, although slight decreases in cortisol level were observed in two studies, which did not appear to be clinically significant.

In vivo, letrozole (**3**) was efficacious in two animal models. One was a postmenopausal hormone-dependent breast cancer mouse model, the other was a dimethylbenz[*a*]-anthracene (DMBA)-induced mammary carcinoma rat model.

In terms of pharmacokinetics, letrozole is well absorbed following 2.5-mg oral administration in healthy postmenopausal women. The maximum plasma concentration of

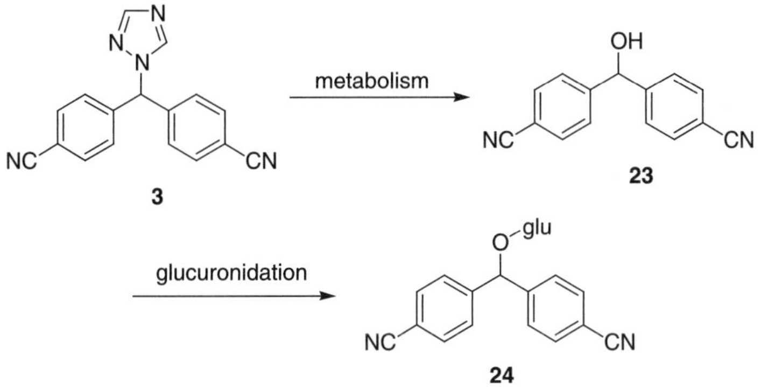

metabolism

3 **23**

glucuronidation

24

Scheme 3.4.

CH_2Cl_2, rt, 15 h

25 **26**

t-BuOK, DMF

3

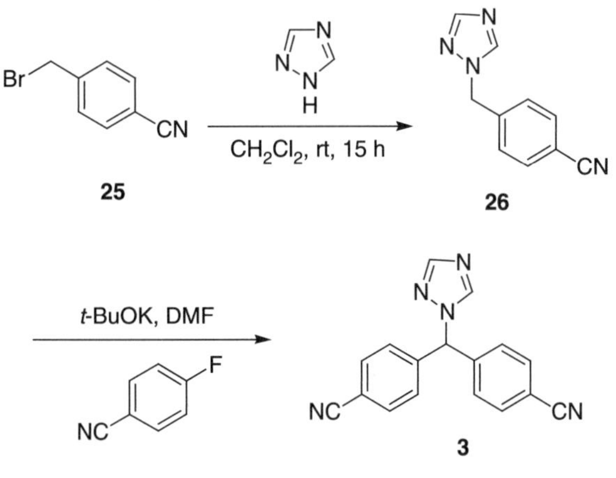

Scheme 3.5.

approximately 115 nmol/L of the drug occurs at 1 h from dosing, that is, $T_{max} = 1$ h. The systemic bioavailability is 99.9% and the volume of distribution is 1.87 L/kg.

The most significant metabolite of letrozole (**3**) is its secondary alcohol metabolite (SAM) **23** (Scheme 3.4). Biotransformation of letrozole is the main elimination mechanism, with the glucuronide conjugate of the secondary alcohol metabolite (**24**) being the prominent species found in urine. However, the total body clearance of letrozole is slow (2.21 L/h). Its elimination half-life is long, at 42 h. Letrozole and its metabolites are excreted mainly via the kidneys.

The synthesis of letrozole (**3**) is quite simple (Scheme 3.5). An S_N2 displacement of 4-bromomethyl-benzonitrile (**25**) with imidazole was carried out in methylene chloride at room temperature to produce adduct **26**. Deprotonation of **26** in DMF using potassium *t*-butoxide was followed by addition of *para*-fluorobenzonitrile. The S_NAr reaction afforded letrozole (**3**) as an oil. In order to make the crystalline drug, a hemisuccinate was prepared, which has a melting point of 149–150°C.

REFERENCES

Bazell, R. (1998). *Her-2, The Making of Herceptin, a Revolutionary Treatment for Breast Cancer*, Random House, New York.

Bhatnagar, A. S. (1990). *J. Steroid Biochem. Mol. Biol.*, 37: 1021–1027.

Bowman, R. M., Steele, R. E., and Browne, L. J. (1990). US 497872 (to Ciba-Gigy).

Buzdar, A. U., Robertson, J. F. R., Eiermann, W., and Nabholtz, J.-M. U. (2002). *Cancer*, 95: 2006–2016.

Buzzetti, F. (1989). US 4,808,616 (to Farmitalia Carlos Erba S.r.l.).

Crotty, S. (2001). *Ahead of the Curve, David Baltimore's Life in Science*, University of California Press, Berkeley.

Demers, L. M. (1994). *J. Steroid Biochem. Mol. Biol.*, 44: 520.

Di Salle, E. and Zaccheo, T. (1992). *Drugs Fut.*, 17: 278–280.

Edwards, P. N. and Large, M. S. (1990). US 4,935,437 (to Imperial Chemical Industries, plc).

Haynes, B. P., Dowsett, M., Miller, W. R., Dixon, J. M., and Bhatnagar, A. S. (2003). *J. Steroid Biochem. Mol. Biol.*, 87: 35–45.

Lamb, H. M. and Adkins, J. C. (1998). *Drugs*, 56: 1125–1140.

Longo, A. and Lambardi, P. DE 3622841 (1987); US 4,876,045 (1989) (to Farmitalia Carlos Erba S.r.l.).

Miller, W. R. and Iqbal, S. (2001). *Expert Opinion of Pharmacotherapy*, 2(6): 975–985.

Mokbel, K. (2003). *Current Medical Res. Opin.*, 19: 683–688.

Plourde, P. V. (1994). *Breast Cancer Res. Treat.*, 30: 103–111.

Plourde, P. V., Dyroff, M., Dowsett, M., Demers, L., Yates, R., and Webster, A. (1995). *J. Steroid Biochem. Mol. Biol.*, 53: 175–179.

Prous, J., Graul, A., and Castañer, J. (1994). *Drugs Fut.*, 19: 335–337.

Prous, J., Mealy, N., and Castañer, J. (1995). *Drugs Fut.*, 20: 30–32.

Saji, S. and Toi, M. (2002). *Expert Opinion on Emerging Drugs*, 7: 303–319.

Sioufi, A., Gauducheau, N., Pineau, V., Marfil, F., Jaouen, A., Cardot, J. M., Godbillon, J., Czendlik, C., Howald, H., Pfister, Ch., and Vreeland, F. (1997). *Biopharm. Drug Dispos.*, 18: 779–789.

Vasella, D. and Slater, R. (2003). *Magic Cancer Bullet, How a Tiny Orange Pill Is Rewriting Medical History*, Harper Business, New York.

4

QUINOLONE ANTIBIOTICS: LEVOFLOXACIN (LEVAQUIN®), MOXIFLOXACIN (AVELOX®), GEMIFLOXACIN (FACTIVE®), AND GARENOXACIN (T-3811)

Chris Limberakis

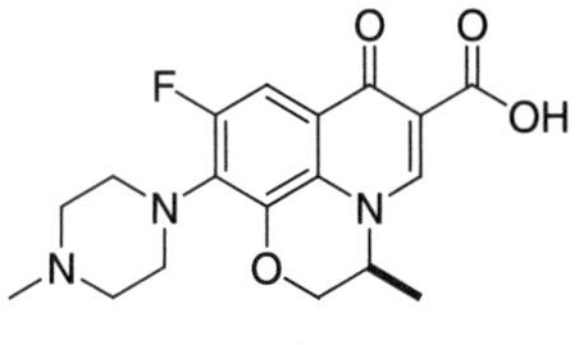

1

USAN: Levofloxacin
Trade name: Levaquin®
Company: Daiichi, Ortho-McNeil (J&J)
Launched: 1997 (USA)
M.W. 361.37 (parent)

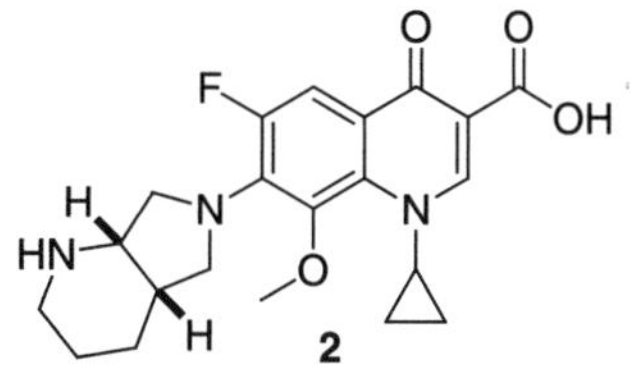

2

USAN: Moxifloxacin hydrochloride
Trade name: Avelox®
Company: Bayer
Launched: 1999 (USA)
M.W. 401.43 (parent)

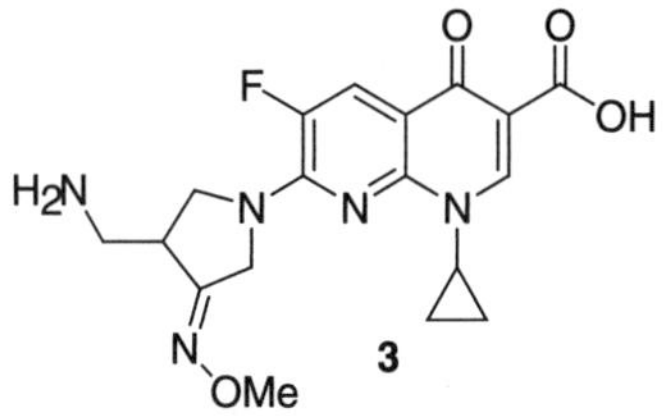

3

USAN: Gemifloxacin mesylate
Trade name: Factive®
Company: LG Life Sciences, Oscient
Launched: 2004 (USA)
M.W. 389.38 (parent)

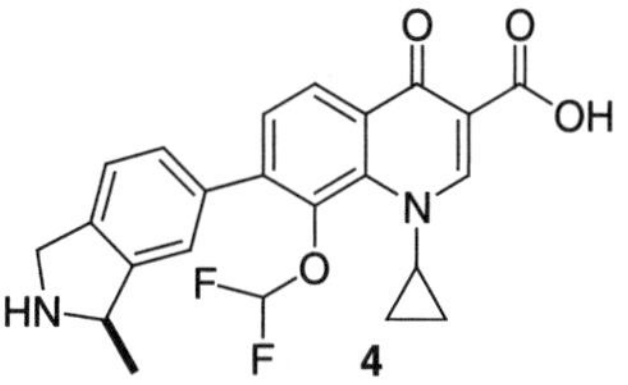

4

USAN: Garenoxacin (T-3811)
Phase III clinical studies completed
NDA submitted February 2006
Company: Toyama, Schering-Plough
M.W. 426.41 (parent)

The Art of Drug Synthesis. Edited by Douglas S. Johnson and Jie Jack Li

4.1 INTRODUCTION

Over the course of the last several decades, the quinolones have evolved into a germane class of antibacterial agents. The genesis of these agents began with the discovery of nalidixic acid in 1962, a relatively poor antibacterial agent, which was limited to urinary tract infections against Gram-negative organisms. However, subsequent generations of quinolones provided enhanced activity against Gram-negative, Gram-positive, and anaerobic bacteria, culminating with the emergence of the fluoroquinolones. Today, quinolones are prescribed for upper and lower respiratory tract infections, gastrointestinal infections, gynecological infections, sexually transmitted diseases, urinary tract infections, prostatitis, bacterial meningitis, and osteomyelitis. Due to the quinolones' importance in combating bacterial infections, extensive reviews have been written (Andriole, 2000a, 2003, 2005; Bryskier, 2005; De Souza, 2005; Hooper and Rubenstein, 2003; Howe and MacGowan, 2004; Mascaretti, 2003; Mitscher, 2005; Ronald and Low, 2003).

In terms of structural classes, Mitscher described seven cores that fall under the quinolonefamily, as shown below in Figure 4.1 (Mitscher, 2005). Of these cores, three templates have produced the most commercially successful drugs. They include the 4-oxo-1,4-dihydroquinolone (**5**), 7-oxo-2,3-dihydro-7H-pyrido-[1,2,3-d,e]-1,4-benzoxazine (**6**), and the 4-oxo-1,4-dihydro-[1,8]-naphthyridine cores (**7**).

The quinolones are also categorized into generations (Fig. 4.2) based on their in vitro activity (Andriole, 2000a, 2005; Ball, 2000; Brighty and Gootz, 2000; Hooper and Rubenstein, 2003; Shams and Evans, 2005). Although there are slight variations in

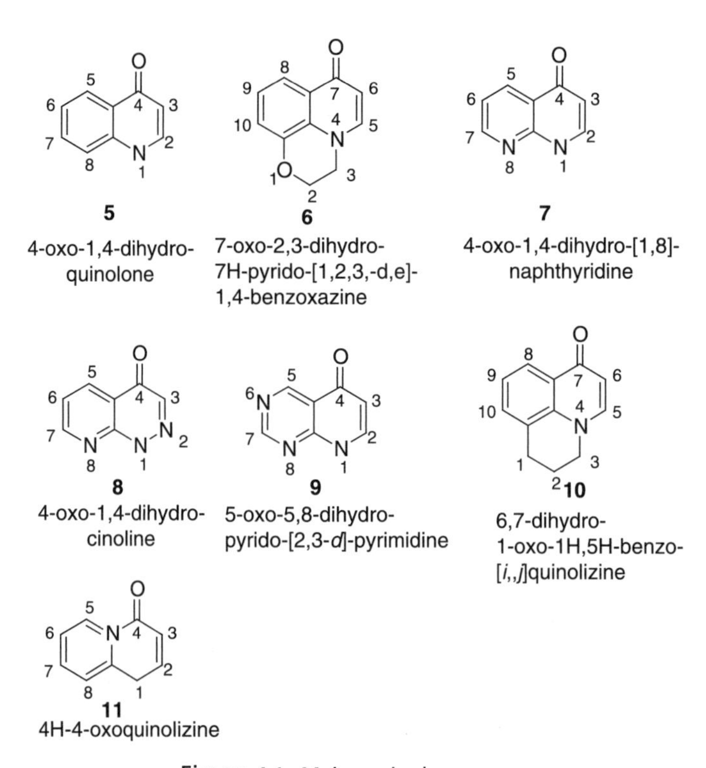

Figure 4.1. Major quinolone cores.

First Generation

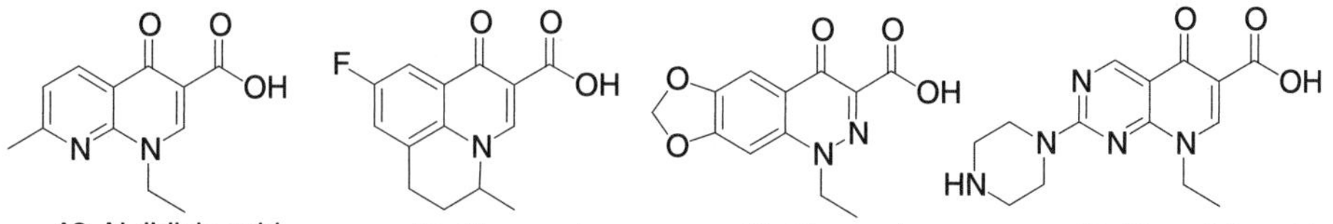

12, Nalidixic acid **13**, Flumequine **14**, Cinoxacin **15**, Pipemidic acid

Second Generation

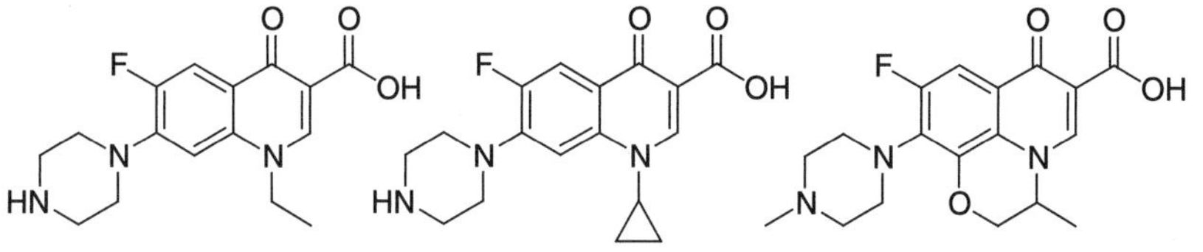

16, Norfloxacin **17**, Ciprofloxacin **18**, Ofloxacin

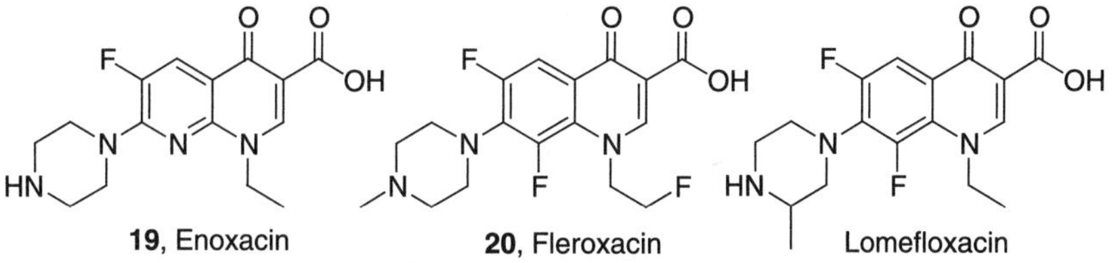

19, Enoxacin **20**, Fleroxacin Lomefloxacin

Figure 4.2. First- and second-generation quinolones.

these generations, this grouping method effectively illustrates the evolution of these drugs. The first generation of quinolones, which included nalidixic acid (**12**), flumequine (**13**), cinoxacin (**14**), and pipemidic acid (**15**) were effective against Gram-negative organisms, but they lacked activity against Gram-positive bacteria and anaerobes.

The late 1970s and early 1980s saw the emergence of the second generation of quinolones, particularly the fluoroquinolones such as norfloxicin (**16**), ciprofloxacin (**17**), ofloxacin (**18**), enoxacin (**19**), fleroxacin (**20**), and lomefloxacin, which exhibited stronger activity against Gram-negative bacteria, albeit moderate activity against Gram-positive. In addition, this generation of agents saw better pharmacokinetic properties with the introduction of the C7-piperdinyl substituent.

These analogs were soon followed by the third generation (Fig. 4.3) including levofloxacin (**1**,(−)ofloxacin), tosufloxacin (**21**), temafloxacin (**22**), sparfloxacin (**23**), grepafloxacin (**24**), and gatifloxacin (**25**). These quinolones exhibited improved activity against the Gram-positive organisms and anaerobes.

Finally, the fourth generation (Fig. 4.3) increased in efficacy against Gram-positive organisms such as *pneumocci* and was very potent against anaerobes. Some of the agents include moxifloxacin (**2**), trovafloxacin (**26**), gemifloxacin (**3**) clinafloxacin (**27**), and sitafloxacin (**28**). More recently, garenoxacin (**4**), a 6 des-F quinolone, has attracted much attention because of its broad-spectrum activity and unique carbon–carbon bond at C7.

Although the availability of these antibacterial agents varies worldwide, over 20 quinolones are currently on the market (Andriole, 2003). For example, the United States has eight quinolones available by prescription (Table 4.1).

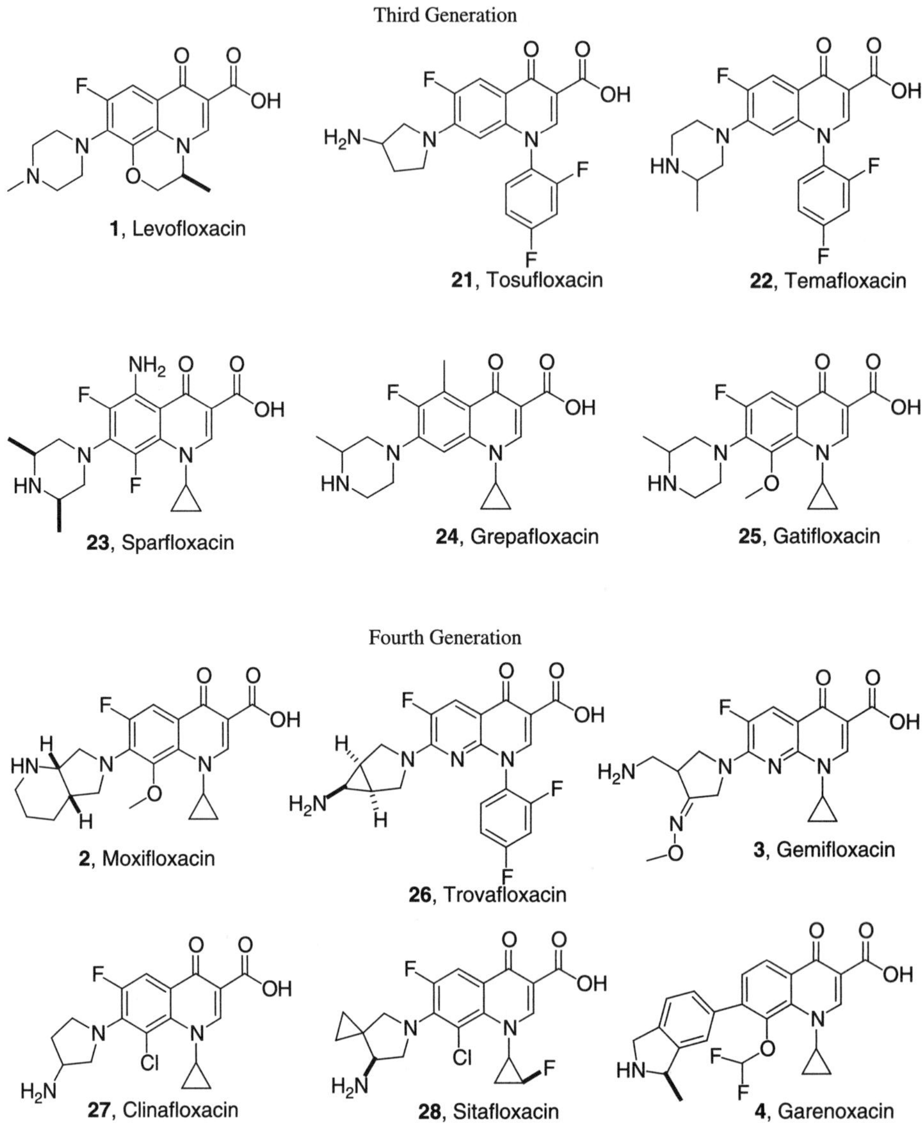

Third Generation

1, Levofloxacin

21, Tosufloxacin

22, Temafloxacin

23, Sparfloxacin

24, Grepafloxacin

25, Gatifloxacin

Fourth Generation

2, Moxifloxacin

26, Trovafloxacin

3, Gemifloxacin

27, Clinafloxacin

28, Sitafloxacin

4, Garenoxacin

Figure 4.3. Third- and fourth-generation quinolones.

Despite major advancements with this class of drugs, some quinolones were plagued with major side effects, causing their withdrawal from the prescription drug sector or the retraction of New Drug Applications (NDAs) (Andriole, 2000b; Leone et al., 2003; Shams and Evans, 2005; Van Bambeke et al., 2005). Some notable examples of marketed quinolones that were removed from the U.S. market examples include temafloxacin (hemolytic-uremic syndrome), sparfloxacin (phototoxicity and QT prolongation), and grepafloxacin (QT prolongation) (Leone et al., 2003). Trovafloxacin (hepatoxicity) was removed from the European market (Leone et al., 2003); however, the FDA has recommended that its use should be limited to life- or limb-threatening infections (Van Bambeke et al., 2005). In addition, the New Drug Applications (NDA) for fleroxacin

TABLE 4.1. Marketed Quinolones in the United States[a]

USAN	Trade Name	FDA Approval	Company
Norfloxacin	Noroxin®	10/1986	Merck
Ciprofloxacin[b]	Cipro®	10/1987	Bayer
Ofloxacin[b]	Floxin®	12/1990	Ortho-McNeil (J&J)
Lomefloxacin	Maxaquin®	2/1992	Pharmacia (Pfizer)
Levofloxacin	Levaquin®	12/1996	Ortho-McNeil (J&J)
Gatifloxacin	Tequin®	12/1999	Bristol-Myers Squibb
Moxifloxacin	Avelox®	12/1999	Bayer
Gemifloxacin[c]	Factive®	4/2003	LG Life Sciences

[a]Approved quinolones as of February 2006; [b]Generic available; [c]Marketed in the United States by Oscient.
Source: U.S. Food and Drug Administration (FDA); website: www.fda.gov/cder

(phototoxicity) (Leone et al., 2003) and clinafloxacin (phototoxicity and hypoglycemia) (Andriole, 2000b; Leone et al., 2003) were retracted. Moreover, in February of 2006, the FDA strengthened the warning label for gatifloxicin (Tequin®) for its existing side effects of hypoglycemia and hyperglycemia (FDA, 2006). In addition, a contraindication for the use of Tequin® in diabetic patients was included. However, the drug remains on the market, because the benefits outweigh the risks.

Although some setbacks have emerged in the last decade, the quinolones are an important class of antibacterial agents in combatting serious Gram-positive, Gram-negative, and anaerobic infections. Based on 2004 global sales, the quinolones garnered $5.752 billion dollars, with levofloxacin generating the highest sales (Fig. 4.4) (IMS, 2005).

4.1.1 Mechanism of Action (Brighty and Gootz, 2000; Drlica and Hooper, 2003; Zhanel et al., 2002)

The quinolones are inhibitors of bacterial DNA gyrase (topoisomerase II) and topoisomerase IV. These two enzymes are essential for the uncoiling of DNA. More specifically, DNA gyrase exists as an A_2B_2 tetramer encoded by *gyrA* and *gyrB* genes, and

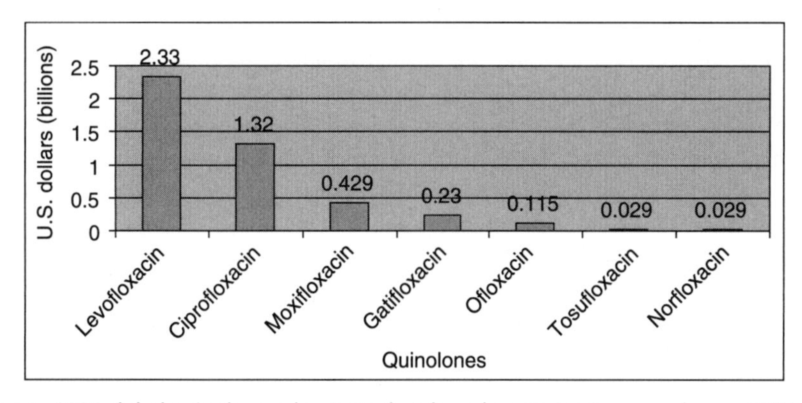

Figure 4.4. 2004 global quinolone sales. Note that, based on Oscient's press releases, 2005 sales of gemifloxacin (Factive®) were approximately 22 million U.S. dollars (Oscient, 2005, 2006). (*Source*: IMS Health, based on 15 leading quinolone products.)

topoisomerase IV (topo IV) is a C_2E_2 tetramer encoded by *par C* and *par E*. In general, DNA gyrase is the target for quinolone inhibition in Gram-negative bacteria such *E. coli.*, and topo IV is the main target in Gram-positive bacteria such as *S. aureus* and *S. pneumo.* However, sparfloxacin and clinafloxacin target DNA gyrase in these Gram-positive bacteria. In addition, moxifloxacin and gatifloxacin target DNA gyrase in *S. pneumo* (Zhanel et al., 2002). It is believed that quinolones bind to complexes formed between DNA and one of the two germane enzymes to afford a quinolone–DNA–enzyme complex. The quinolone complex prevents the possibility of DNA replication. Because this process is reversible, this mode of inhibition is considered bacteriostatic. In other words, there is inhibition of bacterial growth, but the process does not end there. Ultimately, cell death occurs because the DNA ends of the ternary complex fragment, so the net result of quinolone inhibition is bactericidal.

4.1.2 Modes of Resistance (Hooper, 2003)

Like many other antibiotic classes, the quinolones are not immune from bacterial resistance. There are typically two modes of resistance. The major mode of resistance occurs via mutations of an antimicrobial site; that is, mutations of the *gyrA* and *gyrB* genes found in DNA gyrase and modifications of *parC* and *parE* in topo IV. The second mode of resistance results from the lowering of the drug concentration in the target bacterial cell. The drug concentration is reduced via efflux pumps that literally "pump out" the antibacterial agent.

4.1.3 Structure–Activity Relationship (SAR) and Structure–Toxicity Relationship (STR) (Domalaga, 1994; Domalaga and Hagen, 2003; Van Bambeke et al., 2005)

Since the review by Domagala in 1994, little has changed in the SAR and STR of the quinolones; these results are summarized in Table 4.2. Substitution of N-1 appears to be essential for potency control, with the cyclopropyl group appearing to be optimal. Although some success has been achieved with the 2,4-difluorophenyl substituent as evidenced by tosufloxacin, trovafloxacin, and temafloxacin, the latter two have been removed from the market or heavy restrictions have been imposed. At the C(2)-position, hydrogen is optimal. The C3-carboxyl and C4-carbonyl moieties are essential for gyrase binding. Position C5 controls potency, and the general trend in terms of greater potency (i.e., lower MICs) is $NH_2 > OH > CH_3 > H$. However, this site has been also implicated as a source of phototoxicity and genotoxicity. Position 6 controls potency; it has been shown that a fluorine atom at this position increases efficacy, giving rise to the fluoroquinolones. One of the greatest advancements in increased potency as well as improved pharmacokinetics was the introduction of piperidinyl and aminopyrrolidinyl substituents at the C7 position. In general, the piperidinyl substituent produces a higher Gram-negative activity, and the aminopyrrolidinyl delivers enhanced Gram-positive potency. This improvement is exemplified by the major quinolones currently on the market; they all contain one of these two moieties. However, a promising clinical candidate (garenoxacin) has as an isoindoline substituent at C7 via a carbon–carbon connection to the quinolone core. The toxicity associated with C7 substitution is GABA receptor binding (CNS issues). Position 8 is believed to play a role in anaerobic activity and pharmacokinetics. Typically N, OMe, derivatives of OMe, Me, halogens, and H are substituents seen at this position,

TABLE 4.2. SAR and STR Relationships of Quinolones

X = C–H, N, C–OMe, C–Me, C–$OCHF_2$, C–F, C–Cl, etc.

Position	SAR Relationship	STR Relationship
1	Controls potency	Gene toxicity and theophyline interactions
2	No sidechain or small group, because of proximity to gyrase binding	No side effects
3–4	Germane for gyrase binding	Metal chelation site
5	Controls potency for gram (+) bacteria	Influences gene toxicity and phototoxicity
6	Enhances inhibition of DNA gyrase (i.e., fluoroquinolones)	No side effects
7	Influences potency against Gram (+) and Gram (−)-bacteria, spectrum, and ADME	Influences gene toxicity, theophyline and NSAID interactions, and GABA binding
8	Influences anaerobe activity and ADME	Influences gene toxicity and phototoxicity

with levofloxacin containing a cyclic connection. Some of the most potent quinolones have a halogen at C8 such as sparfloxacin (F) and clinafloxacin (Cl); however, these antibacterial agents had significant phototoxicity.

4.1.4 Pharmacokinetics (Dudley, 2003a)

The quinolones have long enjoyed a favorable pharmacokinetic profile. They are well absorbed and distributed in body tissues and fluids. A typical dose is between 100 mg and 1 g. The older agents have half-lives under 3 h, but a more typical value is between 4 and 14 h. Protein binding tends to be low to moderate (15–65%), but there are some exceptions such as nalidixic acid (90%) and garenoxacin (80%) (Howe and MacGowan, 2004). Bioavailability ranges from 55 to 100% (Dudley, 2003b). It has been well established in the literature that the effectiveness of quinolones can be dramatically reduced if the medication is taken with an antacid. Many antacids are salts of divalent and trivalent cations such as Al^{3+}, Ca^{2+}, and Mg^{2+}. In addition, Fe^{3+}, Cu^{2+}, Ni^{2+}, Zn^{2+}, and Ba^{2+} also reduce quinolone activity. All these cations form a chelate with the northeastern sector of the quinolone core to yield **29** (Mitscher, 2005), which dramatically reduces AUCs (area under the plasma or serum concentration-versus-time curve) by 20–98% (Bryskier, 2005; Da Silva et al., 2003). This region of the template, which includes the carbonyl and carboxyl groups, is a critical point of attachment to DNA gyrase (De Souza, 2005).

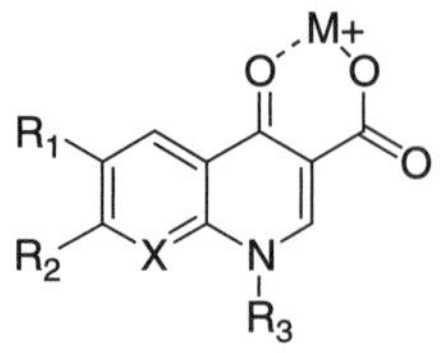

29

4.1.5 Synthetic Approaches

The major synthetic routes to the quinolone cores were recently reviewed (Da Silva et al., 2003; Mitscher, 2005). Several approaches, including the Gould–Jacobs reaction, Grohe–Heitzer cycloacylation, Gerster–Hayakawa syntheses, Chu–Mitscher synthesis, and the Chu–Li syntheses, have been used. The first two strategies will be discussed in this section, and the Gerster–Hayakwa and Chu–Mitscher routes will be addressed in the levofloxacin synthesis section as these strategies directly apply to levofloxacin. Because the Chu–Li syntheses were developed for 9-cyclopropylpyrimidinones, this approach will not be described.

The Gould–Jacobs sequence (Scheme 4.1) commences with an addition–elimination reaction between aniline **30** and substituted ethylenemalonate derivative **31** to yield malonic ester **32**. Subsequent intramolecular cyclization delivers the 4-hydroxy-3-carboalkoxy-quinolone **33**. In the presence of an alkylating agent, **33** is converted to **34**. Saponfication of the ester affords quinolone core **35**.

The Grohe–Heitzer sequence (Scheme 4.2) begins with acylation of malonate derivative **37** with benzoyl chloride **36** to give malonate **38** (Mitscher, 2005). Condensation of the malonate with an ortho-ester in the presence of a dehydrating agent such as acetic anhydride affords enol ether **39**. The enol ether then undergoes an addition–elimination

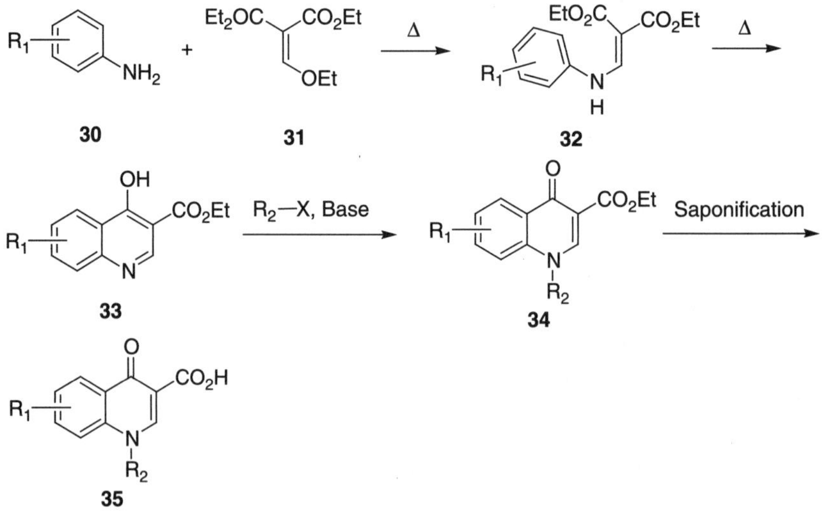

Scheme 4.1. The Gould–Jacobs sequence.

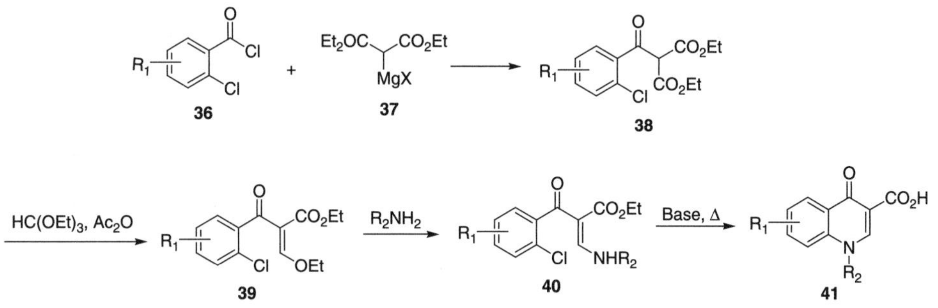

Scheme 4.2. The Grohe–Heitzer sequence.

sequence to afford 3-benzoyl-2-aminoacrylate **40**. Intramolecular addition–elimination delivers the quinolone core **41**.

The remaining portion of this chapter will now concentrate on levofloxacin (LVX), moxifloxacin (MXF), and gemifloxacin (GEM). Although ciprofloxacin (CIP) remains an important quinolone in the battle against bacterial infections, it has been reviewed recently (Li et al., 2004); however, garenoxacin (GAR) will be included in this chapter, because phase III clinical studies have been completed (Hoffman-Roberts et al., 2005), and Schering-Plough submitted an NDA to the FDA in February 2006 (Schering-Plough, 2006).

4.2 LEVOFLOXACIN

Levofloxacin (**1**), the levo-isomer or the (*S*)-enantiomer of ofloxacin, received FDA approval in 1996 (Fish, 2003; Hurst et al., 2002; Mascaretti, 2003; Norrby, 1999; North et al., 1998). The initial approval covered community-acquired pneumonia, acute bacterial exacerbation of chronic bronchitis, acute maxillary sinusitis, uncomplicated skin and skin structure infections, acute pyelonephritis, and complicated urinary tract infections (North et al., 1998). Four years later, the levofloxacin indication list grew to include community-acquired pneumonia caused by penicillin-resistant *Streptococcus pneumoniae*. In addition, in 2002, nosocomial (hospital-acquired) pneumonia caused by methicillin-susceptible *Staphylococcus aureus*, *Pseudomonas aeruginosa*, *Serratia marcescens*, *Haemophilus influenzae*, *Kliebsella pneumoniae*, and *Escherichia coli* was added (Hurst et al., 2002). Finally in 2004, LVX was approved as a post-exposure treatment for individuals exposed to *Bacillus anthracis*, the microbe that causes anthrax, via inhalation (FDA, 2004).

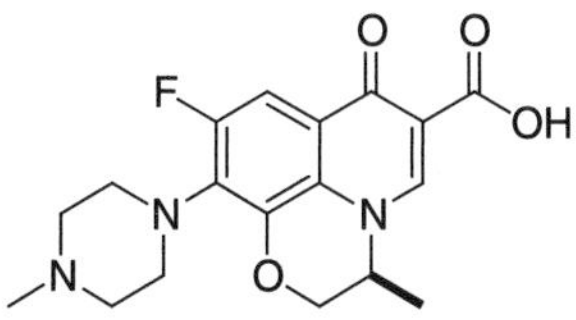

1

Levofloxacin, the (*S*)-enantiomer or (−)-enantiomer of ofloxacin, is approximately 8 to 128 times more potent than the racemic mixture; thus it is believed that the

TABLE 4.3. MIC_{90} Values of Levofloxacin Compared with Other Quinolones

	MIC_{90} (μg/mL)				
Bacterium	LVX	NDX	NFX	CIP	OFX
S. pneumoniae[a]	2	>64	2–16	1–4	2–4
S. aureus[a]	0.25	>64	2	0.5–2	0.5–2
Enterococcus spp.[a]	2–8	>64	8–16	1–8	2–8
H. influenzae[b]	0.03	1	0.25	0.03	0.03
M. catarrhalis[b]	0.06	4	0.25	0.06–0.25	0.12
Neisseria spp.[b]	0.015	0.5	0.03	0.03	0.06
E. coli[b]	0.06–0.25	4–8	0.12–2	0.06–0.025	0.12–0.25
Shigella spp.[b]	0.03	8–16	0.12–0.5	0.12	0.25
P. aeruginosa[b]	4	>64	0.5–2	0.5–2	0.5–4

[a]Gram-positive bacteria; [b]Gram-negative bacteria; LVX, levofloxacin; NDX, nalidixic acid; NFX, norfloxacin; CIP, ciprofloxacin; OFX, ofloxacin.

(*S*)-enantiomer accounts for the activity seen with ofloxacin (Fish, 2003). In Table 4.3, levofloxacin's in vitro activity is compared to quinolones that were on the market when LVX was introduced in 1997 (Howe and MacGowan, 2004). The in vitro activity that will be used through out this chapter is the minimum inhibitory concentration (MIC) required to inhibit growth of 90% of isolates in a given bacterium. This term is often abbreviated as MIC_{90}. Comparison of LVX against the newer fluoroquinolones is reported in the moxifloxacin and gemifloxacin sections.

In terms of pharmacokinetics, LVX has an excellent profile. With an oral dose of 500 mg, LVX has a bioavailability of ≥99%, an AUC range of 41.9–47.7 mg h/mL, C_{max} of 4.5–6.2 mg/mL, a clearance of 10.5–11.9 L/h, and a volume of distribution (V_d) of 1.3 L/kg (Hurst et al., 2002). In terms of protein bound material, only 24–38% is affected (Fish, 2003). Like other fluoroquinolones, there is a 19–44% AUC reduction when co-administered with an aluminum or magnesium antacid or iron sulfate (Qaqish and Polk, 2003). The major metabolite of levofloxacin arises from glucuronidation (Bryskier, 2005).

Before proceeding with the literature syntheses of levofloxacin, it is instructive to describe the synthesis of ofloxacin. The original synthesis of ofloxacin was reported in 1984 by Hayakawa and co-workers with Daiichi Seijaku in Japan (Hayakawa et al., 1984). Hayakawa's strategy was based on the Gerster synthesis, where a quinolone is converted to a carbatricycle, employing a reduction step followed by the Gould–Jacobs sequence, as shown in Scheme 4.3.

Thus, this modification is now called the Gerster–Hayakawa synthesis. In order to obtain the oxo-analog of the Gerster intermediate, Hayakawa and co-workers commenced the synthesis (Scheme 4.4) with selective displacement of the fluorine *ortho* to the nitro

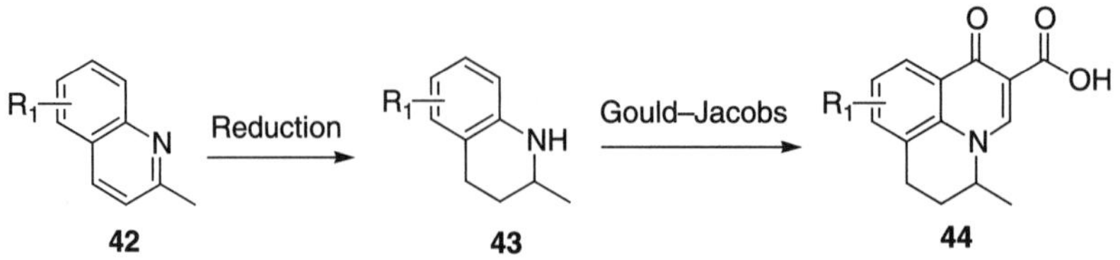

Scheme 4.3.

group in 2,3,4-trifluoro-1-nitrobenzene (**45**) with potassium hydroxide in DMSO to give phenol **46** in 29% yield. Phenol **46** was then alkylated with α-chloroacetone in the presence of potassium iodide to give **47** in moderate yield. Reduction of the nitro group with Raney nickel in the presence of hydrogen followed by concomitant cyclization of the resulting aniline methyl ketone gave requisite benzoxazine **48** in 90% yield. The benzoxazine was then converted to quinolone **50** via the previously described Gould–Jacobs strategy. First, benzoxazine **48** was condensed with diethyl ethoxymethylenemalonate at 145°C to give malonate **49**, which underwent cyclization in the presence of the dehydrating agent polyphosphoric ester (PPE) at 140–145°C to deliver the tricyclic core in 64% yield over two steps. The ethyl ester of **50** was then hydrolyzed in a mixture of hydrochloric acid and acetic acid to carboxylic acid **51** in 94% yield. Selective displacement of the C8 fluorine with *N*-methylpiperizine in DMSO at 100°C delivered (±)-ofloxacin (**17**) in 62% yield.

Before an asymmetric synthesis appeared of levofloxacin (**1**, (−)-ofloxacin), (−)- ofloxacin was isolated via optical, enzymatic, and crystallization resolution of the racemic ofloxacin (**17**) (*Drugs Future*, 1992; Hayakawa et al., 1986, 1991). For instance, tricyclic core **52** was converted to (±)-3,5-dinitrobenzoyl derivative **54** in 75% yield (Scheme 4.5). The enantiomers were then separated via high-performance liquid chromatography (HPLC) with a SUMIPAX OA-4200 column to deliver optically pure benzoyl esters **55a** and **55b** (*Drugs Future*, 1992; Hayakawa et al., 1986, 1991).

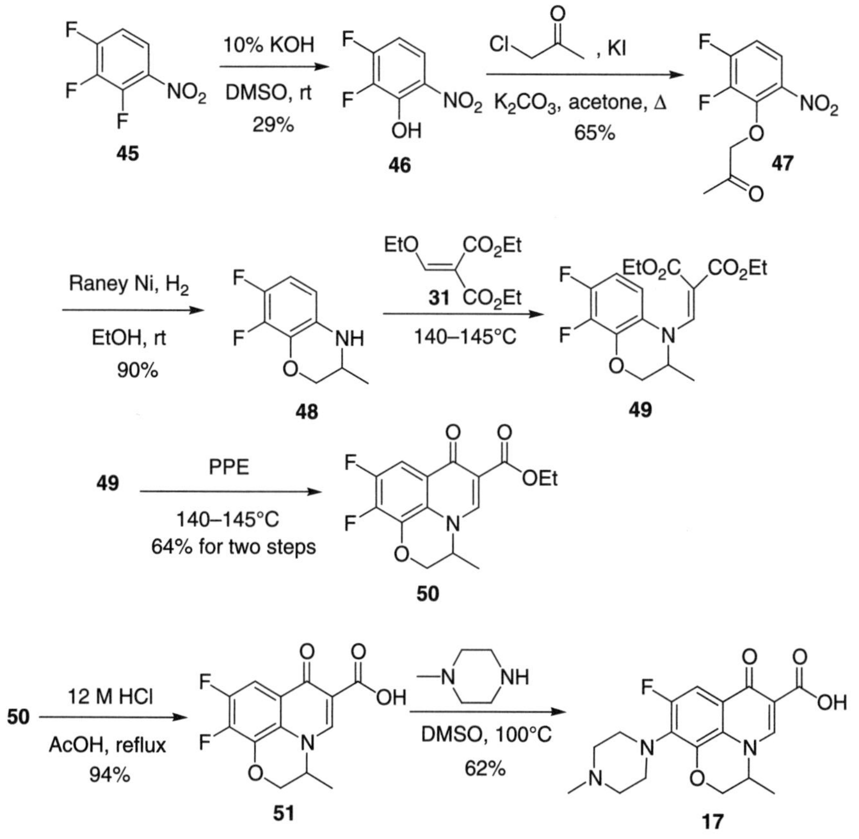

Scheme 4.4.

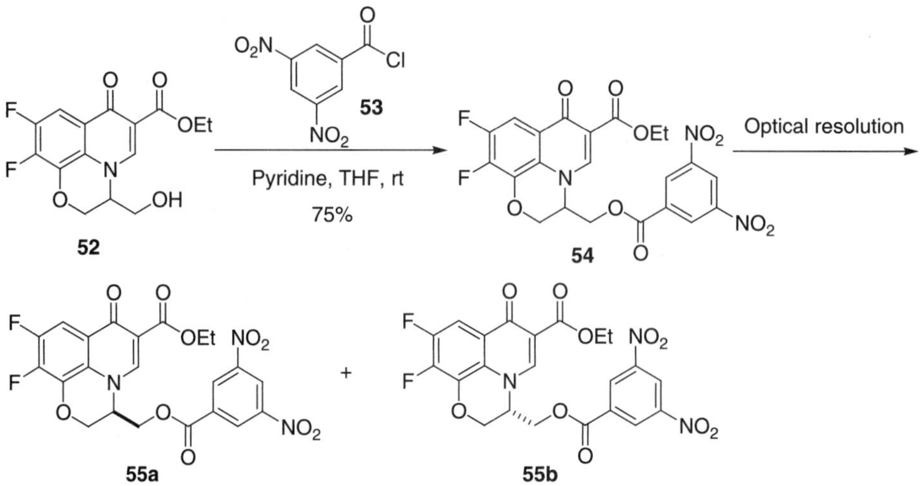

Scheme 4.5.

Benzoyl ester **55a** was then hydrolyzed with aqueous sodium bicarbonate to enantiomerically pure alcohol **56** in 91% yield (Scheme 4.6). The alcohol was converted to iodide **57** with triphenylphosphite methyliodide in excellent yield. Iodide **57** was next reduced under radical conditions, and the resulting core was hydrolyzed with hydrochloric acid in acetic acid to furnish carboxylic acid **58** in 44% yield over two steps. Finally, the 7-fluoroquinolone was allowed to react with *N*-methylpiperazine in DMSO at 130–140°C to deliver (−)-ofloxacin, levofloxacin (**1**), in 52% yield.

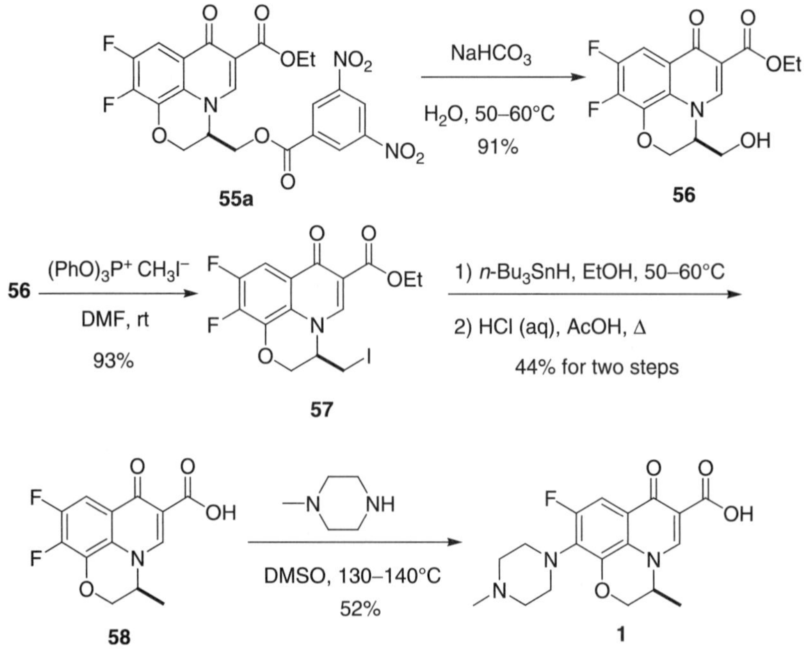

Scheme 4.6.

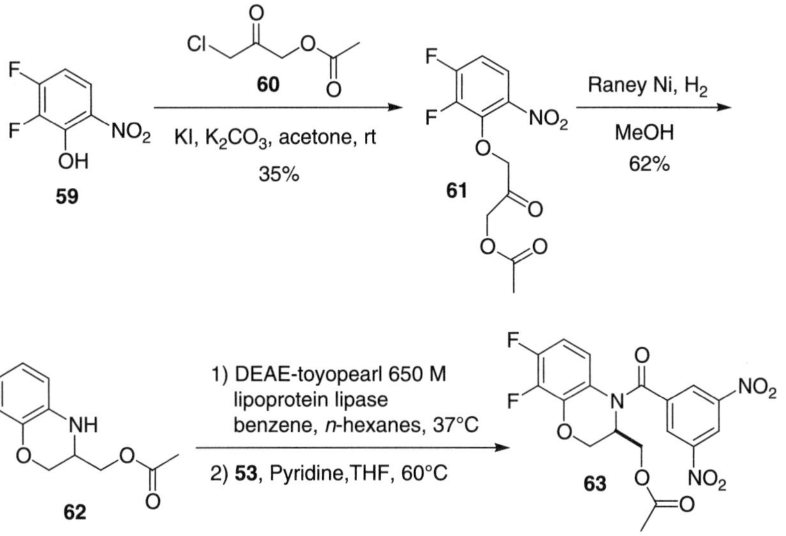

Scheme 4.7.

The second-resolution approach relied on enzymatic resolution of acetate esters **62** (Scheme 4.7) (Hayakawa et al., 1991). The sequence opened with the alkylation of 2,3-difluoro-6-nitrophenol (**59**) with 1-acetoxychloro-2-propane (**60**) to deliver ether **61**. Reduction of the nitro group of **61** gave an intermediate aniline that cyclized to give racemic benzoxazine **62** in 62% yield. A variety of lipases were then examined for the resolution. The best results arose from use of LPL Amano 3, derived from *P. aeruginosa*, which gave a ratio of 73 : 23 in favor of the desired (−)-enantiomer. Benzoylation of the enantiomerically-enriched mixture followed by chromatography of the aryl amides delivered enantiomerically pure **63**.

Hydrolysis of **63** was then achieved with aqueous potassium hydroxide to the hydroxymethylbenzoxazine **64** in 84% yield (Scheme 4.8). Alcohol **64** was converted to

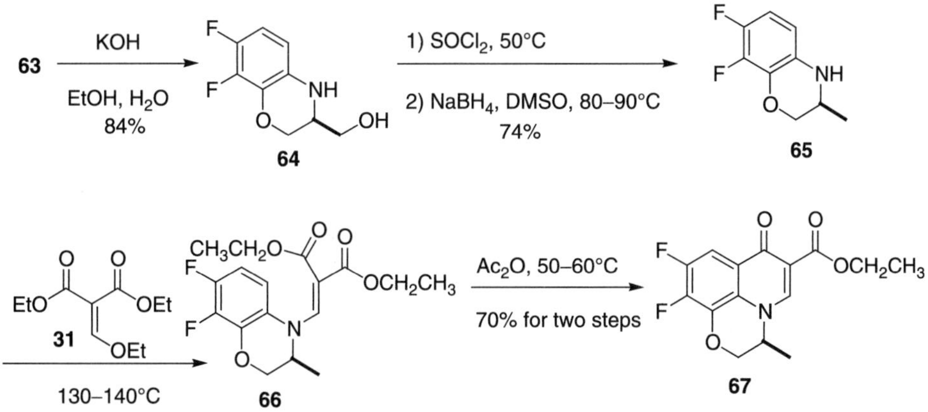

Scheme 4.8.

an alkyl chloride with thionyl chloride and the thus afforded alkyl chloride was reduced with sodium borohydride to deliver methylbenzoxazine **65** in 74% yield over two steps. The methylbenzoxazine **66** was then treated with diethyl ethoxymethylmalonate to afford the methylenemalonate, which was cyclized with acetic anhydride to deliver tricyclic core **67** in 70% over two steps.

Hydrolysis of the ethyl ester proceeded smoothly using hydrochloric acid in acetic acid to give carboxylic acid **69** in 88% yield (Scheme 4.9). Previously, amines were allowed to react with the carboxylic acid core in hot DMSO to deliver the C7 products; however, the difluoroborate **70**, derived from the carboxylic acid **69**, greatly increased the reactivity of the C7 position. Consequently, the displacement of the C7-F with amines was accomplished at lower temperature (Baker et al., 2004; Cecchetti et al., 1996; Domalaga et al., 1993; Ellsworth et al., 2005a,b; Hu et al., 2003). In this event, the carboxylic acid was allowed to react with boron trifluoride to deliver difluroboronate **70** in excellent yield. The thus afforded borate ester reacted with *N*-methylpiperidine in DMSO in the presence of triethylamine at ambient temperature to furnish (−)-ofloxacin (**1**, levofloxacin) in 56% yield.

The third method of resolution occurred via crystallization (Scheme 4.10) (Hayakawa et al., 1991). Benzoxazine **48** was acylated with *S*-(−)-*N*-*p*-toluenesulfonylproline (**71**) to give a mixture of diastereomers. Selective recrystallization of diastereomer **72** followed by hydrolysis of the amide with aqueous sodium hydroxide gave the (*S*)-benzoxazine **73** in 91% yield. Elaboration of **73** could then proceed by the route outlined above.

The first asymmetric synthesis of LVX, (−)-ofloxacin, was performed by Mitscher and co-workers in 1987, and gave rise to the Chu–Mitscher synthesis of LVX (Mitscher et al., 1987). The asymmetric synthesis commenced by using the Grohe–Heitzer strategy (Mitscher, 2005). First a condensation reaction between keto ester **74** and triethyl orthoformate in refluxing acetic anhydride delivered compound **75** (Scheme 4.11). Compound **75** was then allowed to react with (*S*)-(+)-2-amino-1-propanol (**76**) to afford the enantiomerically pure propionate via an addition–elimination reaction in 57% yield over two steps. The LVX core was produced in two steps. First, treatment of the propionate with sodium hydride in DMSO gave bicycle **78** in 59% yield. Subsequent treatment of **78** with potassium hydroxide produced an intermediate alkoxide, which displaced the C8-fluorine to give the LVX core in 70% yield. Selective displacement of the C7-fluorine

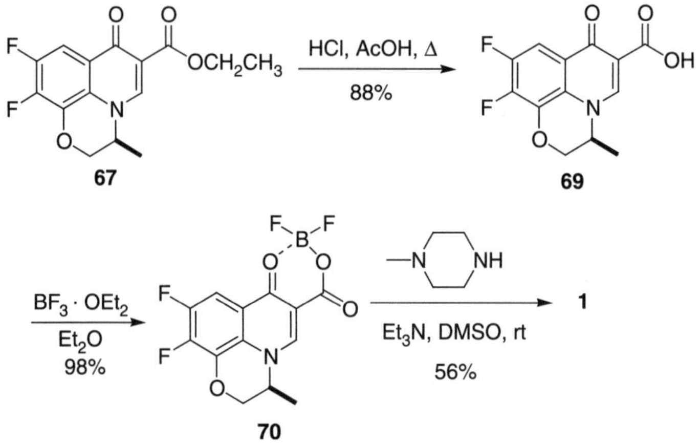

67 → (HCl, $AcOH$, Δ; 88%) → 69

→ ($BF_3 \cdot OEt_2$, Et_2O; 98%) → 70 → (N-methylpiperazine, Et_3N, DMSO, rt; 56%) → 1

Scheme 4.9.

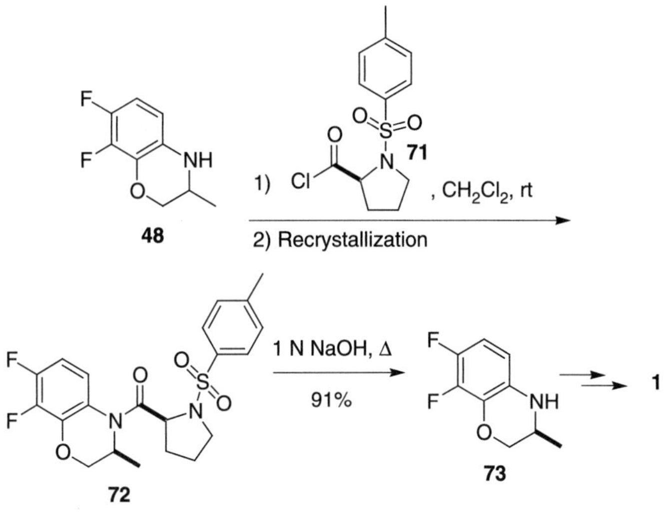

Scheme 4.10.

via an S_NAr reaction with *N*-methylpiperazine in pyridine at 120°C delivered LVX (**1**), (−)-(*S*)-ofloxacin, in 83% yield. In a recent patent application, Teva Pharmaceuticals investigated the effect of polar solvents, which included DMSO, DMA, PGME, and isobutanol, on the C7-fluorine displacement with *N*-methylpiperazine. Although all the solvents worked well, DMSO and DMA delivered the best yields, at 91% and 89%,

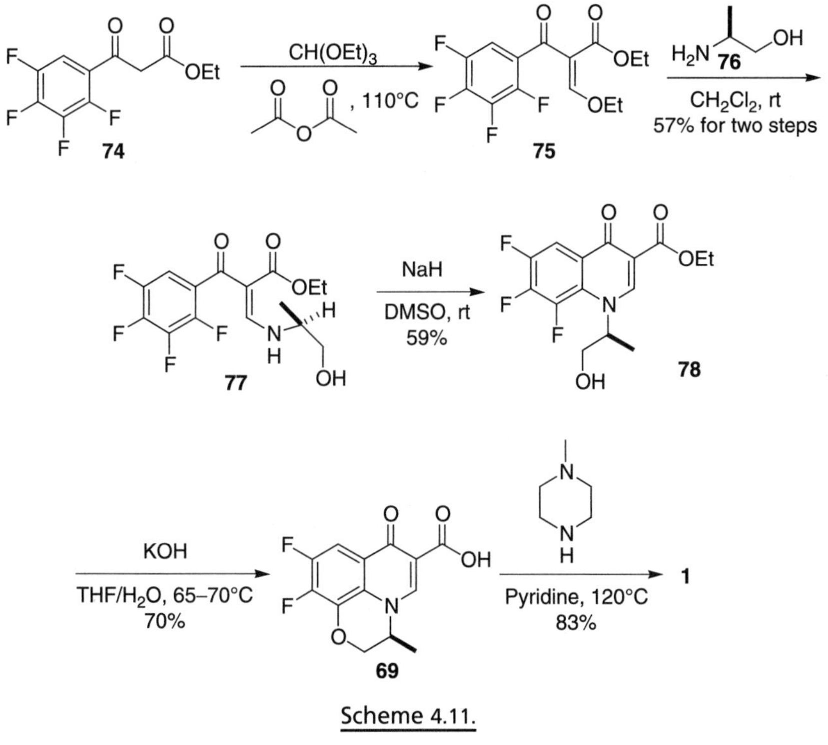

Scheme 4.11.

respectively (Niddam-Hildesheim et al., 2003). Mitscher et al. also showed that the enantiomer, (+)-(*R*)-ofloxacin could be synthesized via this route if (*R*)-(−)-2-amino-1-propanol is substituted for the (*S*)-(+)-2-amino-1-propanol.

Other synthetic approaches to levofloxacin (**1**) stemmed from introducing the asymmetric center at the C3 position of benzoxazine **73**, originally a key intermediate in the synthesis of ofloxacin. Atarshi and colleagues enantioselectively reduced imine **82**, which was synthesized in three steps from methyl ketone **79**, using chiral sodium triacyloxyborohydrides (Scheme 4.12) (Atarashi et al., 1991). They found the asymmetric reduction of **82** with borohydride **A**, derived from (*S*)-*N*-isobutyloxycarbonylproline, gave the enantiomerically enriched 1,4-benzoxazine **73** in high yield with an enantiomeric excess (*ee*) of 95%.

Kim and co-workers also commenced their approach (Scheme 4.13) with ketone **79**, but created the stereogenic center via Baker's yeast reduction to give alcohol **83** in 91% yield with an *ee* of greater than 95% (Kang et al., 1996). Reduction of the nitro group

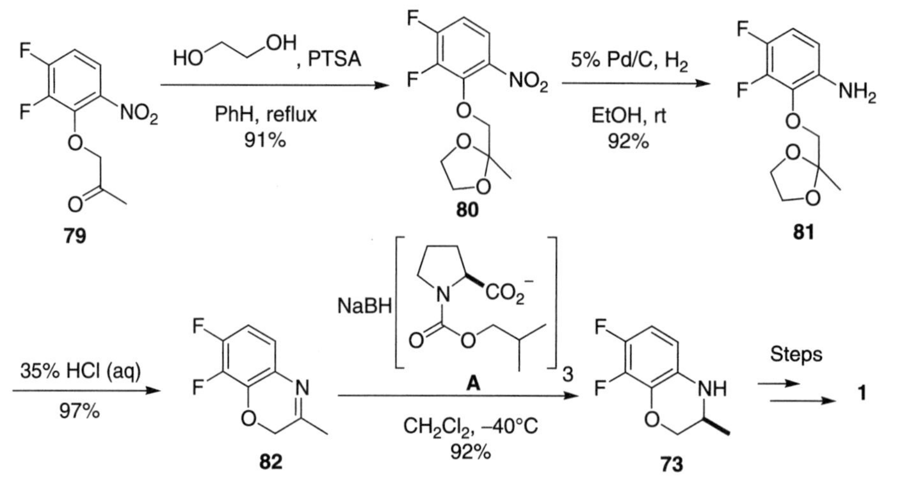

79 → (HO‑CH₂CH₂‑OH, PTSA; PhH, reflux; 91%) → 80 → (5% Pd/C, H_2; EtOH, rt; 92%) → 81 → (35% HCl (aq); 97%) → 82 → (NaBH[A]$_3$; CH_2Cl_2, −40°C; 92%) → 73 → Steps → 1

Scheme 4.12.

79 → (Baker's yeast; MeOH, H_2O, 35°C; 91%, >99% *ee*) → 83 → (Pd/C, H_2; THF, rt; 99%) → 84 → (PPh_3, DEAD, $PhCO_2H$; THF, rt; 95%) → 85 → (KCN; MeOH, rt; 99%, >99% *ee*) → 86 → (PPh_3, DEAD, $ZnCl_2$; CH_3CN, rt; 76%, >99% *ee*) → 73 → Steps → 1

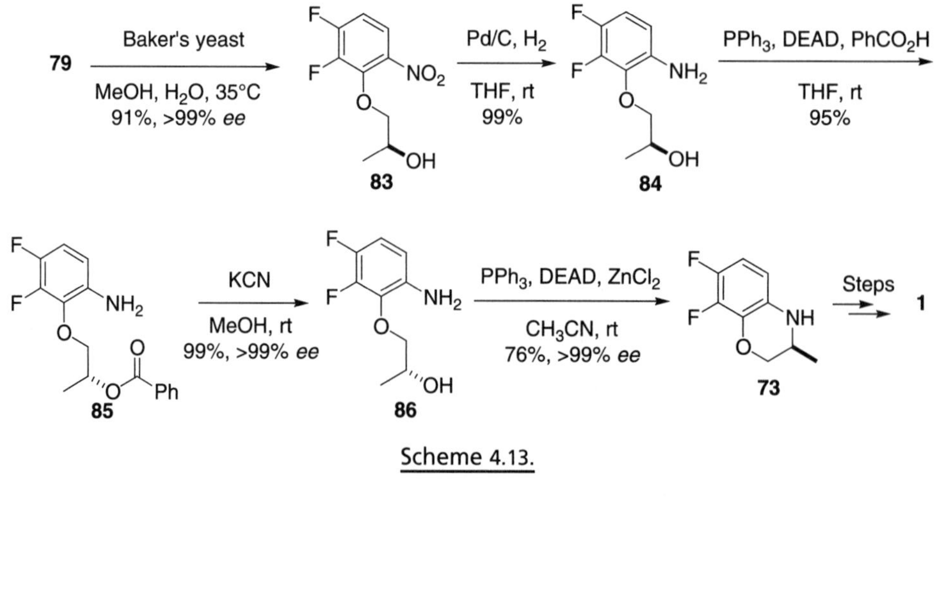

Scheme 4.13.

followed by inversion of the hydroxyl group under Mitsunobu conditions delivered ester **85** in 94% yield over two steps. The alcohol was then unmasked using potassium cyanide to give chiral alcohol **86** without compromising the stereochemistry. The stage was now set for the crucial cyclization to form the tricyclic core. In this event, alcohol **86** was converted to enantiomerically enriched 1,4-benzoxazine in the presence triphenylphosphine, DEAD, and four equivalents of zinc chloride in 75% yield with an *ee* greater then 99%. In the absence of the Lewis acid or if fewer then two equivalents were used, yields were under 25%.

Yang and co-workers introduced chirality in their opening sequence by reacting (*S*)-glycerol acetonide (**87**) with 2,3,4-trifluoro,1-nitrobenzene (**45**) to yield ether **88** (Scheme 4.14) (Yang et al., 1999). Deprotection of the acetonide under standard conditions delivered diol **89**, which was treated with hydrogen bromide in acetic acid to give a mixture

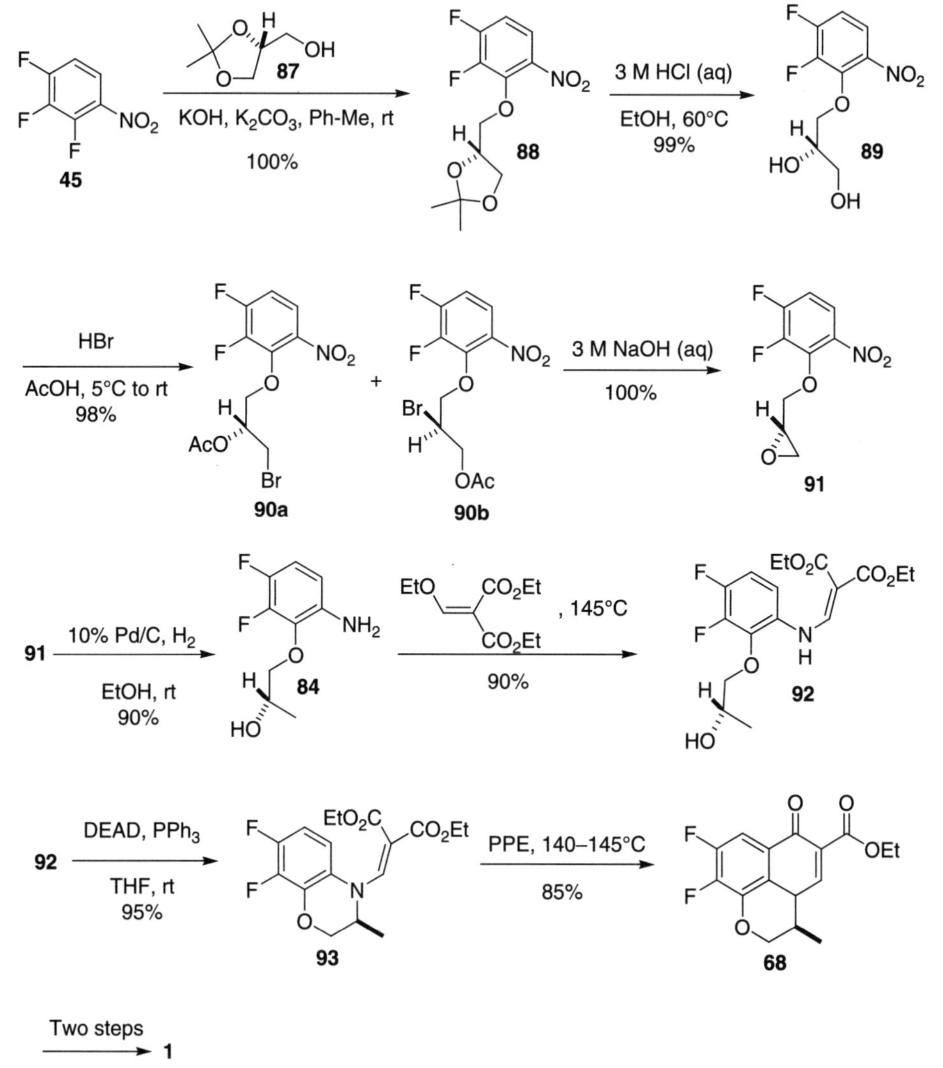

Scheme 4.14.

of acetoxy bromides **90a** and **90b** in high yield. The mixture of acetoxy bromides was then converted to epoxide **91** with aqueous sodium hydroxide in quantitative yield.

A double reduction was achieved under catalytic hydrogenation conditions to open the epoxide and reduce the nitro group to an amino group in 90% yield. The aniline thus afforded was reacted with diethylethoxymethylenemalonate to give **92**. **92** was next cyclized to the 1,4-benzoxazine **93** via a Mitsunobu reaction in the absence of a Lewis acid, unlike Kim's approach (Kang et al., 1996). Completion of the tricycle core was ultimately achieved in PPE at 140–145°C to furnish the LVX core in 85% yield. The core was converted to LVX (**1**) in two precedented steps.

Kim and co-workers described an efficient and high-yielding sequence to the penultimate intermediate of LVX (Kang et al., 1997). Their strategy involved synthesizing a chiral amino acrylate as a 4 : 1 mixture of *Z* to *E* olefin isomers. The reaction sequence

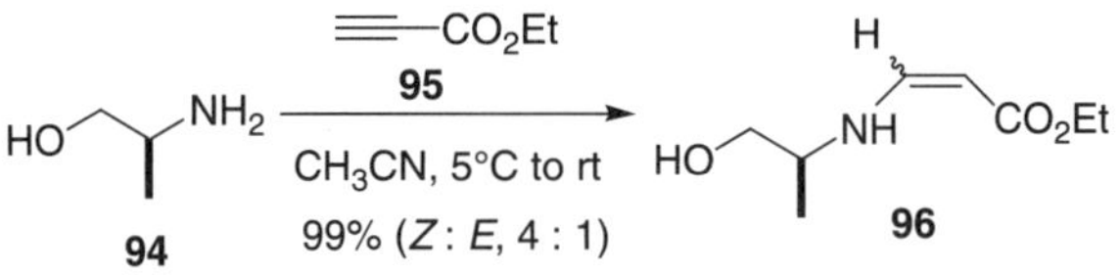

$\equiv$—CO_2Et **95**; CH_3CN, 5°C to rt; 99% (*Z* : *E*, 4 : 1)

94 → **96**

Scheme 4.15.

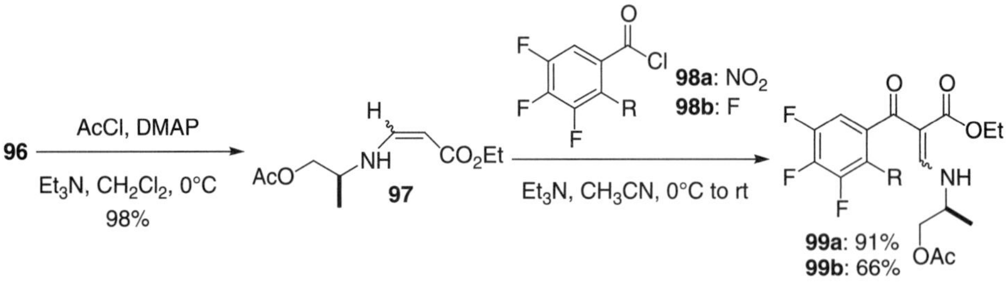

96 → AcCl, DMAP; Et_3N, CH_2Cl_2, 0°C; 98% → **97**

97 + **98a**: NO_2 / **98b**: F; Et_3N, CH_3CN, 0°C to rt → **99a**: 91%, **99b**: 66%

Scheme 4.16.

99a and **99b** → KOH; THF, 0°C to rt → [**100**] → KOH; THF, H_2O, reflux; 92% from **99a**, 90% from **99b** → **68**

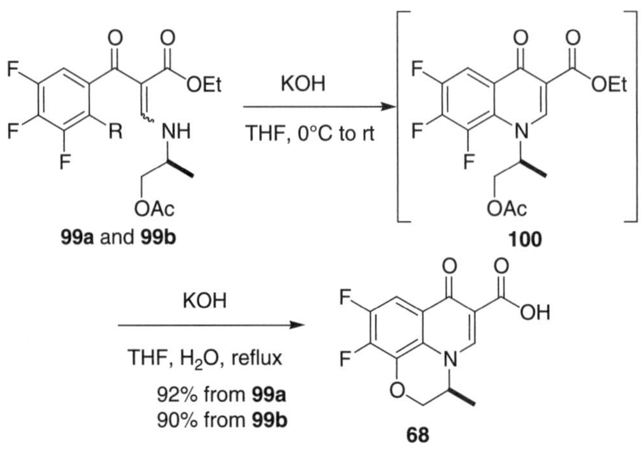

Scheme 4.17.

(Scheme 4.15) began with a Michael addition between (*S*)-2-amino-1-propanol (**94**) and ethyl propiolate (**95**) to afford chiral amino acrylate (**96**) in 99% yield.

Acrylate **96** was acylated with acetyl chloride to deliver the acetate **97** in 98% yield (Scheme 4.16). *C*-Acylation of **97** was achieved at the β-position of the acrylate in the presence of triethylamine in acetonitrile with acid chlorides **98a** and **98b** gave benzoylacrylates **99a** and **99b** in 91% and 66% yield, respectively.

Compounds **99a** and **99b** were smoothly converted to the penultimate intermediate **68** in 92% and 90% yields (Scheme 4.17). First, **99a** and **99b** were treated with potassium hydroxide in tetrahydrofuran (THF) to deliver the intermediate bicycle **100**. Addition of aqueous KOH to **100** and then refluxing the reaction gave **68** in 92% and 90% yields, respectively, to complete a formal total synthesis of levofloxacin (**1**).

4.3 MOXIFLOXACIN

Moxifloxacin (**2**) is an 8-methoxy fluoroquinolone that exhibits broad-spectrum activity against gram-positive and gram-negative microbes as well as anaerobes. The FDA approved this antibacterial agent in 1999 for community-acquired respiratory tract infections, sinusitis, acute exacerbations of chronic bronchitis and pneumonia, and uncomplicated skin and skin structure infections (Barrett, 2000; Blondeau and Hansen, 2001; Caeiro and Iannini, 2003; Keating and Scott, 2004; Miravitlles, 2005). Table 4.4 compares moxifloxacin and other marketed quinolones against microbes implicated in community-acquired respiratory tract infections (CART) and skin and skin structure infections (Keating and Scott, 2004).

2

In terms of pharmacokinetics, moxifloxacin has a high bioavailability, based on once-daily dosing of 400 mg, of 90% (Keating and Scott, 2004). The maximum plasma

TABLE 4.4. MIC_{90} Values of MXF Compared with Other Quinolones Against Bacteria Implicated in CART and Skin/Skin Structure Infections (Keating and Scott, 2004)

	MIC_{90} (μg/mL)				
Organism	MXF	CIP	LVX	GAT	SPX
S. pneumoniae (PR)[a]	0.12–0.25	1	1	0.25–0.5	0.25–0.5
S. pyogenes[a]	0.12–0.25	—	0.5–1	0.25	—
S. aureus (MR)[a]	2–4	>16–32	4–8	4	>8–32
H. influenzae[b]	0.01–0.06	≤0.02–0.03	0.01–0.06	0.02	0.01
K. pneumoniae[b]	0.03–0.25	0.03	0.12	0.06	—
M. catarrhalis[b]	0.03–0.13	≤0.02–0.13	≤0.03–0.13	0.03	0.13

[a]Gram-positive organisms; [b]Gram-negative bacteria; PR, pencillin resistant; MR, methicillin resistant; MXF, moxifloxacin, CIP, ciprofloxacin; LVX, levofloxacin; GAT, gatifloxacin; SPX, sparfloxacin.

concentration (C_{max}) is 3.2–4.5 mg/L and is reached in 1.5 h (Keating and Scott, 2004). The volume of distribution is 2.5–3.5 L/kg, with a clearance of 179 mL/min and a half-life of 11–14 h (Caeiro and Iannini, 2003). Moxifloxacin's AUC is 53.2 μg h/mL (Bryskier, 2005). Protein binding is moderate at 48% (Caeiro and Iannini, 2003). The major paths of metabolism occurring with moxifloxacin include *N*-sulfate (38%) and glucuronide (14%) formation (Keating and Scott, 2004).

The synthesis of the moxifloxacin core (de Souza, 2006; Martel et al., 1997; Seidel et al., 2000) proceeds from a Grohe–Heitzer sequence as described earlier in the chapter. Unlike the traditional Grohe–Heitzer sequence, however, the opening step involved the reaction between acid chloride **101** with the mono potassium salt of malonic acid monoethyl ester (**102**) in the presence of triethylamine to deliver ketoester **103** (Scheme 4.18). Treatment of **103** with ethyl orthoformate furnished acrylate **104**, which reacted with cyclopropyl amine to afford **105**. Cyclization of **105** in the presence of sodium fluoride in DMF gave the moxifloxicin core **106**.

The synthesis of the C7 side-chain began with the formation of anhydride **108** from pyridine-2,3-dicarboxylic acid and acetic anhydride (Scheme 4.19) (de Souza, 2006; Martel et al., 1997; Petersen et al., 1993, 1997; Seidel et al., 2000). The anhydride was then opened with benzylamine to furnish an intermediate amido-acid, which cyclized under the reaction condition to furnish the benzyl succinimide **109**. The pyridyl ring of **109** was reduced under hydrogenation conditions in the presence of 5% Pd/C to afford the dicyclic piperidine **110**. Reduction of the succinimide delivered the octahydro-pyrrolopyridine **111**. The stage was now set to produce the enantiomerically pure side-chain via optical resolution of the tartrate salts of the *cis*-isomers. First, the unwanted *cis*-(*R*,*R*) isomer was removed by forming the crystalline *L*-(+)-tartrate. After basifying the mixture, the desired *cis*-(*S*,*S*) was isolated via crystallization of its *D*-(−)-tartrate salt. The enantiomerically pure *cis*-(*S*,*S*)-octahydropyrrolopyridine **113** was finally isolated after free basing of the salt of **112** and cleavage of the benzyl group.

Displacement of the C7-F of **106** (Scheme 4.20) proceeded via chemoselective attack by the more nucleophilic pyrrolidine nitrogen rather than the piperidine nitrogen to afford the intermediate C7-octahydropyrrolopyridine. Hydrolysis of the ethyl ester with hydrochloric acid ultimately furnished moxifloxacin (**2**).

Like the synthesis of levofloxacin, the coupling between side-chain **113** and the core **106** was achieved using mild conditions (Chava et al., 2005). Ethyl ester **106** was

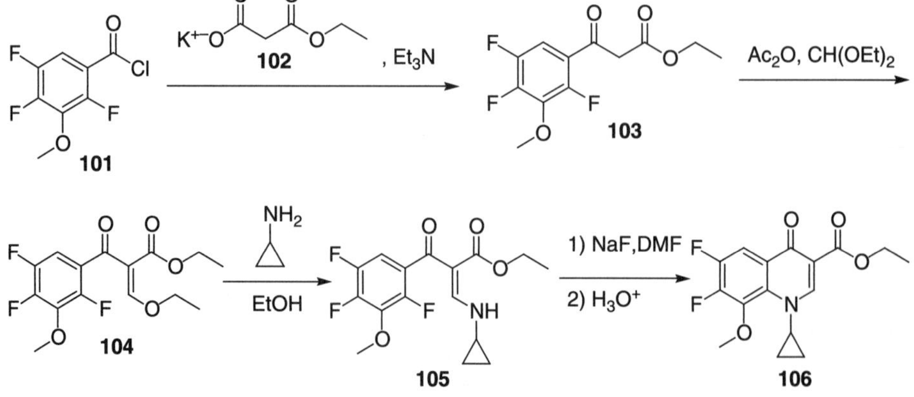

Scheme 4.18.

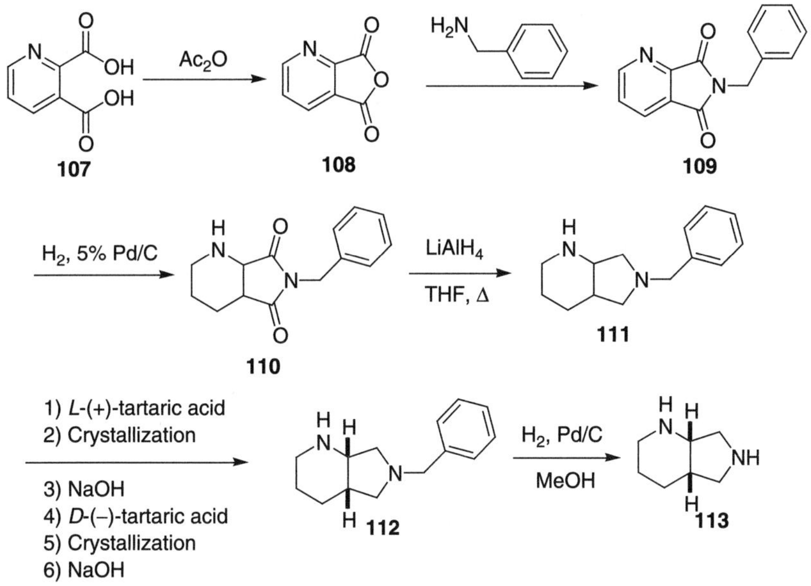

107 → (Ac_2O) → **108** → (H_2N–CH_2Ph) → **109**

→ (H_2, 5% Pd/C) → **110** → ($LiAlH_4$; THF, Δ) → **111**

→ (1) *L*-(+)-tartaric acid; 2) Crystallization; 3) NaOH; 4) *D*-(–)-tartaric acid; 5) Crystallization; 6) NaOH) → **112** → (H_2, Pd/C; MeOH) → **113**

Scheme 4.19.

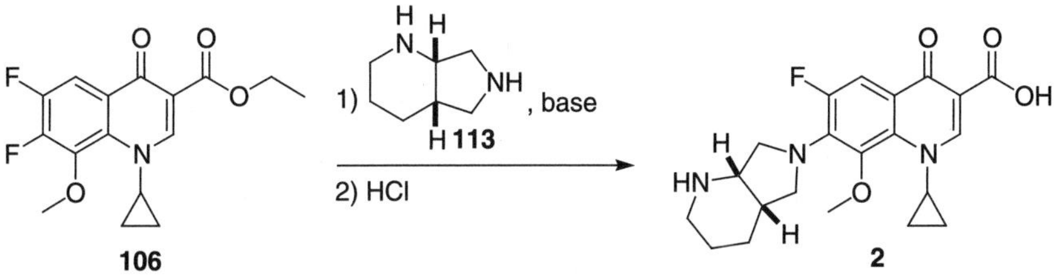

106 → (1) **113**, base; 2) HCl) → **2**

Scheme 4.20.

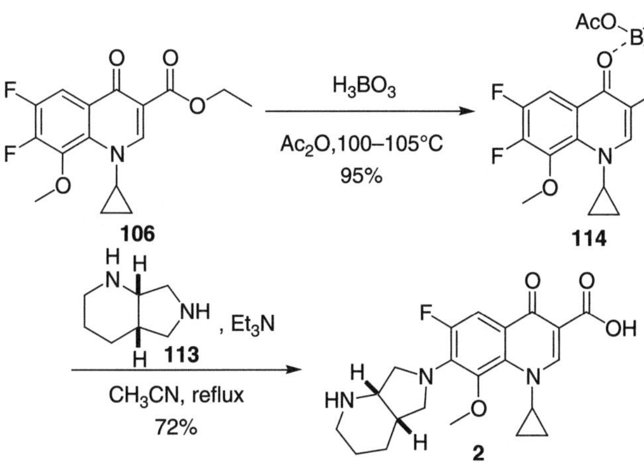

106 → (H_3BO_3; Ac_2O, 100–105°C; 95%) → **114**

→ (**113**, Et_3N; CH_3CN, reflux; 72%) → **2**

Scheme 4.21.

converted to the *bis*-acetoxyboronate in 95% yield using boric acid in hot acetic anhydride (Scheme 4.21). As discussed earlier with difluoroboronates, *bis*-acetoxy boronates (Hashimoto et al., 1996; Miyamoto et al., 1990; Morita et al., 1990) also increase the electrophilicity of the C7 position. Hence, boronate **114** reacted smoothly with **113** in refluxing acetonitrile in the presence of triethylamine to furnish **2** in 72% yield.

4.4 GEMIFLOXACIN

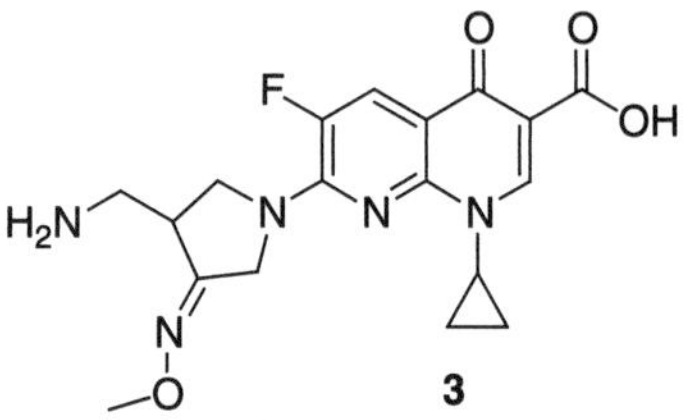

Gemifloxacin (GEM) (**3**), a naphthyridone, was approved in the United States in 2003 for the treatment of acute exacerbations of chronic bronchitis (AECB) and mild community-acquired pneumonia (CAP) (File and Tillotson, 2004). The pathogens most often linked to AECB and CAP are *S. pneumoniae*, *H. influenzae*, and *M. catarrhalis* (only CAP). Gemifloxacin is considered the most bactericidal quinolone with respect to MICs (Appelbaum et al., 2004). Summarized in Table 4.5 are the MIC_{90} values for GEM compared to other marketed quinolones against three important extracellular pathogens, including the Gram-positive *S. pneumoniae* and the Gram-negative *H. influenzae*, and *M. catarrhalis* (Bhavnani and Andes, 2005). Against these organisms GEM has MIC values significantly lower than the other four agents, particularly levofloxacin (LVX) and ciprofloxacin (CIP) (the two oldest of these agents). In addition to the extracellular pathogens, three intracellular microbes have been implicated in pneumonia, including *L. pneumophilia* (nosocomial and community), *C. pneumoniae* (community), and *M. pneumoniae* (community) (Bhavnani and Andes, 2005). Again, GEM showed superior results as compared to CIP, LVX, and GAT.

TABLE 4.5. MIC_{90} of GEM Compared with Other Marketed Quinolones Against Respiratory Microbes Associated with Pneumonia

	MIC_{90} (μg/mL)				
Microbe	GEM	CIP	LVX	GAT	MXF
S. pneumoniae[a]	0.015–0.06	1–4	1–2	0.50	0.12–0.25
H. influenzae[b]	≤0.004–0.06	0.008–0.03	0.015–0.12	0.015–≤0.03	<0.03–0.06
M. catarrhalis[b]	0.008–0.03	≤0.016–≤0.5	≤0.03–0.06	≤0.03–0.03	0.03–0.125
L. pneumophilia[b]	0.016	0.03	0.016	—	0.016
C. pneumoniae[c]	0.25	—	1	—	1
M. pneumoniae[c]	0.125	—	0.5	—	—

[a]Gram-positive bacteria; [b]Gram-negative bacteria; [c]Other organisms; GEM, gemifloxacin; CIP, ciprofloxacin; LVX, levofloxacin; GAT, gatifloxacin; MXF, moxifloxacin.

In terms of pharmacokinetics, with a single daily dose of 320 mg, GEM has a bioavailability (%F) of 71% (Yoo et al., 2004), an AUC of 9.30 mg h/L (File and Tillotson, 2004), a half-life of 6.65 h (File and Tillotson, 2004), peak plasma concentration (C_{max}) of 1.48–1.6 mg/L (File and Tillotson, 2004), T_{max} of 1 h (File and Tillotson, 2004), a volume of distribution of 4.97 L/kg (Bhavnani and Andes, 2005), and a clearance of 151 mL/min (Bhavnani and Andes, 2005). Protein binding is moderate at 55–73% (Bryskier, 2005). As with other fluoroquinolones, its bioavailabilty (%F) is decreased significantly in the presence of antacids (Bryskier, 2005). The primary observed metabolites include the *E*-isomer (4–6%), the acyl glucuronide (2–6%), and *N*-acetyl gemifloxacin (2–5%) (Yoo et al., 2004).

The naphthyridine core of GEM **122** has not been reported in the papers concerning it; however, the standard Grohe–Heitzer sequence described earlier in the chapter will produce the required template (Bouzard et al., 1992; Matsumoto et al., 1985; Mich et al., 1987; Sanchez et al., 1988; Shin et al., 2004). The reaction sequence opened with hydrolysis of arylnitrile **115** to afford carboxylic acid **116** in high yield (Scheme 4.22) (Matsumoto et al., 1985). The carboxylic acid was then converted to the acid chloride, which was treated with diethyl ethoxymagnesiummalonate to yield an intermediate nicotinoylmalonate **118**. This intermediate underwent decarboxylation in the presence of

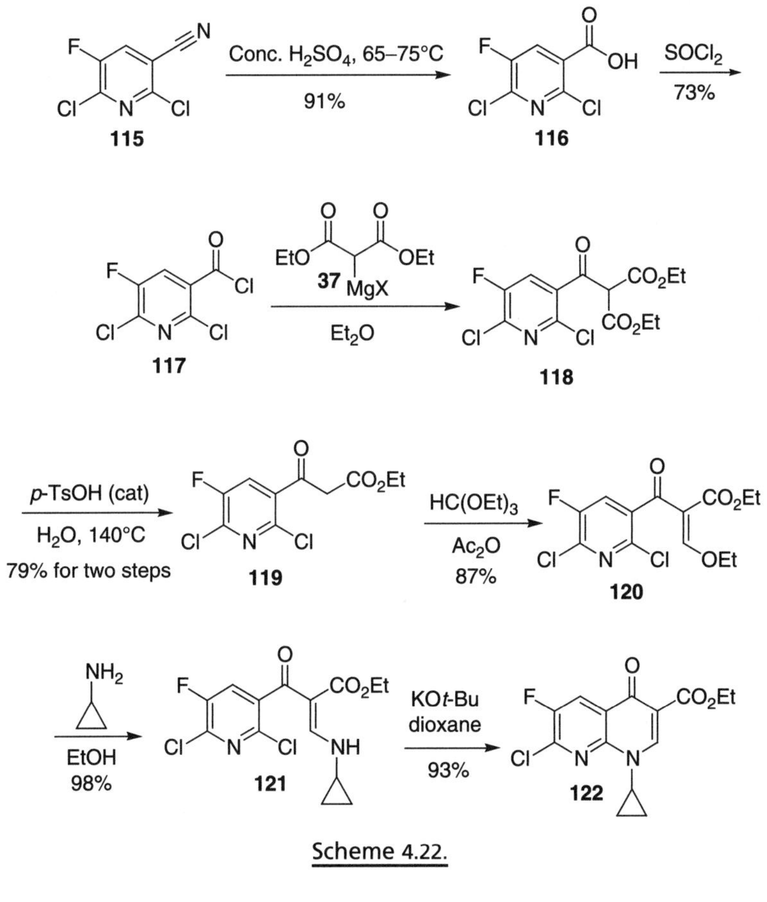

115 → (Conc. H_2SO_4, 65–75°C, 91%) → 116 → ($SOCl_2$, 73%) → 117 → (37 MgX, Et_2O) → 118 → (p-TsOH (cat), H_2O, 140°C, 79% for two steps) → 119 → ($HC(OEt)_3$, Ac_2O, 87%) → 120 → (cyclopropylamine, EtOH, 98%) → 121 → (KO*t*-Bu, dioxane, 93%) → 122

Scheme 4.22.

p-toluenesulfonic acid to deliver the keto ester **119** in 79% yield. The ethoxyacrylate **120** was produced in high yield by treating **119** with ethyl orthoformate; the acrylate afforded thereby was treated with cyclopropyl amine to give aminoacrylate **121** in 98% yield. The aminoacrylate was then cyclized efficiently to the naphthyridine core **122** with a catalytic amount of potassium *tert*-butoxide in 93% yield.

The synthesis of the GEM C7 side-chain commenced with a 1,4-conjugate addition reaction between ethylglycine hydrochloride **123** and acrylonitrile in the presence of sodium hydroxide to deliver acrylonitrile adduct **124** in 48% yield (Scheme 4.23) (Graul and Castañer, 1998; Hong et al., 1997a,b,c, 1998; Hong, 2001; Jeong, 1995). The secondary amine was easily converted to the BOC carbamate, and the resulting product was cyclized to α-cyano ketone **125** with sodium ethoxide in 92% yield over two steps. Numerous attempts to selectively reduce the nitrile under catalytic hydrogenation conditions delivered the desired product, albeit in low yield. Thus, a two-step reduction approach was undertaken. First, **125** was treated with sodium borohydride to give the secondary alcohol in quantitative yield. Subsequent reduction of the cyano group of **126** with lithium aluminum hydride and protection of the resulting amine gave crucial intermediate **127** in 83% yield.

As the side-chain contains a methyloxime group, alcohol **127** was oxidized to the ketone using the Parikh–Doering oxidation (Scheme 4.24). Ketone **128** was then transformed exclusively to methyloxime **129** under standard conditions, using methoxylamine hydrochloride in the presence of sodium bicarbonate, in 88% yield. This assignment was based on a literature thermodynamic argument and derivatization by converting oxime **129** to the corresponding dicyclic pyrazoline. Cleavage of the BOC group was achieved using acetyl chloride in refluxing methanol to give the dihydrochloride salt **130** in 94% yield.

The first manufacturing route of the GEM side-chain relied on α-cyanoketone **125**; however, the number of chemical steps from **125** to the final side-chain was reduced by one step (Noh et al., 2004a). The sequence began with a selective hydrogenation with Raney nickel followed by double bond migration to enamine **131** (Scheme 4.25). The amino functionality of **131** was then monoprotected, and the double bond was reduced under hydrogenation conditions to afford pyrrolidine-3-one **133**. Treatment of **133** with methoxylamine yielded methoxyoxime **129**. Deprotection of the carbamate functionality was achieved with methanesulfonic acid to afford the C7-side-chain as the *bis*-methansulfonate salt.

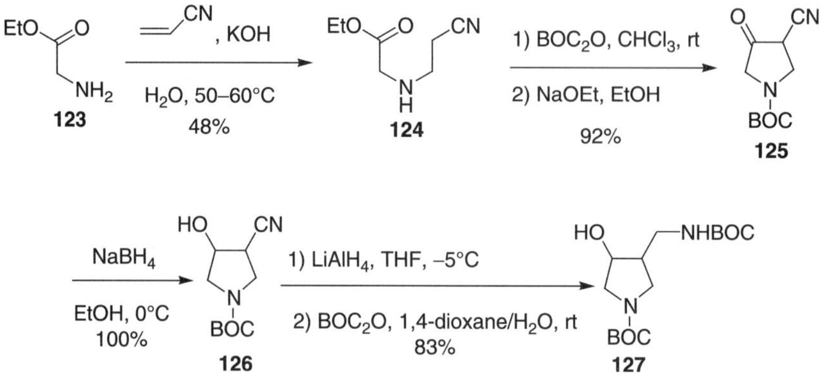

Scheme 4.23.

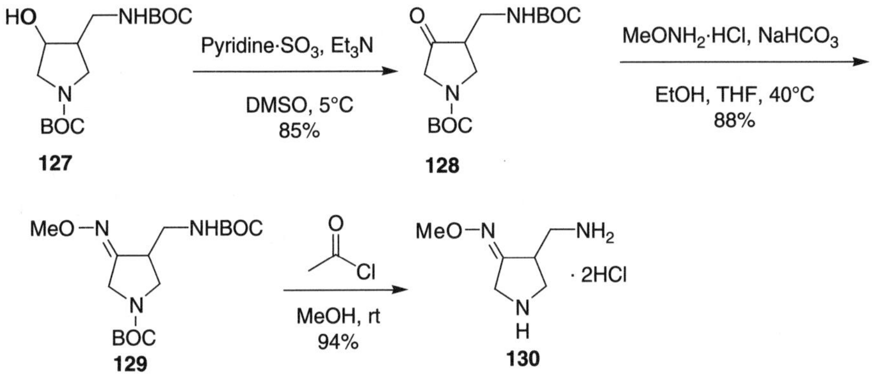

127 → 128: Pyridine·SO_3, Et_3N; DMSO, 5°C; 85%

128 → 129: $MeONH_2$·HCl, $NaHCO_3$; EtOH, THF, 40°C; 88%

129 → 130 (·2HCl): acetyl chloride; MeOH, rt; 94%

Scheme 4.24.

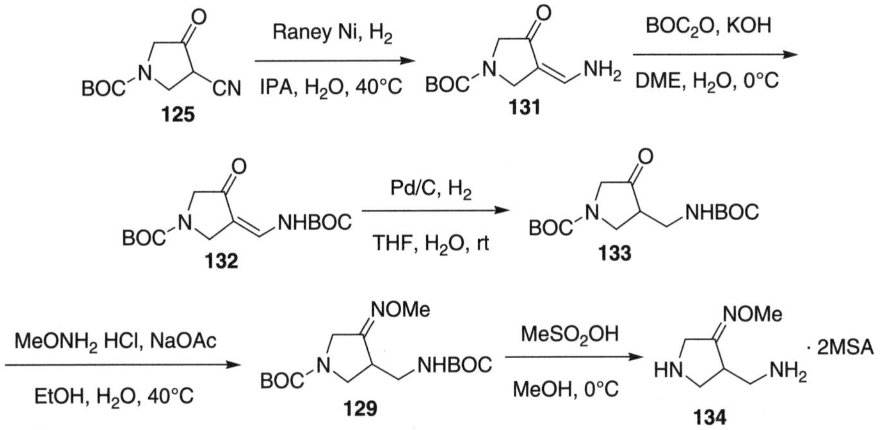

125 → 131: Raney Ni, H_2; IPA, H_2O, 40°C

131 → 132: BOC_2O, KOH; DME, H_2O, 0°C

132 → 133: Pd/C, H_2; THF, H_2O, rt

133 → 129: $MeONH_2$ HCl, NaOAc; EtOH, H_2O, 40°C

129 → 134 (·2MSA): $MeSO_2OH$; MeOH, 0°C

Scheme 4.25.

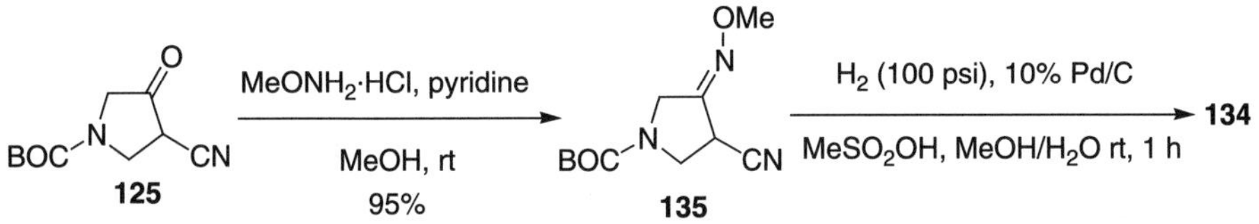

125 → 135: $MeONH_2$·HCl, pyridine; MeOH, rt; 95%

135 → 134: H_2 (100 psi), 10% Pd/C; $MeSO_2OH$, MeOH/H_2O rt, 1 h

Scheme 4.26.

122 → 3: **130**, DBU; CH_3CN, rt; 85%

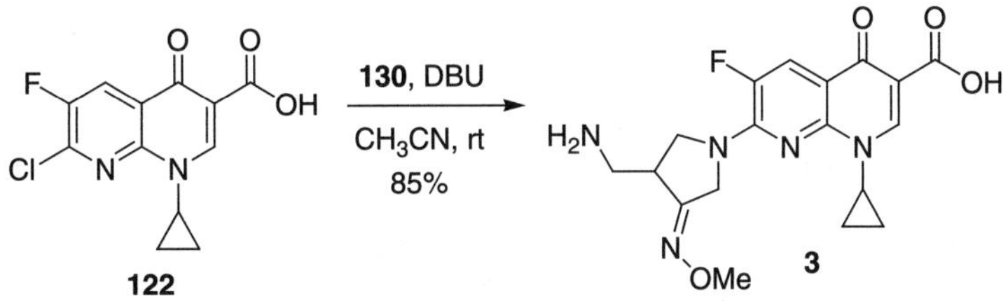

Scheme 4.27.

Recently, the manufacturing route was streamlined to three steps from the α-cyano ketone **125** (Scheme 4.26) (Noh et al., 2004b). First the α-cyano pyrrolidinone was converted to methyl oxime **135** in 95% yield. After an extensive survey of catalysts, solvents, and acids, selective hydrogenation of the cyano group was achieved using 10% palladium/carbon in methanol in the presence of methanesulfonic acid to deliver **134** in high yield (Hashimoto et al., 1996). The authors noted that the chemoselective hydrogenation could also be performed in anhydrous methanol; however, other solvents such as DME, ethanol, isopropyl alcohol, and THF required the addition of varying amounts of water.

The coupling reaction between **122** and **130** proceeded smoothly in the presence of DBU in acetonitrile at ambient temperature to give gemifloxacin **3** in 85% yield (Scheme 4.27).

4.5 GARENOXACIN (T-3811): A PROMISING CLINICAL CANDIDATE

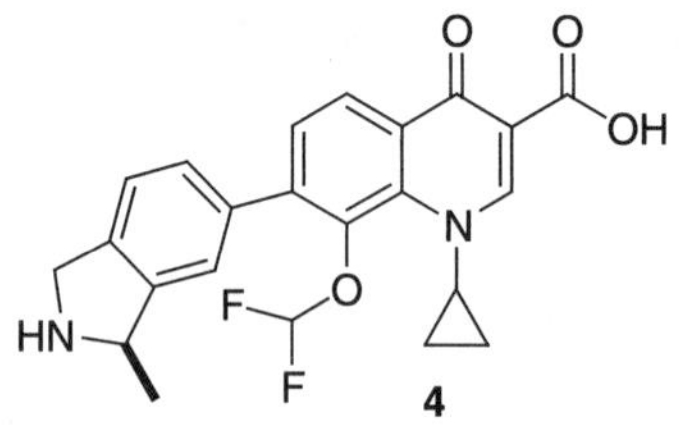

Garenoxacin (**4**) is a promising novel des-F(6) quinolone that is currently in phase III clinical trials with a broad range of activity against Gram-positive and Gram-negative bacteria, and anaerobes (Table 4.6) (Christiansen et al., 2004; Graul et al., 1999; Noviello et al., 2003). In February 2006, the U.S. FDA accepted Schering-Plough's NDA for review (Schering-Plough, 2006).

In terms of chemical structure, garenoxacin contains an isoindoline moiety at C7 via a sp^2C-sp^2C bond, unlike most quinolones, which have a sp^2C-N bond. This brief section will only discuss the incorporation of the C7 side-chain (Hayashi et al., 2002) and not the

TABLE 4.6. MIC_{90} of GAR Compared with Other Quinolones Against Gram-Positive and Gram-Negative Bacteria

	MIC_{90} (μg/mL) (Christiansen et al., 2004)					
Bacterium	GAR	CIP	LVX	GAT	MXF	GEM
MRSA[a]	2	32	8	—	2	—
ORSA[a]	4	>2	>4	>4	—	4
S. pneumoniae[a]	0.06	2	1	5	0.25	≤0.03
S. pyogenes[a]	0.12	1	1	0.25	—	≤0.03
H. influenzae[b]	≤0.03	≤0.25	≤0.03	≤0.03	≤0.03	≤0.03
M. catarrhalis[b]	0.03	≤0.25	0.06	≤0.03	0.06	≤0.03

[a]Gram-positive; [b]Gram-negative; MRSA, methicillin-resistant *Staphylococcus aureus*; ORSA, oxacillin- resistant *Staphylococcus aureus*; GAR, garenoxacin; CIP, ciprofloxacin; LVX, levofloxacin; GAT, gatifloxacin; MXF, moxifloxacin; GEM, gemifloxacin.

synthesis of the core **136** (Graul et al., 1999; Todo et al., 2000).

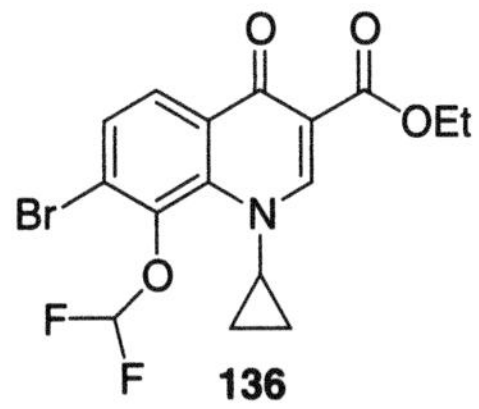

136

Two strategies have been disclosed concerning the incorporation of the (*R*)-methylisoindoline side-chain **138**, which is derived from racemic methylisoindoline **137** via two chemical steps and chromatographic separation of diastereomers (Graul et al., 1999), to the garenoxacin core (Graul et al., 1999; Hayashi et al., 2002). These approaches include a Suzuki reaction and Stille coupling (Scheme 4.28).

In the Suzuki strategy **138** was treated with trityl chloride to afford amine **139** in 84% yield (Scheme 4.29) (Todo et al., 2000). Metal–halogen exchange of 5-bromoisoindoline **139** was achieved using *n*-butyl lithium at −65°C to deliver an intermediate organolithium species that was trapped with triisopropylborate to afford the boronic acid upon work-up. Further treatment of the boronic acid with diethanolamine gave diethanolamine boronic ester **140** (Hayashi et al., 2002). Presumably, the diethanolamine boronic ester was formed for stability reasons, as a given diethanolamine boronic ester tends to be less

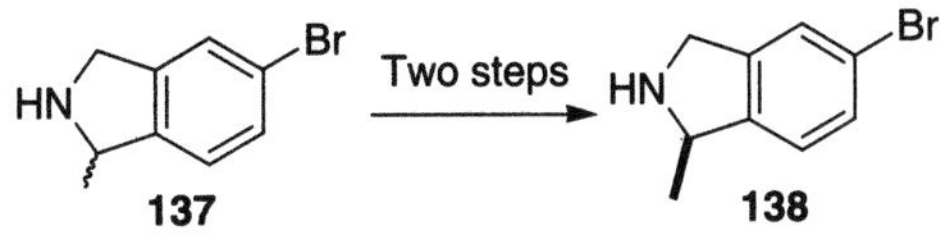

Two steps

137 **138**

Scheme 4.28.

138 → Ph_3CCl, Et_3N, CH_2Cl_2, 84% → **139**

139 → 1) *n*-BuLi, THF, −65°C; 2) *i*-$(PrO)_3B$; 3) $HN(CH_2CH_2OH)_2$, 87% → **140**

140 → 1) AcOH, DME, H_2O, rt; 2) **136**, Na_2CO_3, $PdCl_2(PPh_3)_2$, reflux, 92% → **141**

141 → HCl, EtOH, H_2O, rt, 86% → **4**

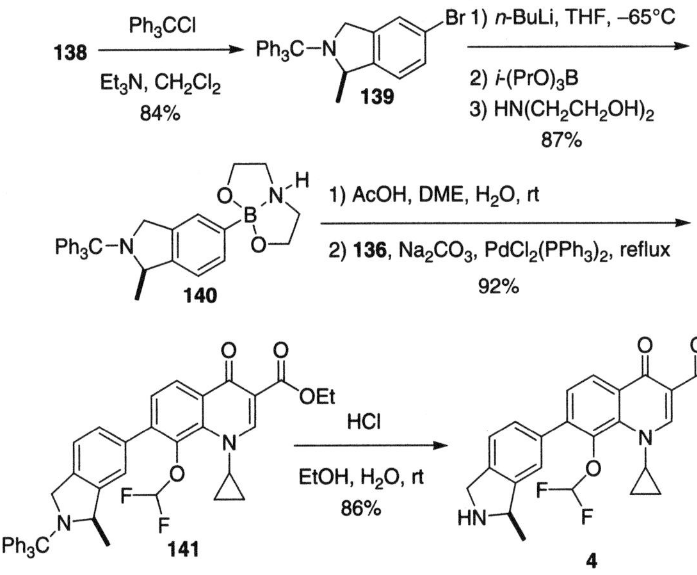

Scheme 4.29.

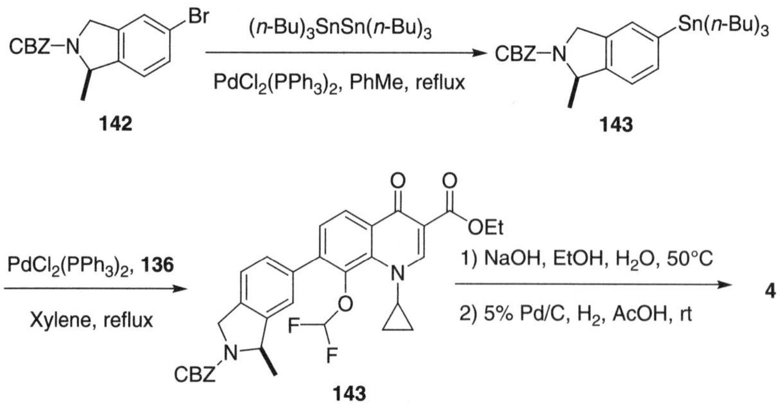

Scheme 4.30.

susceptible to hydrolysis and air oxidation than its parent boronic acid (Hall, 2005). Boronate **140** was treated with AcOH in dimethoxyethane and water; the resulting borate species reacted smoothly with bromoquinolone **136** in the presence of palladium *bis*(triphenylphosphine) dichloride [$Pd(PPH_3)_2Cl_2$] and sodium carbonate to give **141** in 92% yield. Exposure of **141** to hydrochloric acid cleaved the trityl group and hydrolyzed the ethyl ester to deliver garenoxacin (**4**) in 86% yield (Hayashi et al., 2002).

Alternatively, **138** can be converted to CBZ carbamate **142** using standard conditions (Scheme 4.30). This 5-bromoisoindoline was transformed to stannane **143** in one step with bistributyltin in the presence of a palladium catalyst. The stannane was treated with **136** under Stille conditions to afford coupled product **144**. The resulting ester was hydrolyzed with sodium hydroxide, and the CBZ group was removed under hydrogenolysis conditions to deliver garenoxacin (**4**) (Hayashi et al., 2002).

REFERENCES

Andriole, V. T. (Ed.) (2000a). *The Quinolones*, Academic Press, San Diego.

Andriole, V. T. (2000b). In *The Quinolones*, Andriole, V. T. (Ed.) Academic Press, San Diego, p. 3.

Andriole, V. T. (2003). In *Antibiotic and Chemotherapy Antiinfective Agents and their Use in Therapy*, Finch, R. G. et al. (Eds) Churchill Livingstone, Edinburgh, (29): 349–373.

Andriole, V. T. (2005). *Clin. Infect. Dis.*, 41: S113–S119.

Appelbaum, P. C., Gillespie, S. H., Burley, C. J., and Tillotson, G. S. (2004). *Int. J. Antimicrob. Agents*, 23: 533–546.

Atarashi, S., Tsurumi, H., Fujiwara, T., and Hayakawa, I. (1991). *J. Heterocyclic Chem.*, 28: 329–331.

Baker, W. R., Cai, S., Dimitroff, M., Fang, L., Huh, K. K., Ryckman, D. R., Shang, X., Shawar, R. M., and Therrien, J. H. (2004). *J. Med. Chem.*, 47: 4693–4709.

Ball, P. (2000). *J. Antimicrob. Chemother.*, 46: 17–24.

Barrett, J. F. (2000). *Curr. Opin. Invest. Drugs*, 1: 45–51.

Bhavnani, S. M. and Andes, D. R. (2005). *Pharmacotherapy*, 25: 717–740.

Blondeau, J. M. and Hansen, G. T. (2001). *T. Expert. Opin. Pharmacother.*, 2: 317–335.

Blondeau, J. M. and Missaghi, B. (2004). *Expert Opin. Pharmacother.*, 5: 1117–1152.

Bouzard, D., Di Cesare, P., Essiz, M., Jacquet, J. P., Ledoussal, B., Remuzon, P., Kessler, R. E., and Fung-Tomc, J. (1992). *J. Med. Chem.*, 35: 518–525.

Brighty, K. E. and Gootz, T. D. (2000). *The Quinolones.* In Andriole, V. T. (Ed.) Academic Press, San Diego, (2): 36–43.

Bryskier, A. (2005). In *Antimicrobial Agents: Antibacterials and Antifungals*, Bryskier, A. (Ed.) ASM Press, Washington, D.C., (26): 668–788.

Bryskier, A. (2005). In *Antimicrobial Agents: Antibacterials and Antifungals*, Bryskier, A. (Ed.) ASM Press, Washington, D.C., (26): 750.

Bryskier, A. (2005). In *Antimicrobial Agents: Antibacterials and Antifungals*, Bryskier, A. (Ed.) ASM Press, Washington, D.C., (26): 759.

Caeiro, J.-P. and Iannini, P. B. (2003). *Expert Rev. Anti-infect. Ther.*, 1: 363–370.

Cecchetti, V., Fravolini, A., Lorenzini, M. C., Tabarrini, O., Terni, P., and Xin, T. (1996). *J. Med. Chem.*, 39: 436–445.

Chava, S., Gorantla, S. R., Vasireddy, U. R., and Dammalapati, V. L. N. (2005). WO2005012285A1.

Christiansen, K. J., Bell, J. M., Turnidge, J. D., and Jones, R. N. (2004). *Antimicrob. Agents Chemother.*, 48: 2049–2055.

Curran, T. T. (2005). In *Reactions in Heterocyclic Chemistry*, Li, J.-J. (Ed.) Wiley & Sons, Inc., Hoboken, NJ, 423–436.

Da Silva, A. D., De Almeida, M. V., De Souza, M. V. N., and Couri, M. R. C. (2003). *Curr. Med. Chem.*, 10: 21–39.

De Souza, M. V. N. (2005). *Mini-Rev. Med. Chem.*, 5: 1009–1017.

De Souza, M. V. N. (2006). *Recent Pat. Anti-infect. Drug Discovery*, 1: 33–44.

Domagala, J. M. (1994). *J. Antimicrob. Chemother.*, 33: 685–706.

Domagala, J. M. and Hagen, S. E. (2003). In *Quinolone Antimicrobial Agents*, Hooper, D. C. and Rubenstein, E. (Eds) ASM Press, Washington, D.C., (1): 3–18.

Domagala, J. M., Hagen, S. E., Joannides, T., Kiely, J. S., Laborde, E., Schroeder, M. C., Sesnie, J. A., Shapiro, M. A., Suto, M. J., and Vanderroest, S. (1993). *J. Med. Chem.*, 36: 871–882.

Drlica, K. and Hooper, D. C. (2003). In *Quinolone Antimicrobial Agents*, Hooper, D. C. and Rubenstein, E. (Eds) ASM Press, Washington, D.C., (8): 19–40.

Drlica, K. and Malik, M. (2003). *Curr. Topics Med. Chem.*, 3: 249–282.

Drugs Fut. (1992). 17: 559.

Dudley, M. (2003a). In *Quinolone Antimicrobial Agents*, Hooper, D. C. and Rubenstein, E. (Eds) ASM Press, Washington D.C., (2): 115–132.

Dudley, M. (2003b). In *Quinolone Antimicrobial Agents*, Hooper, D. C. and Rubenstein, E. (Eds) ASM Press, Washington D.C., (2): 117.

Dudley, M. (2003c). In *Quinolone Antimicrobial Agents*, Hooper, D. C. and Rubenstein, E. (Eds) ASM Press, Washington D.C., (2): 136.

Ellsworth, E. L., Limberakis, C., and Taylor, C. B. (2005a). WO 2005026145A2.

Ellsworth, E. L., Taylor, C. B., Murphy, S. T., Rauckhorst, M. R., Starr, J. T., Hutchings, K. M., Limberakis, C. L., and Hoyer, D. W. (2005b) WO 2005049602 A1.

FDA (2004). www.fda.gov/cder (Drugs@FDA): FDA letter to J&J dated November 24, 2004 concerning Levaquin®.

FDA (2006). www.fda.gov/cder; News item: March 7, 2006.

File, T. M., Jr. and Tillotson, G. S. (2004). *Expert. Rev. Anti. Infect. Ther.*, 2: 831–843.

Fish, D. N. (2003). *Expert Rev. Anti-infect. Ther.*, 1: 371–387.

Graul, A. and Castañer, J. (1998). *Drugs Fut.*, 23: 1199–1204.

Graul, A., Rabasseda, X., and Castañer, J. (1999). *Drugs Fut.*, 24: 1324–1331.

Hall, D. G. (2005). *Boronic Acids: Preparation and Applications in Organic Synthesis and Medicine*. Wiley-VCH Verlag GmbH, Weinheim, (1): 17–18.

Hashimoto, K., Okaichi, Y., Nomi, D., Miyamoto, H., Bando, M., Kido, M., Fujimura, T., Furuta, T., and Minamikawa, J. (1996). *Chem. Pharm. Bull.*, 44: 642–645.

Hayakawa, I., Atarashi, S., Imamura, M., Yokohama, S., Higashihashi, N., Sakano, K., and Ohshima, M. (1991). US Patent 5053407.

Hayakawa, I., Atarashi, S., Yokohama, S., Imamura, M., Sakano, K., and Furukawa, M. (1986). *Antimicrob. Agents Chemother.*, 29: 163–164.

Hayakawa, I., Hiramitsu, T., and Tanaka, Y. (1984). *Chem. Pharm. Bull.*, 32: 4907–4913.

Hayashi, K., Takahata, M., Kawamura, Y., and Todo, Y. (2002). *Arzneim.-Forsch./Drug Res.*, 52: 903–913.

Hoffman-Roberts, H. L., Babcock, E. C., and Mitropoulos, I. F. (2005). *Expert Opin. investig. Drugs*, 14: 973–995.

Hong, C. Y. (2001). *Farmaco*, 56: 41–44.

Hong, C. Y., Kim, Y. K., Chang, J. H., Kim, S. H., Choi, H., Kwak, J. H., Nam, D. H., Jeong, Y. N., Oh, J. I., and Kim, M. Y. (1997a). US 5633262.

Hong, C. Y., Kim, Y. K., Chang, J. H., Kim, S. H., Choi, H., Nam, D. H., Kwak, J. H., Jeong, Y. N., Oh, J. I., and Kim, M. Y. (1997b). US 5698570.

Hong, C. Y., Kim, Y. K., Chang, J. H., Kim, S. H., Choi, H., Nam, D. Y., Kim, Y. Z., and Kwak, J. H. (1997c) *J. Med. Chem.*, 40: 3584–3593.

Hong, C. Y., Kim, Y. K., Kim, S. H., Chang, J. H., Choi, H., Nam, D. H., Kim, A. R., Lee, J. H., and Park, K. S. (1998). US 5776944.

Hooper, D. C. (2003). In *Quinolone Antimicrobial Agents*, Hooper, D. C. and Rubenstein, E. (Eds) ASM Press, Washington D.C., (3): 41–67.

Hooper, D. C. and Rubenstein, E. (Eds) (2003). *Quinolone Antimicrobial Agents*, ASM Press, Washington, D.C.

Howe, R. and MacGowan, A. (2004). In *Infectious Diseases*, Cohen, J. and Powderly, W. G. (Eds) Mosby, Edinburgh, (198): 1827.

Howe, R. and MacGowan, A. (2004). In *Infectious Diseases*, Cohen, J. and Powderly, W. G. (Eds) Mosby, Edinburgh, (198): 1827–1833.

Howe, R. and MacGowan, A. (2004). In *Infectious Diseases*, Cohen, J. and Powderly, W. G. (Eds) Mosby, Edinburgh, (198): 1833.

Hu, X. E., Kim, N. K., Gray, J. L., Almstead, J.-I. K., Seibel, W. L., and Ledoussal, B. (2003). *J. Med. Chem..*, 46: 3655–3661.

Hurst, M., Lamb, H. M., Scott, L. J., and Figgitt, D. P. (2002). *Drugs*, 62: 2127–2167.

IMS Health World Review (2005). www.imshealth.com.

Jeong, Y. N. and Oh, J. I. (1995). EP 0688772A1.

Kang, S. B., Ahn, E. J., Kim, Y., and Kim, Y. H. (1996). *Tetrahedron Lett.*, 37: 9317–9320.

Kang, S. B., Park, S., Kim, Y. H., and Kim, Y. (1997). *Heterocycles*, 45: 137–145.

Keating, G. M. and Scott, L. J. (2004). *Drugs*, 64: 2347–2377.

Leone, R., Venegoni, M., Motola, D., Moretti, U., Piazzetta, V., Cocci, A., Resi, D., Mozzo, F., Velo, G., Burzilleri, L., Montanaro, N., and Conforti, A. (2003). *Drug Safety*, 26: 109–120.

Li, J.-J., Johnson, D. S., Sliskovic, D. R., and Roth, B. D. (2004). *Contemporary Drug Synthesis*, Wiley & Sons, Inc., Hoboken NJ, (7): 75–83.

Martel, A. M., Leeson, P. A., Castañer, J. (1997). *Drugs Fut.*, 22: 109–113.

Mascaretti, O. A. (2003). *Bacteria versus Antibacterial Agents an Integrated Approach*, ASM, Washington, D.C., 303–304.

Mascaretti, O. A. (2003). In *Bacteria versus Antibacterial Agents An integrated Approach*, ASM Press, Washington, D.C., (23): 295–309.

Matsumoto, J., Nakamura, S., Miyamoto, T., and Uno, H. (1985). EP Appl. 0132845A2.

Mich, T. F., Sanchez, J. P., Domagala, J. M., and Trehan, A. K. (1987). US 4663457.

Miravitlles, M. (2005). *Expert. Opin. Pharmacother.*, 6: 283–293.

Mitscher, L. A. (2005). *Chem. Rev.*, 105: 559–592.

Mitscher, L. A., Sharma, P. N., Chu, D. T. W., Shen, L. L., and Pernet, A. G. (1987). *J. Med. Chem.*, 30: 2283–2286.

Miyamoto, H., Ueda, H., Otsuka, T., Aki, S., Tamaoka, H., Tominaga, M., and Nakagawa, K. (1990). *Chem. Pharm. Bull.*, 38: 2472–2475.

Morita, S., Otsubo, K., Uchida, M., Kawabata, S., Tamaoka, H., and Shimizu, T. (1990). *Chem. Pharm. Bull.*, 38: 2027–2029.

Niddam-Hildesheim, V., Gershon, N., Amir, E., and Wizel, S. (2003). WO 03028664A2.

Noh, H. K., Lee, J. S., Kim, Y., Hwang, G., Chang, J. H., Shin, H., Nam, D. H., and Lee, K. H. (2004a). *Org. Proc. Res. Dev.*, 8: 781–787.

Noh, H. K., Lee, J. S., Kim, Y., Hwang, G., Chang, J. H., Shin, H., Nam, D. H., and Lee, K. H. (2004b). *Org. Proc. Res. Dev.*, 8: 788–795.

Norrby, S. R. (1999). *Exp. Opin. Pharmacother.*, 1: 109–119.

North, D. S., Fish, D. N., and Redington, J. J. (1998). *Pharmacotherapy*, 18: 915–935.

Noviello, S., Ianniello, F., Leone, S., and Esposito, S. (2003). *J. Antimicrob. Chemother.*, 52: 869–872.

Oscient (2005, 2006). www.oscient.com; Oscient quarterly financial press releases: March 6, 2006; November 2, 2005; August 2, 2005; May 2, 2005.

Petersen, U., Krebs, A., Schenke, T., Philipps, T., Grohe, K.-D., Bremm, K.-D., Endermann, R., and Metzger, K. G. (1993). DE 4208792A1.

Petersen, U., Schenke, T., Krebs, A., Grohe, K., Schriewer, M., Haller, I., Metzger, K. G., Endermann, R., and Zeiler, H.-J. (1997). US 5607942.

Qaqish, R. and Polk, R. E. (2003). In *Quinolone Antimicrobial Agents*, Hooper, D. C. and Rubenstein, E. (Eds) ASM Press, Washington, D.C., (7): 136.

Ronald, A. R. and Low, D. E. (Eds) (2003). *Fluoroquinolone Antibiotics*, Birkhäuser, Boston.

Sanchez, J. P., Domagala, J. M., Hagen, S. E., Heifetz, C. L., Hutt, M. P., Nichols, J. B., and Trehan, A. K. (1988). *J. Med. Chem.*, 31: 983–991.

Schering-Plough News Release February 13, 2006. www.schering-plough.com.

Seidel, D., Conrad, M., Brehmer, P., Mohrs, K., and Petersen, U. (2000). *J. Labelled Cpd. Radiopharm.*, 43: 795–805.

Shams, W. E. and Evans, M. E. (2005). *Drugs*, 65: 949–991.

Shin, H. I., Rim, J. G., Lee, K. S., Kim, Y. S., Nam, D. H., Chang, J. H., Oh, C. Y., and Ham, W. H. (2004). *J. Label. Compd. Radiopharm.*, 47: 779–786.

Todo, Y., Hayashi, K., Takahata, M., Watanabe, Y., and Narita, H. (2000). US 6025370.

Van Bambeke, F., Michot, J.-M., Van Eldere, J., and Tulkens, P. M. (2005). *Clin. Microbiol. Infect.*, 11: 256–280.

Walsh, C. (2003). *Antibiotics: Actions, Origins, and Resistance*, ASM Press, Washington, D.C., (5): 71–77.

Yang, Y.-S., Ji, R.-Y., and Chen, K.-X. (1999). *Chinese J. Chem.*, 17: 539–544.

Yoo, B. K., Triller, D. M., Yong, C.-S., and Lodise, T. P. (2004). *Ann Pharmacother.*, 38: 1226–1235.

Zhanel, G. G., Ennis, K., Vercaigne, L., Walkty, A., Gin, A. S., Embil, J., Smith, H., and Hoban, D. J. (2002). *Drugs*, 62: 13–59.

5

TRIAZOLE ANTIFUNGALS: ITRACONAZOLE (SPORANOX®), FLUCONAZOLE (DIFLUCAN®), VORICONAZOLE (VFEND®), AND FOSFLUCONAZOLE (PRODIF®)

Andrew S. Bell

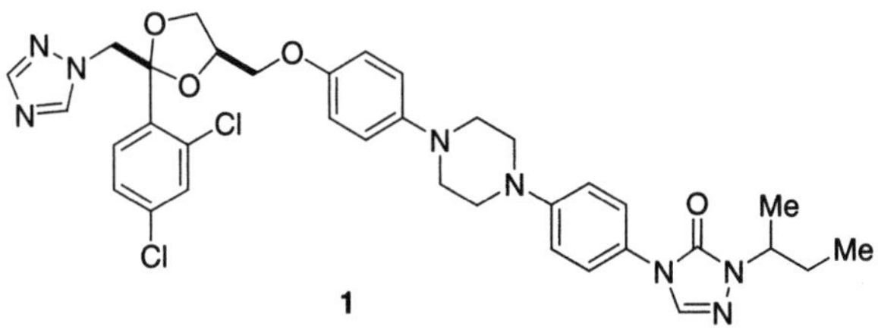

1

USAN: Itraconazole
Trade name: Sporanox®
Johnson & Johnson
Launched: 1988
M.W. 705.65

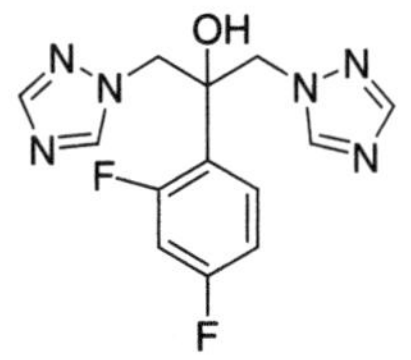

2

USAN: Fluconazole
Trade name: Diflucan®
Pfizer
Launched: 1988
M.W. 306.28

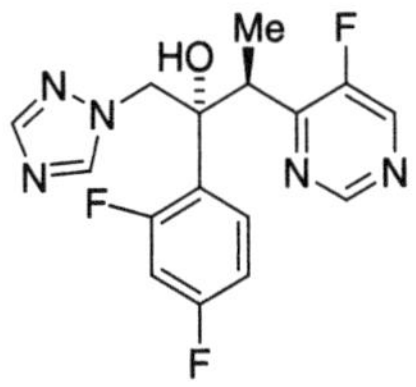

3

USAN: Voriconazole
Trade name: Vfend®
Pfizer
Launched: 2002
M.W. 349.32

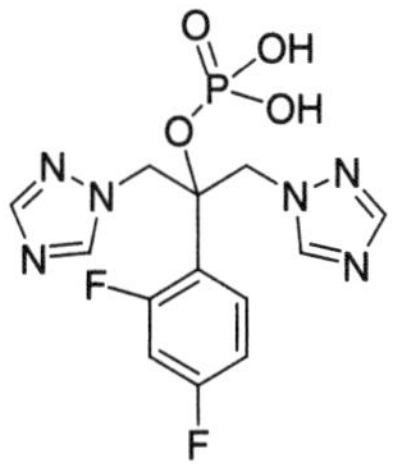

4

USAN: Fosfluconazole
Trade name: Prodif®
Pfizer
Launched: 2003
M.W. 386.26

The Art of Drug Synthesis. Edited by Douglas S. Johnson and Jie Jack Li

5.1 INTRODUCTION

Over the last fifty years, as chemotherapeutics for the treatment of cancer, bacterial infections, and to prevent transplant rejections have improved, there has been a concomitant increase in the number of patients suffering from fungal infections. In the past, most fungal infections were superficial, affecting only the skin, hair, or nails. However, as medical science has progressed in other disease areas, many more patients are immunocompromised, which has resulted in a greater incidence of systemic fungal infections.

Although there are some 300,000 different fungal species, only 600 are pathogenic to humans and of these, 20 account for >99% of all infections. Unfortunately, these fungi differ widely from each other, often displaying widely different susceptibilities to drug treatments. In addition, unlike bacteria, which are prokaryotes, fungi are eukaryotes, and consequently, make use of the same enzymes as humans; hence, selectivity over the corresponding mammalian target can be problematic. For example, terbinafine (**5**, Lamisil®), an inhibitor of squalene epoxidase (SE), is highly selective for isozymes from certain fungi and is a weak inhibitor of the mammalian homologue (Fig. 5.1). Unfortunately, it is also a weak inhibitor of SE from *Candida* spp., the most common systemic pathogen. In contrast, the natural product, amphotericin B (**6**), which acts through binding to ergosterol, is active against most fungi, but is also toxic to mammalian cells, resulting in a low therapeutic index.

The most widely used class of antifungal drugs is known as the azoles, based on their common feature, an imidazole or triazole ring. The first compounds in this class were discovered by the Janssen group in the late 1960s, who found that phenethylimidazoles possessed antifungal activity. They subsequently discovered that their leads were acting through inhibition of lanosterol-14α-demethylase (CYP51), a key enzyme in the biosynthesis of ergosterol (Scheme 5.1), a major component of fungal cell membranes. This enzyme catalyses the oxidative removal of a specific methyl group from lanosterol through a cytochrome-P450-dependent mechanism. The azoles act through coordination to the heme group, preventing coordination of the oxygen required to initiate oxidation. Fortuitously, the azole drugs are selective for CYP51 inhibition over the human CYP3A4, the major metabolizing enzyme.

Based on their initial leads, Janssen launched a series of imidazole antifungals (Fig. 5.2), miconazole (**7**), econazole (**8**), and ketoconazole (**11**), which were complemented by clotrimazole (**10**, Bayer) and tioconazole (**9**, Pfizer). With the exception of ketoconazole, all were characterized by potent antifungal activity but very low systemic exposure; hence, they were limited to topical application for skin or nail infections.

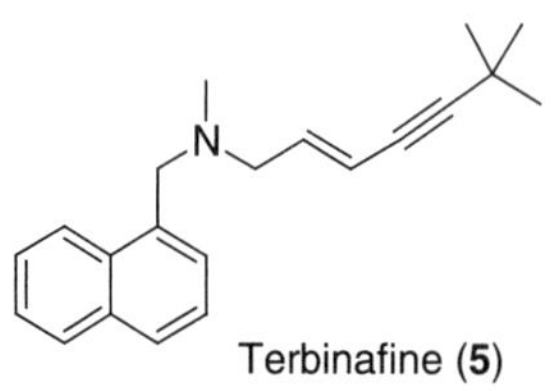

Terbinafine (**5**) 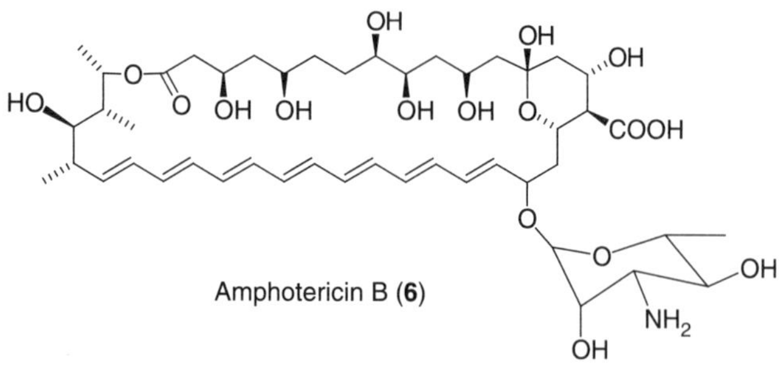 Amphotericin B (**6**)

Figure 5.1. Structures of terbinafine (**5**) and amphotericin B (**6**).

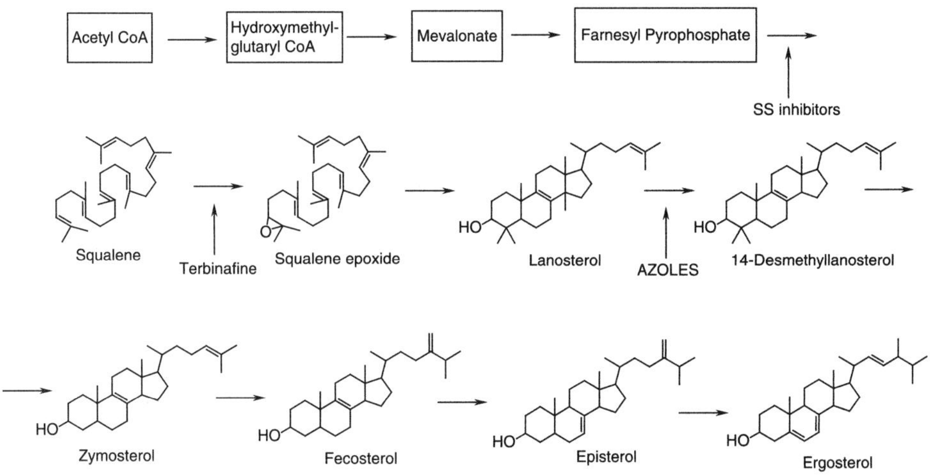

Scheme 5.1. Biosynthetic pathway to Ergosterol showing points of inhibition by drug classes.

Ketoconazole was the first azole to demonstrate activity against systemic pathogens, but it is a potent inhibitor of hepatic CYP3A4, resulting in inhibition of its own metabolism. As a consequence, the half-life of ketoconazole is both dose- and patient-dependent.

In an attempt to improve on the pharmacokinetic profile of the first-generation agents, researchers at Janssen and Pfizer independently discovered that the compounds bearing a 1-linked triazole in place of the imidazole substituent were also potent inhibitors of lansterol demethylase, but were much less prone to metabolism. As a consequence, the triazoles, as they became known, are able to demonstrate broad-spectrum antifungal activity against systemic infections such as candidosis, cryptococcosis, and aspergillosis, which became much more prevalent in the 1980s, particularly in patients suffering from AIDS.

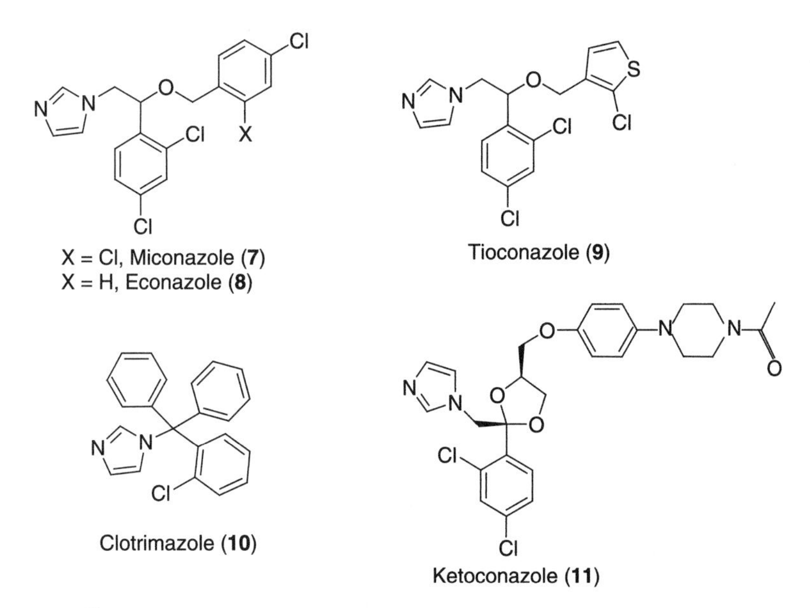

Figure 5.2. Structures of imidazole antifungal drugs.

5.2 SYNTHESIS OF ITRACONAZOLE (SPORANOX®)

The discovery pathway to itraconazole (**1**) (Heeres and Backx, 1980, 1981; Heeres et al., 1988) can be traced back through the earlier Janssen antifungal drugs from miconazole in a series of publications (Heeres et al., 1979, 1983, 1984). The ethyl spacer between the azole and phenyl rings was quickly found to be optimal, as was 2,4-dichlorosubstitution on the aryl ring. The strategy of elaboration of the remaining side-chain from ethers (miconazole, econazole) to aryloxymethyldioxolanes (ketoconazole) was also continued in the synthesis of itraconazole, albeit with the requisite azole installed at the first stage. Consequently, the route to itraconazole is relatively lengthy and linear.

Compared to ketoconazole, itraconazole is much less active in in vitro tests, possibly due to its low solubility in the test medium. However, it is still active at low concentrations (0.1–1 μg/mL) against dermatophytes (skin infections) and against the three main fungal pathogens (*Candida albicans, Cryptococcus neoformans*, and *Aspergillus fumigatus*) at 0.1 μg/mL. An entirely different picture is seen in animal infection models where itraconazole is much more potent than ketoconazole. Due to its wide spectrum of activity, itraconazole is used to treat a wide range of infections in man. Clinical trials demonstrated that high cure rates could be achieved in fingernail and toenail onchomycosis (200 mg/day for 3 months), dermatophytosis (100 mg/day for 2–4 weeks) and vaginal candidiasis (400 mg single dose).

As a consequence of its high lipophilicity and low aqueous solubility, gastric acidity is also required for itraconazole absorption (Haria et al., 1996). It is best absorbed when administered with food, although there is considerable interpatient variability. The oral bioavailability or itraconazole from a 100-mg solution dose was 55%. Like ketoconazole, its bioavailability and half-life are dose-dependent, indicating saturable metabolism. Once absorbed, itraconazole is highly plasma protein bound (99.8%) and widely distributed (10.7 L/kg). Although itraconazole is widely metabolized, it does produce an active metabolite by hydroxylation of the triazolone side-chain.

The synthesis of itrazonazole (Scheme 5.2) started by ketalization of 2,4-dichloroacetophenone with glycerine under mildly acidic condition in refluxing benzene (Heeres et al., 1979). The crude product was then brominated at 40°C to give **10** in 91% yield for the two steps. The primary alcohol was protected as the benzoate by treatment with benzoyl chloride in pyridine, enabling isolation of a crystalline solid in 50% yield. Treatment of the bromide with the sodium salt of 1,2,4-triazole (generated in situ from triazole and sodium hydride) in DMSO at 130°C gave a mixture of the regioisomeric triazole derivatives, which were saponified using a mixture of aqueous sodium hydroxide and dioxane (Heeres et al., 1983). The isomers were separated by chromatography, with the major product (approximately 10 : 1) being the desired 1-substituted triazole isomer.

Following activation of the alcohol through mesylation with mesyl chloride in pyridine (87%), a second nucleophilic displacement with the sodium salt of *N*-acetyl-4-hydroxyphenylpiperazine at 80°C, again generated in situ with sodium hydride in DMSO, provided the triazole analog of ketoconazole in 64% yield (Heeres et al., 1984). Deacetylation was achieved with sodium hydroxide in *n*-butanol under refluxing conditions (70%) before attachment of the final phenyl ring by nucleophilic aromatic substitution using 4-chloronitrobenzene under mildly basic conditions in DMSO at 120°C. Catalytic hydrogenation over platinum on carbon in ethylene glycol monomethyl ether at 50°C gave the corresponding poorly soluble aniline derivative, which required the crude reaction mixture to be heated to avoid filtration of the product

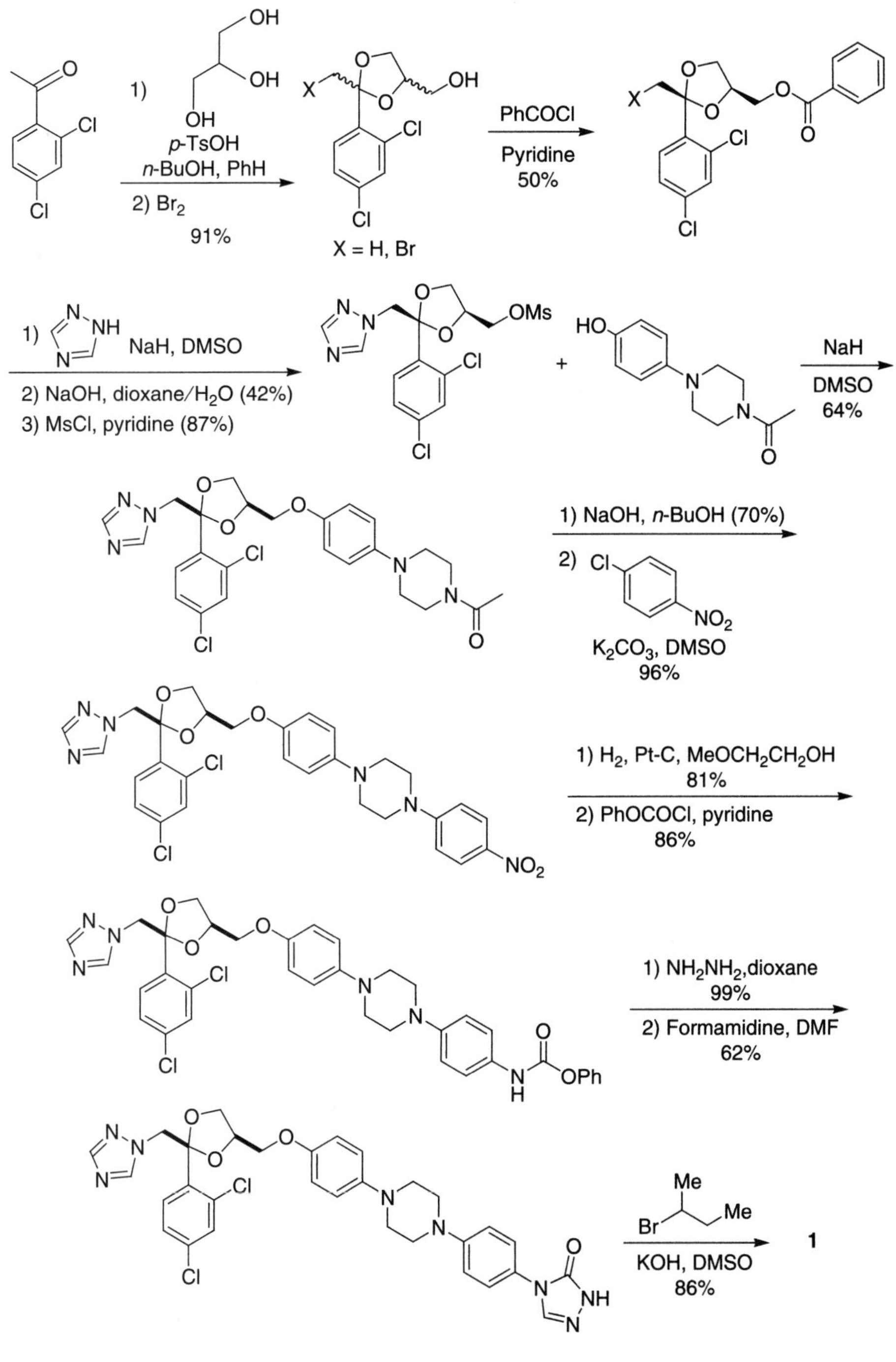

1) p-TsOH, n-BuOH, PhH
2) Br_2
91%

X = H, Br

PhCOCl
Pyridine
50%

1) NaH, DMSO
2) NaOH, dioxane/H_2O (42%)
3) MsCl, pyridine (87%)

NaH
DMSO
64%

1) NaOH, n-BuOH (70%)
2) K_2CO_3, DMSO
96%

1) H_2, Pt-C, $MeOCH_2CH_2OH$
81%
2) PhOCOCl, pyridine
86%

1) NH_2NH_2, dioxane
99%
2) Formamidine, DMF
62%

KOH, DMSO
86%

Scheme 5.2. Synthesis of itraconazole (1).

along with the catalyst. Carbamoylation of the aniline derivative with phenyl chloroformate in a mixture of chloroform and pyridine gave an activated carbamate, which was treated with hydrazine to yield the semicarbazide in 86% yield over two steps. Condensation of the semicarbazide with formamidine acetate in DMF at 130°C for 3 h gave the cyclized triazolone (62% yield). Subsequent alkylation with 2-bromobutane was achieved using powdered potassium hydroxide in DMSO to give itraconazole (**1**) as a crystalline solid from toluene.

5.3 SYNTHESIS OF FLUCONAZOLE (DIFLUCAN®)

In contrast to the evolutionary discovery process that lead to itraconazole, the discovery of fluconazole (Narayanaswami and Richardson, 1983; Richardson, 1982) appears more revolutionary. The Pfizer team decided to target compounds with in vivo activity from the outset, deliberately focusing on compounds with a good ADME profile at the expense of in vitro potency (Richardson et al., 1988). Having reviewed the profiles of each of the known series of antifungal imidazole derivatives, they concentrated their efforts on the tertiary alcohol series, as examples often gave activity in animal infection models. However, when pharmacokinetic studies indicated that the imidazole ring was still susceptible to metabolism, a number of alternative heterocyclic replacements were investigated. Only the 1-linked 1,2,4-triazole offered any encouragement in vivo, despite being less active than the corresponding imidazole derivative in vitro. In continuation of the focus on pharmacokinetic properties, the hexyl side-chain of the lead compound was replaced by a second 1,2,4-triazole, on the assumption that the resulting derivatives would have low lipophilicity. It was hoped that this would be consistent with high blood levels and, due to reduced protein binding, higher free drug levels. This hypothesis turned out to be true and the series of *bis*-triazoles were found to be highly active in animal infection models. Fluconazole, the 2,4-difluorophenyl analog, emerged as the only compound to combine good aqueous solubility, a long half-life, and an excellent safety profile.

Fluconazole is effective against oropharyngeal and esophageal candidiasis when used orally once daily either as a treatment or prophylactically in patients with AIDS or undergoing cancer therapy. It is also effective in patients with cryptococcal meningitis, especially as a maintenance therapy. Vaginal candidiasis can be treated using a single 50-mg dose. Fluconazole is available for both oral and i.v. administration, particularly for seriously ill patients, with few side-effects.

Fluconazole (Goa and Barradell, 1995) is very well absorbed, reaching peak plasma concentrations in 1–2 h with virtually complete bioavailability following oral administration. Neither food nor gastric acidity has any effect. Its volume of distribution is low (0.7–0.8 L/kg), approximating to body water, with very low plasma protein binding (12%). Fluconazole also crosses membranes easily such that CSF : plasma ratios of 0.5–0.9 are achieved. There is very little evidence for metabolism of fluconazole, which is excreted mainly through the renal elimination route (60–90% of dose). However, there is extensive tubular reabsorption, resulting in a long plasma elimination half-life of 30–40 h, enabling once-daily dosing for systemic candidoses and a single-dose therapy for vaginal candidiasis.

In keeping with the simplicity of its structure, fluconazole can be synthesized in two steps from commercially available starting materials (Scheme 5.3). Metallation of 1-bromo-2,4-difluorobenzene with butyllithium in ether gave the corresponding aryllithium,

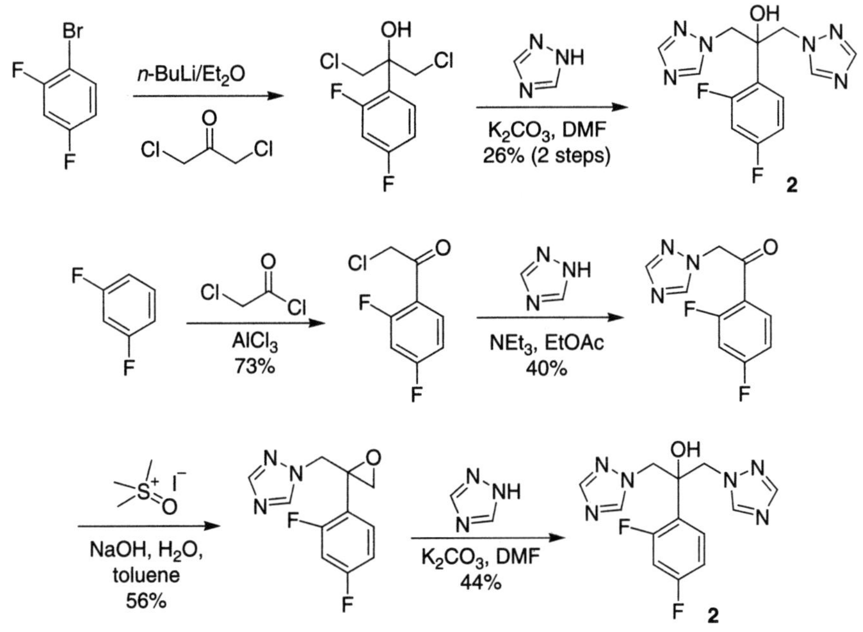

Scheme 5.3. Synthesis of fluconazole (2).

which was trapped with 1,3-dichloroacetone. The crude alcohol intermediate was reacted with 1,2,4-triazole in DMF in the presence of potassium carbonate at 70°C to give a mixture of regioisomeric triazoles from which the desired *bis*-1-linked triazole could be isolated by chromatography in 26% overall yield.

An alternative larger scale chromatography-free four-step route has also been described in the patent literature (Narayanaswami and Richardson, 1983; Richardson, 1982). Aluminum-chloride-catalysed chloroacetylation of 1,3-difluorobenzene using chloroacetyl chloride without solvent at 50°C afforded 1′-chloro-2,4-difluoroacetophenone in 73% yield. Displacement of the chloride with 1,2,4-triazole using triethylamine as base in refluxing ethyl acetate gave a modest yield (40%) of the desired regioisomeric triazolylketone as the hydrochloride salt. Corey epoxidation using the sulfoxonium ylide [generated in situ from trimethylsulfoxonium iodide under phase transfer conditions (cetrimide/toluene/aqueous sodium hydroxide)] at 60°C provided the epoxide, which was isolated as its mesylate salt in 56% yield. The epoxide ring was opened with a second equivalent of 1,2,4-triazole using potassium carbonate as base to give a mixture of triazole isomers. The minor, undesired isomer is even more water soluble than fluconazole and consequently, can be removed by a simple water wash of the crude product solution in chloroform. Fluconazole (**2**) was isolated as a crystalline free base in 44% yield from isopropanol.

5.4 SYNTHESIS OF VORICONAZOLE (VFEND®)

Although clinical experience with fluconazole demonstrated excellent efficacy against infections from two of the main human fungal pathogens, *Candida albicans* and *Crypotococcus neoformans*, it is poorly effective against infections caused by the third major pathogen, *Aspergillus fumigatus*. It was subsequently found that fluconazole is a weaker inhibitor of the target enzyme, lanosterol-14α-demethylase, from *Aspergillus* compared to its potency against the corresponding *Candida* enzyme (Ballard et al., 1990).

Pfizer scientists found that introduction of a methyl group adjacent to one of the triazole rings of fluconazole increased potency against *Aspergillus fumigatus*, while retaining potency against the other fungal pathogens. Replacement of the triazole ring adjacent to the branch point with six-membered heterocycles (Ray and Richardson, 1991, 1994, 1998) provided compounds with broad-spectrum in vitro activity, and surprisingly, a fungicidal mechanism of action against *Aspergillus* spp. The optimal compound, in the series, voriconazole (Dickinson et al., 1996), features a 5-fluoro-4-pyrimidinyl substituent. Voriconazole is a single enantiomer, whose opposite enantiomer is at least 500-fold less active. The difference in activity of fluconazole and voriconazole against CYP51 in *Aspergillus* has been rationalized as a difference in binding mode (Gollapudy et al., 2004).

Like fluconazole, voriconazole (Muijsers et al., 2002) is polar, with moderate aqueous solubility (0.5 mg/mL), resulting in rapid absorption (maximum concentration achieved in less than 2 h) and oral bioavailability (96%). Moderate food effects have been observed. Distribution is wide with a steady-state volume of 4.6 L/kg and moderate binding to plasma proteins (58%).

Owing to its inferior pharmacokinetic profile, voriconazole is dosed twice daily; i.v. doses of 6 mg/kg are used on the first day, followed by 200 mg orally or continued i.v. dosing at 4 mg/kg. Voriconazole is recommended for the treatment of adults with invasive aspergillosis and can be used for rare infections caused by *Fusarium* spp. and *Scedosporium apiospermum*, where treatment with other agents has failed. Its primary use is in immunocompromised patients with progressive, life-threatening infections.

The synthetic route to voriconazole (Scheme 5.4) also demonstrates its evolution from fluconazole, as both make use of the same 1-linked 1,2,4-triazolyl-substituted acetophenone. For voriconazole, this required condensation is with a carbon nucleophile based on 4-ethyl-5-fluoropyrimidine (Butters et al., 2001). A variety of substituted ethylpyrimidine analogs were investigated in this reaction under different metallation conditions. These analogs were all prepared starting from 5-fluorouracil, initially by chlorination with phosphorous oxychloride at 95°C in the presence of *N*, *N*-dimethylaniline, followed by quenching into aqueous hydrochloric acid (to avoid hydrolysis of the chlorines), to give the corresponding 2,4-dichloropyrimidine in an optimized yield of 95%. Introduction of the ethyl group could be achieved via selective displacement of the 4-chloro substituent of the pyrimidine ring with the sodium salt of diethyl methylmalonate, followed by decarboxylation in refluxing HCl and acetic acid to give 2-chloro-4-ethyl-5-fluoropyrimidine (**12**). Alternatively, addition of ethyl magnesium bromide to 2,4-dichloro-5-fluoropyrimidine gave the unstable dihydropyrimidine analog. Oxidation back to the aromatic system could be achieved using a variety of oxidants, but iodine was found to be optimal, especially when added to the crude Grignard addition product without work-up. This two-step one-pot procedure provided pyrimidine **13** in an overall yield of 75%.

Although 2,4-dichloro-6-ethyl-5-fluoropyrimidine (**13**) contained the correct carbon skeleton for condensation with the acetophenone, only self-condensation products were obtained following metallation with LDA. A reduction in electrophilicity was achieved by removal of the 2-chloro substituent via a three-step sequence involving hydrolysis of the more reactive 4-chloro substituent, followed by hydrogenation of the 2-chlorine and rechlorination with phosphorous oxychloride. The overall yield for the three steps resulting in 4-chloro-6-ethyl-5-fluoropyrimidine (**14**) was 72%.

The key condensation of anions derived from the various ethylpyrimidines was the subject of extensive investigation (Table 5.1). Depending on the substitution pattern on

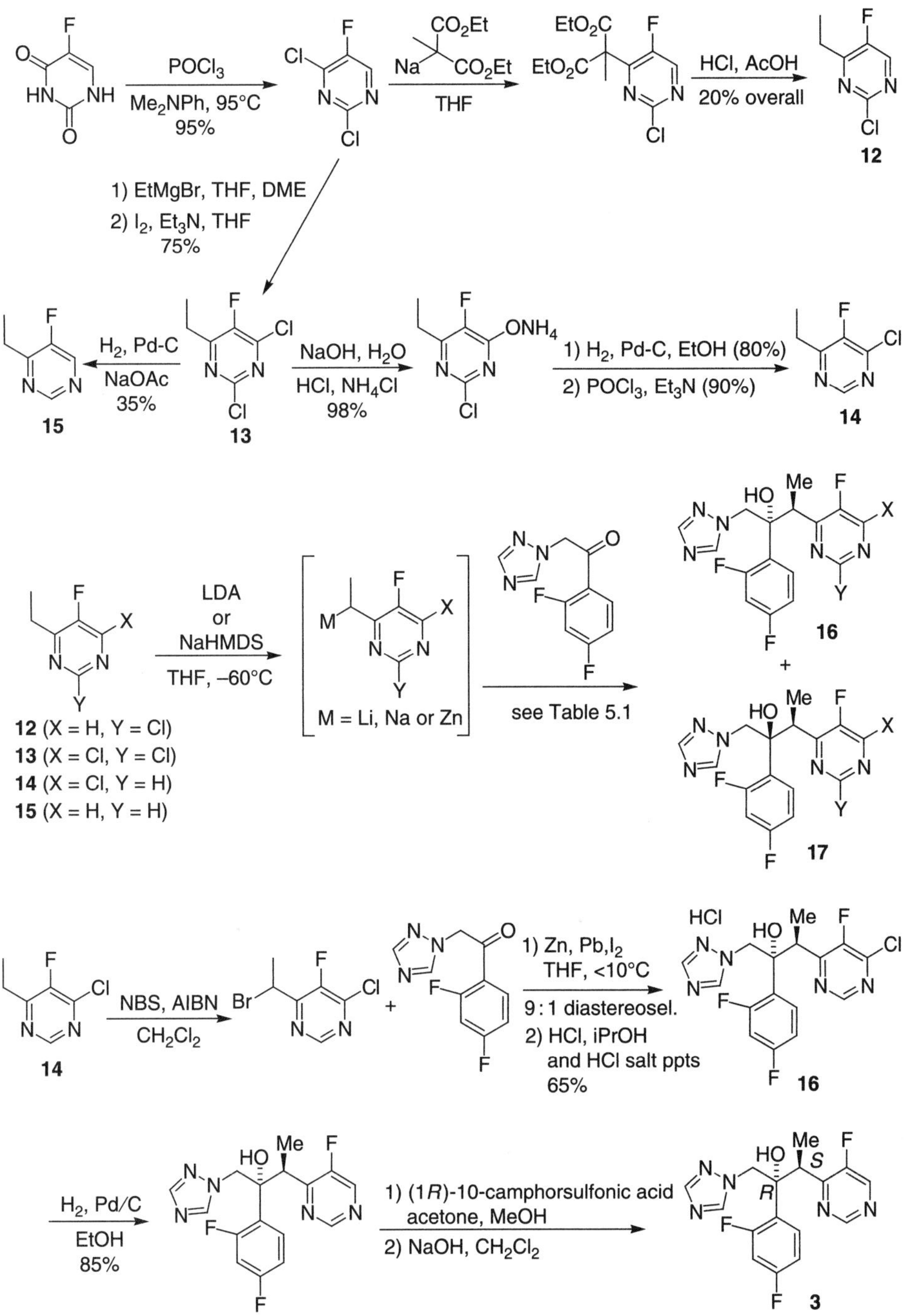

Scheme 5.4. Synthesis of voriconazole (**3**).

the pyrimidine ring, self-condensation competed with addition to the triazolylketone, which was also susceptible to deprotonation. In an additional complication, the diastereoselectivity of the condensation also depended on the heterocyclic substitution pattern. The optimal balance of these competing reactions was achieved with

TABLE 5.1. Variation in Yields of Diastereomeric Products (Ratio of **16** : **17**) from Various Metallated Pyrimidines

X, Y	M = Li	M = Na	M = Zn
H, Cl (**12**)	0	8% (1 : 1)	11% (1.2 : 1)
Cl, Cl (**13**)	0	0	9.9% (1.1 : 1)
Cl, H (**14**)	50% (1.2 : 1)	25% (1 : 3)	71.5% (2 : 1)
H, H (**15**)	0	30% (1 : 2)	71% (1 : 3)

4-chloro-6-ethyl-5-fluoropyrimidine, which gave a 50% yield of a 55 : 45 mixture of diastereoisomeric aldol products (**16** : **17**) following deprotonation with LDA. Alternative bases such as NaHMDS or KHMDS gave a poorer diastereoselectivity.

As the relatively low yield of the condensation was ascribed to competing deprotonation, a number of alternative counterions were also investigated. The optimal was found to be zinc, generated from the corresponding bromoethylpyrimidine derivative (prepared by radical bromination of the ethyl derivative with NBS). Thus, condensation of 6-(1-bromoethyl)-4-chloro-5-fluoropyrimidine with the ketone using activated zinc in refluxing THF gave a 50% yield of a 2 : 1 mixture of desired:undesired diastereoisomers (**16** : **17**). Improvements in diastereoselectivity to 12 : 1 could be achieved by running the reaction at 10°C.

Following isolation of the desired diastereoisomer by partial crystallization, hydrogenation of the extraneous chlorine was carried out under standard palladium on carbon-catalysed conditions to give racemic voriconazole. Fortunately, resolution was readily achieved using (1*R*)-10-camphorsulfonic acid, yielding a highly enantiopure salt from which the desired voriconazole free base could be regenerated. Voriconazole (**3**) is obtained as a crystalline solid from isopropanol.

5.5 SYNTHESIS OF FOSFLUCONAZOLE (PRODIF®)

Although fluconazole is the most aqueous soluble of all of the azoles discussed so far, intravenous solutions achieve a maximum concentration of 2 mg/mL. As many seriously ill patients require doses of >400 mg, high-volume infusions are required. In order to avoid this problem, a number of prodrugs of fluconazole were investigated (Green and Stephenson, 1997; Green et al., 2003). Strategically, derivatization of the tertiary hydroxyl was preferred, because it avoided the creation of a chiral center. As high aqueous solubility was a key objective, efforts rapidly focused on phosphate ester prodrugs, because they are often very soluble, especially in salt forms. Thus, it was unsurprising that the disodium salt of the phosphate ester of fluconazole, known as fosfluconazole, was found to be soluble in water at >300 mg/mL.

Although fosfluconazole is some 25-fold less active than fluconazole in vitro, it is rapidly hydrolyzed by phosphatases in vivo and has similar efficacy in animal models and in patients, where it is 97% bioavailable (Hegde and Schmidt, 2005). The prodrug has a half-life of 2.3 h in patients, and the vast majority of the dose is excreted as fluconazole: Consequently, fosfluconazole has a similar clinical profile to that of fluconazole.

A number of synthetic routes to fosfluconazole (**4**) have been described (Scheme 5.5) (Bentley et al., 2002). The initial small-scale route used conditions described by Fraser–Reid involving treatment of fluconazole (**2**) with dibenzyl diisopropylphosphoramidate in

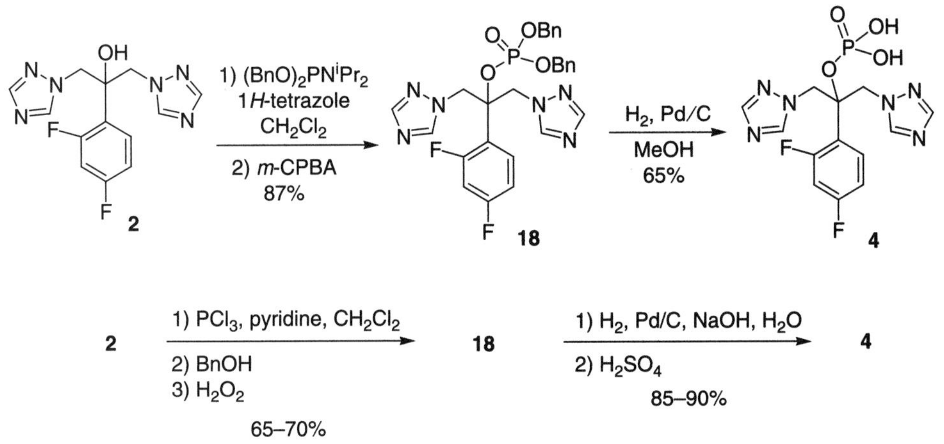

1) $(BnO)_2PN^iPr_2$, 1*H*-tetrazole, CH_2Cl_2; 2) *m*-CPBA, 87% (2 → 18); H_2, Pd/C, MeOH, 65% (18 → 4)

2 → 18: 1) PCl_3, pyridine, CH_2Cl_2; 2) BnOH; 3) H_2O_2; 65–70%

18 → 4: 1) H_2, Pd/C, NaOH, H_2O; 2) H_2SO_4; 85–90%

Scheme 5.5. Synthetic routes to fosfluconazole (**4**).

the presence of tetrazole in dichloromethane to give the phosphate, which was oxidized in situ with *m*-chloroperoxybenzoic acid (*m*-CPBA) to afford **18** in 87% overall yield. Hydrogenolysis of the benzyl groups over palladium on carbon in methanol gave the free diacid **4** in 65% yield. The initial route was improved by replacement of the thermally labile tetrazole and *m*-CPBA with 1,2,4-triazole and hydrogen peroxide, respectively. However, the limited commercial availability of the Fraser–Reid reagent prompted a reassessment of the route. It was subsequently found that sequential treatment of fluconazole (**2**) in dichloromethane/pyridine with phosphorus trichloride, benzyl alcohol, and hydrogen peroxide gave the desired phosphate ester **18** in 65–70% yield. As the final acid product is relatively insoluble in methanol, the final hydrogenolysis is optimally carried out in aqueous sodium hydroxide. After filtration of the catalyst, fosfluconazole (**4**) can be isolated in 88% yield by filtration, following acidification of the product solution with sulfuric acid.

REFERENCES

Ballard, S. A., Ellis, S. W., Kelly, S. L., and Troke, P. F. (1990). *Journal of medical and veterinary mycology: bi-monthly publication of the International Society for Human and Animal Mycology*, 28: 335–344.

Bentley, A., Butters, M., Green, S. P., Learmonth, W. J., MacRae, J. A., Morland, M. C., O'Connor, G., and Skuse, J. (2002). *Org. Process Res. Dev.*, 6: 109–112.

Butters, M., Ebbs, J., Green, S. P., MacRae, J., Morland, M. C., Murtiashaw, C. W., and Pettman, A. J. (2001). *Org. Process Res. Dev.*, 5: 28–36.

Dickinson, R. P., Bell, A. S., Hitchcock, C. A., Narayanaswami, S., Ray, S. J., Richardson, K., and Troke, P. F. (1996). *Bioorg. Med. Chem. Lett.*, 6: 2031–2036.

Goa, K. L. and Barradell, L. B. (1995). *Drugs*, 50: 658–690.

Gollapudy, R., Ajmani, S., and Kulkarni, S. A. (2004). *Bioorg. Med. Chem.*, 12: 2937–2950.

Green, S. and Stephenson, P. T. (1997). WO 97/28169.

Green, S., Stephenson, P. T., and Murtiashaw, C. (2003). W. US 2003/0144250.

Haria, M., Bryson, H. M., and Goa, K. L. (1996). *Drugs*, 51: 585–620.

Heeres, J. and Backx, L. J. J. (1980). EP 6711 (to Janssen Pharmaceutica NV Belg.).

Heeres, J. and Backx, L. J. J. (1981). US 4267179.

Heeres. J., Hendrickx. R., and Van Cutsem. J. (1983). *J. Med. Chem.*, 26: 611–613.

Heeres, J., Backx, L. J. J., and Van Cutsem, J. (1984). *J. Med. Chem.*, 27: 894–900.

Heeres, J., Backx, L. J. J., Mostmans, J. H., and Van Cutsem, J. (1979). *J. Med. Chem.*, 22: 1003–1005.

Heeres, J., Backx, L. J. J., Thijssen, J. B. A., and Knaeps, A. G. (1988). US 4791111.

Hegde, S. and Schmidt, M. (2005). *Ann. Rep. Med. Chem.*, 40: 457–458.

Muijsers, R. B. R., Goa, K. L., and Scott, L. J. (2002). *Drugs*, 62: 2655–2664.

Narayanaswami, S. and Richardson, K. (1983). EP 0096569.

Ray, S. J. and Richardson, K. (to Pfizer Limited UK). EP 0440372 (1991); US 5278175 (1994); US 5773443 (1998).

Richardson, K., Cooper, K., Marriott, M. S., Tarbit, M. H., Troke, P. F., and Whittle, P. J. (1988). *Annals New York Acad. Sci.*, 544: 4–11.

Richardson, K. (1983). US 4404216 Patent GB 2099818 (to Pfizer Limited UK).

6

NON-NUCLEOSIDE HIV REVERSE TRANSCRIPTASE INHIBITORS

Arthur Harms

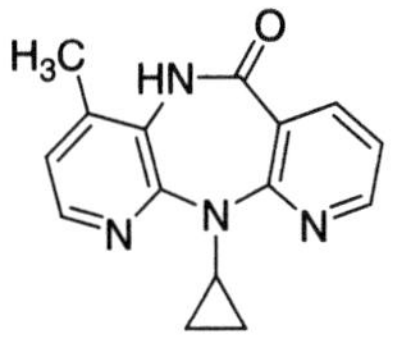

1

USAN: Nevirapine
Trade name: Viramune®
Boehringer Ingelheim
Launched: 1996
M.W. 266.30

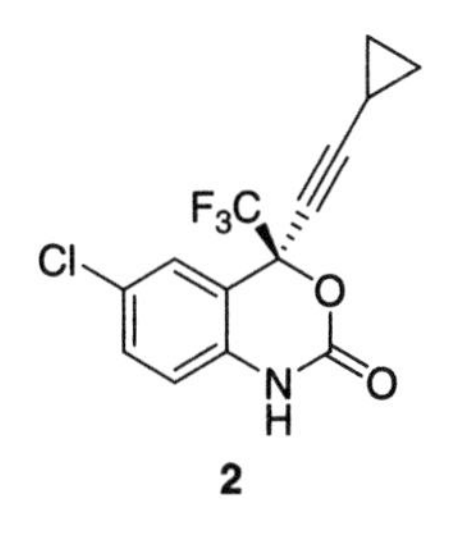

2

USAN: Efavirenz
Trade name: Sustiva®, Stocrin®
Bristol-Myers Squibb/Merck
Launched: 1998
M.W. 315.67

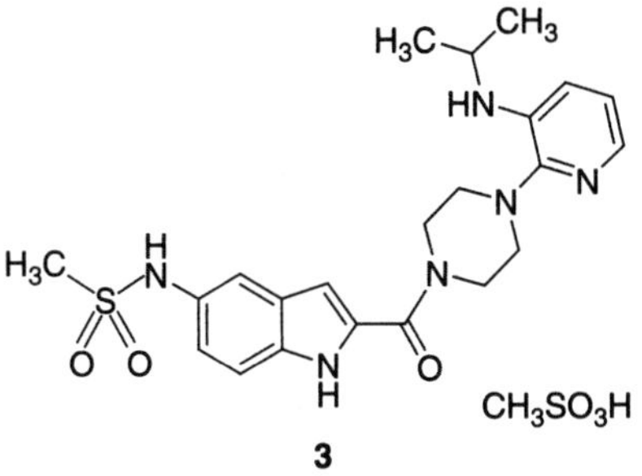

3

CH_3SO_3H

USAN: Delavirdine mesylate
Trade name: Rescriptor®
Pfizer
Launched: 1997
M.W. 456.56 (parent)

The Art of Drug Synthesis. Edited by Douglas S. Johnson and Jie Jack Li

6.1 INTRODUCTION

Nearly 40 million people are infected with the human immunodeficiency virus (HIV). Over half of those infected reside in sub-Saharan Africa. Worldwide during 2004, it is estimated that nearly 14,000 people a day were infected. Human immunodeficiency virus type 1 is the primary etiological source for the acquired immunodeficiency syndrome (AIDS). Fortunately, people infected with HIV are leading longer and more productive lives due to the availability of more effective therapies. Better medicines have evolved due to the efforts of scientists worldwide who find targets and compounds that inhibit the virus life-cycle. The current treatment for HIV infection is via a drug cocktail that usually includes a protease inhibitor (PI), a nucleoside reverse transcriptase inhibitor (NRTI), and a non-nucleoside reverse transcriptase inhibitor (NNRTI).

The reverse transcriptase enzyme (RT) is the primary enzyme responsible for the conversion of the viral single-strand RNA to the double-strand DNA. The reverse transcriptase enzyme is a component of the virion and is encoded by the *pol* gene. The RT is manufactured in the HIV-infected cells as a *gag–pol* fusion polyprotein. The RT is not the only enzyme necessary for the translation of RNA to DNA. The other enzymes for this conversion include RNA-dependent DNA polymerase, DNA-dependent DNA polymerase, and RNase H (Gilboa and Mitra, 1978; Prasad and Gogg, 1990). The reverse transcriptase enzyme has a high error rate (1 in 2000 bases), which produces higher incidents of mutation. Some of these mutations make the virus resistant to NNRTI treatment.

To date, over 30 structurally different classes of NNRTIs have been published (Balzarini, 2004; De Clercq, 1998; Hajós et al., 2000). Three compounds (nevirapine, delavirdine, and efavirenz) have been approved by the U.S. FDA. Many of the latest drug candidates result from a publication of X-ray crystal structures of nevirapine and R86183 (Breslin et al., 1995, 1999; Ho et al., 1995; Kukla et al., 1991; Pauwels et al., 1994; White et al., 1991) with RT, which reveals how these compounds interact with RT (Arnold et al., 1996; Ding et al., 1995; Esnouf et al., 1995; Kohlstaedt et al., 1992; Smerdon et al., 1994). The similarities between the crystal structures of nevirapine and R86183 (Fig. 6.1) are striking. Each compound binds in a hydrophobic pocket that is close but separate from the natural substrate site. The crystal structure reveals a butterfly-like shape for R86183 and nevirapine. For R86183, the butterfly morphology is a result of the interaction of the dimethylallyl group with the side-chains of tyrosine at positions 181, 188, and 229. The chlorophenyl group interacts with Leu100, Lys101, and Tyr318. The allosteric pocket is occupied similarly with nevirapine. The 4-methylpyridyl occupies the same space as the dimethylallyl group, and the unsubstituted pyridyl group occupies

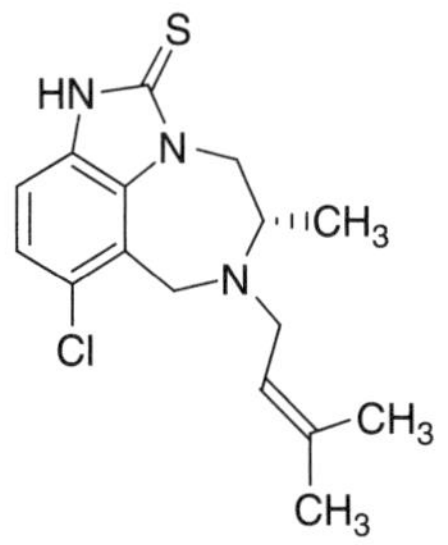

S
HN
N
CH_3
N
Cl
CH_3
CH_3

Figure 6.1. R86183.

the space where the chlorophenyl group resides (Nanni et al., 1993; Tantillo et al., 1994). The thiourea NH group of R86183 forms a critical hydrogen bond with the carbonyl oxygen from Lys101. However, nevirapine lacks that hydrogen bond. The same type of interactions can be observed between efavirenz and delavirdine in the enzyme pocket. Inhibitor binding at the allosteric site changes the three-dimensional shape of the active site necessary for transcription of the viral RNA resulting in inhibition of the enzyme.

As mentioned earlier, NNRTI therapy gives rise to mutants and drug resistance. The general trends have shown that mutations create a larger allosteric site. As a result, the NNRTI can no longer stay in the site, so the RT can continue to transcribe RNA. Many of the amino-acid residues surrounding this binding site are highly variable, including many of those identified in connection with non-nucleoside drug resistance. Also located in this same region are a number of other amino acids that are more highly conserved, particularly aspartic acid residues (Asp110, Asp185, Asp186). Mutation of these aspartic residues inactivates RT (Larder et al., 1987). The mutations associated with nevirapine, efavirenz, and delavirdine are discussed in their respective sections in this chapter.

6.2 SYNTHESIS OF NEVIRAPINE

Nevirapine (**1**) was invented and synthesized by scientists at Boehringer Ingelheim Pharmaceuticals, Inc. Nevirapine is a tricyclic dipyridodiazepinone and a specific noncompetitive inhibitor of reverse transcriptase (Hargrave et al., 1991; Merluzzi et al., 1990). Nevirapine does not inhibit the reverse transcriptase from simian immunodeficiency virus or feline leukemia virus, nor does it inhibit calf thymus DNA polymerase α or human DNA polymerase α, β, γ, or δ. This compound also has no effect on HIV-1 protease, human plasmin renin, and HSV-1 ribonucleotide reductase enzymes. With regard to inhibiting HIV-1 replication, nevirapine inhibits HIV-1 replication in CD_4^+ T-cell cultures (c8166) (Salahuddin, 1983). The IC_{50} against HIV-1 (IIIb) is 40 nM, with a maximum inhibition of 100% as determined by inhibition of p24 production. Viability of c8166 cells is determined by means of a tetrazolium salt (MTT) metabolic assay (Denizot and Lang, 1986). This assay shows 50% toxicity of nevirapine at 321,000 nM, resulting in an 8025 in vitro selectivity ratio (ED_{50}/IC_{50}). In addition, nevirapine shows no cytotoxic effects on human bone marrow colonies including erythroid burst-forming units and colony-forming units of granulocyte, erythroid, macrophage megakarocyte, and granulocyte macrophage at concentrations up to 37,500 nM. The NRTI inhibitor, AZT, is a more potent inhibitor ($IC_{50} = 6$ nM); however, AZT is more cytotoxic ($CC_{50} = 66{,}000$ nM) in the same assays. In in vivo assays in cynomolgus monkeys and chimpanzees, the plasma levels remained between 35 and 600 times the IC_{50} during an 8-h period after a single oral dose of 20 mg/kg of body weight. Nevirapine can also treat HIV, which may reside in the central nervous system. In rodent and primate models, the ratio of nevirapine in plasma versus that in the brain is 0.8 to 1.0, which is observed using tissue distribution studies following oral distribution.

Most of the viral resistance of RT to nevirapine is based on the mutation found in position 181. In this mutation, cysteine is substituted for tyrosine (Y181C) (Richman et al., 1991, 1994). The Y181C mutant is less sensitive to nevirapine than the wild-type enzyme ($IC_{50} = 2.6$ μM vs. $IC_{50} = 0.08$ μM). This mutation is also less sensitive to other NNRTIs (Dueweke et al., 1993; Nunberg et al., 1991). Other mutations known to occur from nevirapine treatment are K103N, V106A, G190A, and Y188L.

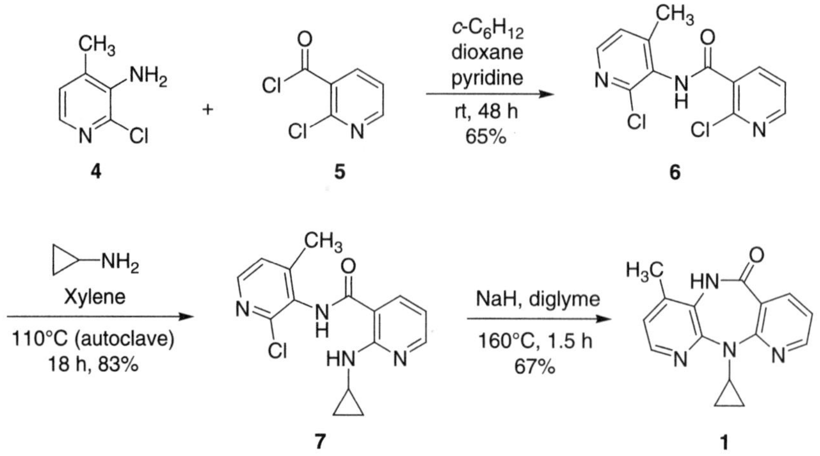

Scheme 6.1.

Nevirapine is given orally at doses of one 200-mg tablet per day for 14 days followed by two 200-mg tablets for one day. Nevirapine is always given as a part of the AIDS cocktail, which may contain one or two of the following protease inhibitors, Reyataz® (atazanavir), Invirase® (saquinavir), Kaletra® (lopinavir and ritonavir), Crixivan® (indinavir), Agenerase® (amprenavir), or Lexiva® (fosamprenavir), and sometimes with AZT or other NRTIs. In terms of pharmacokinetics, nevirapine has a half-life of 56 h. Women and patients with higher CD4 counts are at increased risk of liver problems while taking nevirapine.

Synthesis (Hargrave et al., 1991) of nevirapine uses the acylation of 3-amino-2-chloro-4-methylpyridine (**4**) (Grozinger et al., 1995) with 2-chloronicotinoyl chloride (**5**) to provide 2,2′-dichloro amide **6** (Scheme 6.1). Activation of the 2′-chlorine from the adjacent carbonyl group facilitates displacement by cyclopropylamine to give **7**. Displacement of the 2-chlorine is observed if electron-withdrawing groups are present on the other pyridine ring. To affect ring closure, the dianion of **7** was formed and then heated under reflux to provide nevirapine (**1**). When dimethylformamide was used, lower yields of less pure product were observed.

A pharmacokinetic study shows that CYP3A4 is responsible for the oxidative metabolism of nevirapine, resulting in five major metabolites. The major metabolite of nevirapine from the liver microsomes of humans, rats, monkeys, and dogs results from the hydroxylation of the 4-methyl substituent (compound **8**) (Cheeseman et al., 1993; Grozinger et al., 2000).

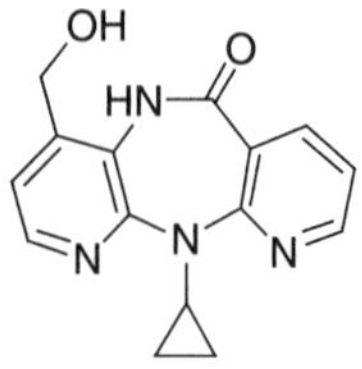

8

6.3 SYNTHESIS OF EFAVIRENZ

Efavirenz (**2**) was discovered by Merck scientists for the treatment of HIV infection (Young et al., 1995, 1996). The IC_{95} of efavirenz against the wild-type laboratory adapted strains and clinical isolates ranges from 1.7 to 25 nM in lymphoblastoid cell lines, peripheral blood mononuclear cells, and macrophage/monocyte cultures. Efavirenz was tested for its ability to inhibit the spread of single and double mutants. Based on the results observed in their RT assay, substitutions at A98G, K101E, V106A, V108I, V179E, and Y181C did not show any resistance as much as substitutions at positions L100I and K103N (efavirenz was approximately six times less effective in relation to the wild-type virus). The double mutants K103N-Y181C and L100I-K103N are eight to twenty times less sensitive to efavirenz. In the cell culture assay, mutants V108I, V179E, and Y181C have comparable sensitivity to the wild-type virus (IIIb, MN, and RFH). However, mutants A98G, K101E, and V106A are four to fifteen times less sensitive. Single mutants L100I and K103N have significantly less sensitivity (30–60 times less sensitive). Double mutant L100I–K103N is greater than 8000 times less sensitive. In clinical trials, 90% of the virus from patients whose viral load rebounded after the initial response of load reduction had the K103N mutation. This mutation is observed in both the monotherapy, in vitro selection experiments (Bacheler, 1999) and in clinical trials (Bacheler et al., 1998). Approximately four months following the appearance of the K103N mutation, double mutations K103N–V108I or K103N–P225H are observed in a large number of samples.

Results from testing against a variety of polymerase enzymes show that efavirenz is inactive up to 300 μM for a 50% inhibition (Young et al., 1995). The polymerase enzymes studied were Moloney murine leukemia virus RT, human DNA polymerases α, β, and γ, *Escherichia coli* RNA polymerase, and the Klenow fragment. Cytotoxicity studies in their primary cells and in a T-cell line reveal that efavirenz has a selectivity index of 80,000.

The establishment of the stereocenter in efavirenz provides a challenging goal for the synthetic chemist (Pierce et al., 1998; Thompson et al., 1995). The synthesis starts by treating 4-chloroaniline with pivaloyl chloride under biphasic conditions to provide the desired amide **10** (Scheme 6.2). Ortho metallation as directed by the amide is accomplished with two equivalents of *n*-butyllithium (or *n*-hexyllithium) in tetramethylethylene diamine (TMEDA) and MTBE. The resulting dianion is quenched with ethyl trifluoroacetate to provide pivaloylamide ketone **11** (Fuhrer and Gschwend, 1979). The amide is hydrolyzed in situ to provide the trifluoroketone hydrate hydrochloride **12**, which crystallizes from the reaction mixture (>98% pure).

The free base **13** is obtained by stirring with sodium acetate in MTBE. Benzylation by treatment with a mild acid and *p*-methoxybenzyl alcohol provides **14** (Emert et al., 1977; Henneus et al., 1996). The initial conditions for the asymmetric addition of the lithium acetylide to the trifluoroketone appear in an earlier Merck paper (Thompson et al., 1993, 1996). Optimization of these conditions, which include some elaborate NMR studies (Thompson et al., 1998) and key scale-up experiments, provides a reliable and scaleable procedure to install the stereocenter in high yield, purity, and enantioselectivity (Scheme 6.3). *n*-Butyllithium (or *n*-hexyllithium, minimum four equivalents) is added to a solution of (1*R*,2*S*)-*N*-pyrrolidinylnorephedrine (Corey and Cimrich, 1994) (two equivalents) and cyclopropylacetylene (two equivalents) at −10°C and the reaction is allowed to warm to 0°C. These conditions are critical to establish the chiral complex that is responsible for the high enantioselectivity. This solution is cooled below −50°C, and trifluoroketone **14** in THF is added and stirred for about 1 h at this temperature before

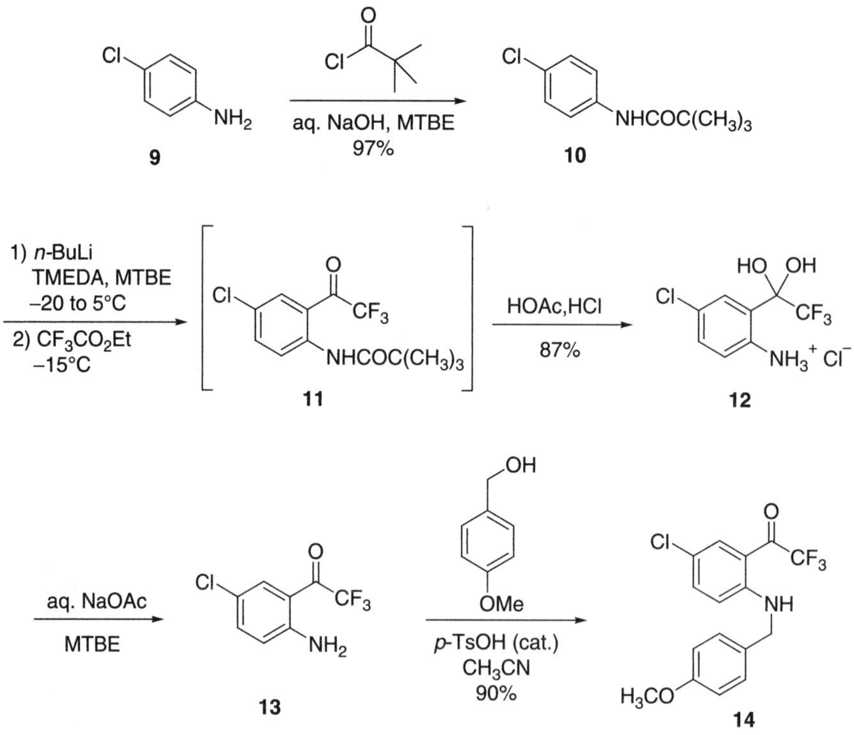

Scheme 6.2.

quenching with aqueous citric acid. The organic layer is exchanged to toluene and the product is crystallized by the addition of heptane.

Removal of the *p*-methoxybenzyl group is accomplished by treatment with dichlorodicyanoquinone (DDQ), which forms quantitatively aminal **17** in an 11.5:1 diastereomeric ratio (Yu and Levy, 1984). The solution is treated with sodium methoxide in methanol, which decomposes the aminal into the desired amine **18** and *p*-methoxybenzaldehyde. Due to the difficulties in separating the *p*-methoxybenzaldehyde from amino alcohol **18**, the aldehyde is reduced in situ with sodium borohydride. The amino alcohol **18** is crystallized from the reaction mixture after neutralization with acetic acid. Additional recrystallization provides the desired amino alcohol **18** in 94% yield.

The final step of the synthesis is to complete the benzoxazinone ring of efavirenz. The most direct and economically desirable route utilizes phosgene (THF–heptane, 0–25°C). After aqueous sodium bicarbonate work-up, efavirenz was crystallized in excellent yields (93–95%) with excellent chemical and optical purities (>99.5%, >99.5% *ee*). When methyl chloroformate is used, the two-step process does not perform as well as phosgene because the residual intermediate carbamate is not easily removed from the desired product and the yield (83%) is therefore lower than the phosgene method. The procedure utilizing the *p*-nitrophenylcarbamate intermediate is conducted in one pot (However, the reaction does not proceed to completion if run at initially high pH (pH > 11) so $KHCO_3$ followed by KOH was used. In this case, ring closure is rapid and K-nitrophenylate competes with **18** for nitrophenyl chloroformate (forming nitrophenyl carbonate as a by-product). The use of $KHCO_3$ ensures pH < 8.5 and complete formation of the

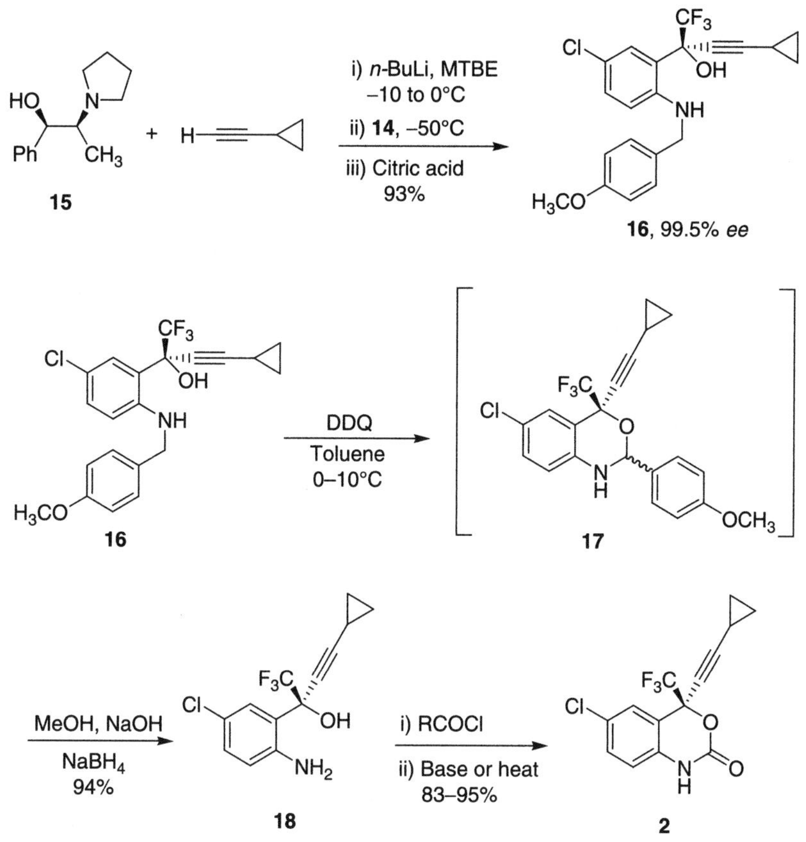

HO
N
Ph
CH3
15
+ H
i) n-BuLi, MTBE
−10 to 0°C
ii) 14, −50°C
iii) Citric acid
93%
CF3
Cl
OH
NH
H3CO
16, 99.5% ee
Cl
CF3
OH
NH
H3CO
16
DDQ
Toluene
0–10°C
F3C
Cl
O
N
H
OCH3
17
MeOH, NaOH
NaBH4
94%
F3C
Cl
OH
NH2
18
i) RCOCl
ii) Base or heat
83–95%
F3C
Cl
O
N
H
O
2

Scheme 6.3.

carbamate before addition of KOH to effect ring closure.) The work-up is quite simple. The organic phase is washed with brine and the solvent is switched to isopropanol. The product is crystallized by the addition of water, which provides the desired efavirenz in 94% yield (>99.5% purity), with no trace of *p*-nitrophenylcarbamate.

Merck scientists made changes to this process, resulting in fewer equivalents of acetylide and chiral ligand without the use of *N*-protection (Chen et al., 1998). In a general procedure, a solution of the chiral ligand (1.5 equivalents) is treated with either dimethyl or diethylzinc at 0°C to room temperature (Scheme 6.4). This mixture is stirred for 1 h followed by the addition of an additive. Most of the additives studied are either carboxylic acids or alcohols. The range of *ee* values from the different additives is 71.6% (2,2-dimethylpropanoic acid) to 96% (2,2,2-trichloroethanol). This solution is mixed

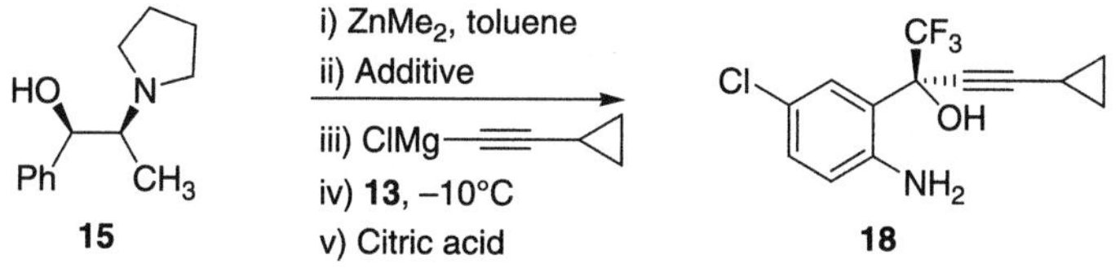

HO
N
Ph
CH3
15
i) ZnMe2, toluene
ii) Additive
iii) ClMg
iv) 13, −10°C
v) Citric acid
CF3
Cl
OH
NH2
18

Scheme 6.4.

with a solution of the chloromagnesium acetylide (1.2 equivalents). The mixture is stirred for about 30 min and then cooled to -10°C. A solution of trifluoroketone **13** is added and the mixture is allowed to stir for 7 h. The optimized conditions are employed on a kilogram scale to provide 95.2% yield of efavirenz that is 99.8% pure and 99.3% *ee*.

Efavirenz is given orally at a dose of one 600-mg tablet per day. As an important part of the AIDS cocktail, efavirenz is administered with the same protease inhibitors and NRTIs as described in the nevirapine section. The half-life of efavirenz after a single dose is 52–76 h, and multiple doses lower the half-life to 40–55 h. Treatment with efavirenz has been associated with the development of serious psychiatric side-effects, including severe depression, suicidal thoughts, aggressive behavior, and paranoid and manic reactions.

The major isozymes that metabolize efavirenz are CYP3A4 and CYP2B6. In fact, after the administration of efavirenz, the CYP3A4 isozyme is upregulated, which is the same isozyme responsible for the metabolism of protease inhibitors. As a result, the amount of protease inhibitor may need to be increased. The major primary metabolite is a result of hydroxylation at C-7 (compound **19**).

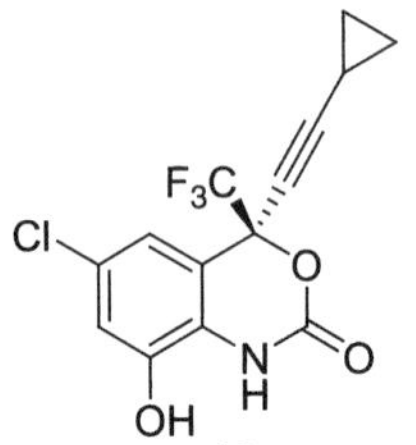

F_3C Cl O N H O OH

19

6.4 SYNTHESIS OF DELAVIRDINE MESYLATE

Delavirdine mesylate is a member of the *bis*(heteroaryl)piperazine (BHAP) class of non-nucleoside HIV-1 reverse transcriptase inhibitors (Adams et al., 1998; Romero et al., 1993; Romero, 1994). This class of compounds was discovered by Upjohn scientists from a computer-directed dissimilarity analysis of the Pharmacia & Upjohn chemical library to select compounds for screening against HIV-1 RT. The result of the in vitro assay (Deibel et al., 1990) is an IC_{50} of 0.260 μM, which is comparable to AZT. In accordance with the previous NNRTIs, delavirdine is a noncompetitive inhibitor of reverse transcriptase, and has a synergistic effect with nucleoside transcriptase and protease inhibitors (Chong et al., 1994).

The antiviral effect of delavirdine against multiple laboratory adapted strains and clinical isolates of HIV-1 is measured in cell culture using various cell lines [MT-2 cells and peripheral blood mononuclear cells (PBMC)] (Dueweke et al., 1993a). In the MT-2 cells infected with HIV-1 (IIIb) with a MOI (multiplicity of infection) of 0.001 (as determined by measuring levels of syncytia formation), the ED_{50} values for delavirdine and AZT are 10 nM and 100 nM, respectively. The CC_{50} for MT-2 cells is >10 μM for delavirdine and 10 μM for AZT. Under the same MOI with HIV (IIIb), comparable results are observed in the PBMC cell line. Using this assay, the ED_{50} of delavirdine is 0.1–1 μM, and the ED_{50} for AZT is 1 μM.

As the NNRTIs are structurally diverse and yet bind to RT at a common site, the similar occurrence of resistance-conferring mutations is not surprising. As a consequence, the effectiveness of other NNRTIs may be compromised by the emergence of HIV-1 variants caused by a previous NNRTI therapy (Sardana et al., 1992). Experiments were performed in which HIV-1 strains (JR-CSF or MF) are cultured in human lymphocytes in the presence of partially inhibitory concentrations of delavirdine (Dueweke et al., 1993b). These conditions yield mutants that are 100-fold resistant. In order to determine what mutation(s) occurs, PCR (polymerase chain reaction) amplification and DNA sequence analysis of the RT coding region were applied and indicated that mutation P236L had occurred. Mutations at amino acids 181 or 183, which have been associated with resistance to other NNRTIs, are not detected.

In order to measure the effect of other mutated HIV (IIIb) RTs, oligonucleotide-directed mutagenesis is conducted (Sheehan et al., 1961). Both the Y181C and K103N mutations possess resistance to delavirdine and nevirapine. However, delavirdine has a greater inhibitory effect than nevirapine. As an example, the IC_{50} for delavirdine against the K103N mutant was 8.3 μM. The other NNRTIs studied fail to achieve 50% inhibition at 60 μM (the highest concentration examined). The role of position 181 appears to be more important in the binding of nevirapine and the other NNRTIs tested than in the case of delavirdine.

Additional cell culture experiments are conducted to confirm the sensitization observed with the mutant RTs. Lower concentrations of nevirapine (5–20-fold) are

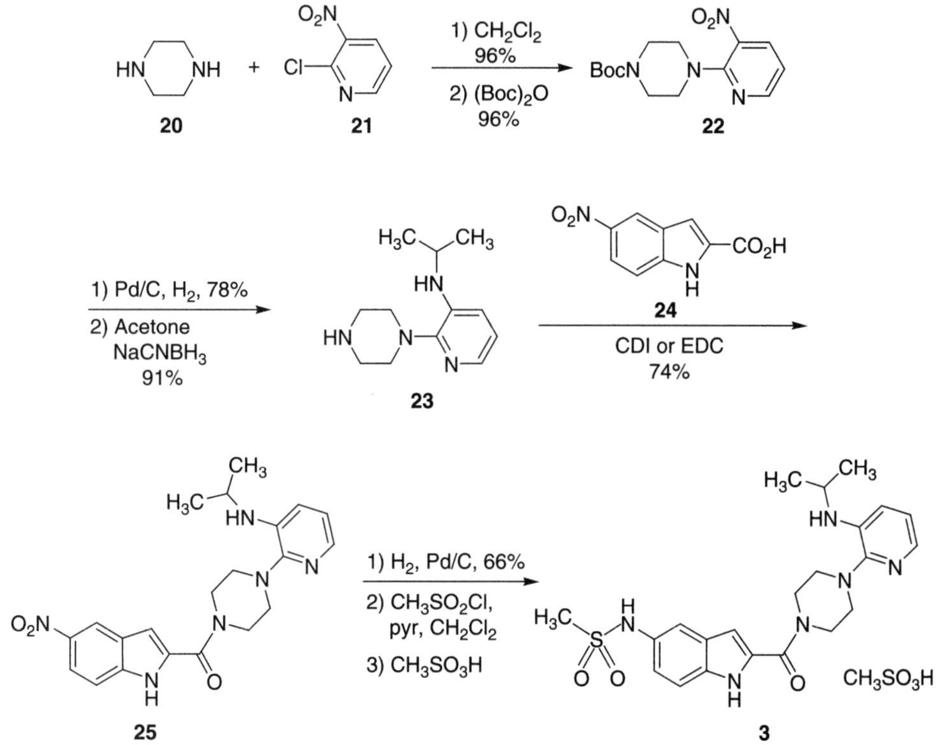

20 + 21 → 1) CH_2Cl_2, 96%; 2) $(Boc)_2O$, 96% → 22

→ 1) Pd/C, H_2, 78%; 2) Acetone, $NaCNBH_3$, 91% → 23

23 + 24 → CDI or EDC, 74% → 25

25 → 1) H_2, Pd/C, 66%; 2) CH_3SO_2Cl, pyr, CH_2Cl_2; 3) CH_3SO_3H → 3 (CH_3SO_3H)

Scheme 6.5.

required to completely inhibit viral replication of delavirdine-resistant HIV-1 variants as compared to wild-type HIV-1, substantiating the above results (Sheehan et al., 1961).

Synthesis of delavirdine begins with the addition of piperazine (**20**) to chloropyridine **21** (Scheme 6.5). The nitro group is reduced and the resultant amine undergoes reductive amination with acetone to provide pyridylpiperazine **23**. Coupling of **23** with 6-nitroindole-2-carboxylic acid (**24**) is accomplished using either 1-ethyl-3-(dimethylamino)propylcarbodiimide (EDC) (Desai and Straniello, 1993; Sheehan et al., 1961) or 1,1′-carbonyldiimidazole (CDI) (Morton et al., 1988; Romero et al., 1994) to give amide **25**. The nitro group is reduced under standard palladium on carbon hydrogenation conditions. The resulting amine is then sulfonylated with methanesulfonyl chloride to provide delavirdine, which is then transformed to delavirdine mesylate (**3**).

Delavirdine mesylate is given orally at doses of two 200-mg tablets, three times a day. The half-life of delavirdine mesylate is approximately 6 hs. Severe, life-threatening skin reactions have been associated with the use of delavirdine. Skin rashes appear in about 25% of patients. Delavirdine may cause serious and/or life-threatening side-effects if taken with certain other medications, including astemizole (Hismanal®), terfenadine (Seldane®), midazolam (Versed®), triazolam (Halcion®), ergotamine (Ergostat®, Ergomar®, and others), dihydroergotamine (D.H.E. 45), nifedipine (Procardia®, Adalat®), sildenafil (Viagra®), quinidine (Cardioquin®, Quinaglute®, Quinidex®, and others), and others. Delavirdine mesylate is administered as part of the AIDS cocktail with protease inhibitors and NRTIs.

Investigation of the in vitro metabolism of delavirdine is accomplished using mouse, rat, dog, monkey, rabbit, and human liver microsomes. The primary metabolite observed is the *N*-dealkylated delavirdine **26**. Another primary metabolite observed is the hydroxylation of the pyridine ring at C-6′ (compound **27**). The primary metabolism is by CYP3A4 and also CYP2D6. Delavirdine reduces the activity of CYP3A4, thereby inhibiting its own metabolism.

26 **27**

REFERENCES

Adams, W. J., Aristoff, P. A., Jensen, R. K., Morozowich, W., Romero, D. L., Schinzer, W. C., Tarpley, W. G., and Thomas, R. C. (1998). *Pharmaceutical Biotechnology*, 11: 285.

Arnold, E., Das, K., Ding, J., Yadav, P. N. S., Hsiou, Y., Boyer, P. L., and Hughes, S. H. (1996). *Drug Des. Discovery*, 13: 29.

Bacheler, L. T. (1999). *Drug Resist. Updates*, 2: 56.

Bacheler, L., Weislow, O., Snyder, S., Hanna, G., D'Aquila, R. and the SUSTIVA Resistance Study Team (1998). Abstract presented at the 12th World AIDS Conference.

Balzarini, J. (2004). *Curr. Top. Med. Chem.*, 4: 921.

Breslin, H. J., Kukla, M. J., Kromis, T., Cullis, E., De Knaep, F., Pauwels, R., Andries, K., De Clercq, E., Janssen, M. A. C., and Janssen, P. A. J. (1999). *Bioorg. Med. Chem.*, 7: 2427.

Breslin, H. J., Kukla, M. J., Ludovici, D. W., Mohrbacher, R., Ho, W., Miranda, M., Rodgers, J. D., Hitchens, T. K., Leo, G., Gauthier, D. A., Ho, C. Y., Scott, M. K., De Clercq, E., Pauwels, R., Andries, K., Janssen, M. A. C., and Janssen, P. A. J. (1995). *J. Med. Chem.*, 38: 771.

Cheeseman, S. H., Hattox, S. E., McLaughlin, R. A., Koup, R. A., Andrews, C., Bova, C. A., Pav, J. W., Roy, T., Sullivan, J. L., and Keirns, J. J. (1993). *Antimicrobial Agents Chemother.*, 37: 178.

Chen, C. Y., Tillyer, R., and Tan, L. (1998). WO 98/51676.

Chong, K. T., Pagano, P. J., and Hinshaw, R. R. (1994). *Antimicrob. Agents Chemother.*, 38: 288.

Corey, E. J. and Cimprich, K. A. (1994). *J. Am. Chem. Soc.*, 116: 3151.

De Clercq, E. (1998). *Antiviral Res.*, 38: 153.

Deibel, M. R., McQuade, T. J., Brunner, D. P., and Tarpley, W. G. (1990). *AIDS Res. Hum. Retroviruses*, 6: 329.

Denizot, F. and Lang, R. (1986). *J. Immunol. Methods*, 89: 271.

Desai, M. C. and Straniello, L. M. S. (1993). *Tetrahedron Lett.*, 34: 7685.

Ding, J., Das, K., Moereels, H., Koymans, L., Andries, K., and Janssen, P. A. J. (1995). *Nat. Struct. Biol.*, 2: 407.

Dueweke, T. J., Poppe, S. M., and Romero, D. L. (1993a). *Antimicrob. Agents Chemother.*, 37: 1127.

Dueweke, T. J., Pushkarskaya, T., Poppe, S. M., Swaney, S. M., Zhao, J. Q., and Chen, I. S. (1993b). *Proc. Natl. Acad. Sci. USA*, 90: 4713.

Dueweke, T. J., Pushkarskaya, T., Poppe, S. M., Swaney, S. M., Zhao, J. Q., Chen, I. S. Y., Stevenson, M., and Tarpley, W. G. (1993). *Proc. Natl. Acad. Sci. USA*, 90: 4713.

Emert, J., Goldenberg, M., Chiu, G. L., and Valeri, A. (1977). *J. Org. Chem.*, 42: 2012.

Esnouf, R., Ren, J., Ross, C., James, Y., Stammers, D., and Stuart, D. (1995). *Struct. Biol.*, 2: 303.

Fuhrer, W. and Gschwend, H. W. (1979). *J. Org. Chem.*, 44: 1133.

Gilboa, E. and Mitra, S. W. (1978). *Cell*, 18: 93.

Grozinger, K. G., Byrne, D. P., Nummy, L. J., Ridges, M. D., and Salvagno, A. (2000). *J. Heterocyclic Chem.*, 37: 229.

Grozinger, K. G., Fuchs, V., Hargrave, K. D., Mauldin, S., Vitous, J., Campbell, S., and Adams, J. (1995). *J. Heterocyclic Chem.*, 32: 259.

Hajós, G., Riedl, Z., Molnár, J., and Szab, D. (2000). *Drugs Future*, 25: 47.

Hargrave, K. D., Proudfoot, J. R., Grozinger, K. G., Cullen, E., Kapadia, S. R., Patel, U. R., Fuchs, V. U., Mauldin, S. C., Vitous, J., Behnke, M. L., Klunder, J. M., Pal, K., Skiles, J. W., McNeil, D. W., Rose, J. M., Chow, G. C., Skoog, M. T., Wu, J. C., Schmidt, G., Engel, W. W., Eberlein, W. G., Saboe, T. D., Campbell, S. J., Rosenthal, A. S., and Adams, J. (1991). *J. Med. Chem.*, 34: 2231.

Henneus, C., Boxus, T., Tesolin, L., Pantano, G., and Merchand-Brymaert, J. (1996). *Synthesis*, 495.

Ho, W., Kukla, M. J., Breslin, H. J., Ludovici, D. W., Grous, P. P., Diamond, C. J., Miranda, M., Rodgers, J. D., Ho, C. Y., De Clercq, E., Pauwels, R., Janssen, M. A. C., and Janssen, P. A. J. (1995). *J. Med. Chem.*, 38: 794.

Kohlstaedt, L. A., Wang, J., Friedman, J. M., Rice, P. A., and Steitz, T. A. (1992). *Science*, 256: 1783.

Kukla, M. J., Breslin, H. J., Pauwels, R., Feddle, C. L., Miranda, M., Scott, M. K., Sherrill, R. G., Raeymaekers, A., Van Gelder, J., Andries, K., Janssen, M. A. C., De Clercq, E., and Janssen, P. A. J. (1991). *J. Med. Chem.*, 34: 746.

Larder, B. A., Purifoy, D. J. M., Powell, K. L., and Darby, G. (1987). *Nature*, 327: 716.

Merluzzi, V. J., Hargrave, K. D., Labadia, M., Grozinger, K., Skoog, M., Wu, J. C., Shih, C.-K., Eckner, K., Hatton, S., Adams, J., Rosenthal, A. S., Faanes, R., Eckner, R. J., Koup, R. A., and Sullivan, J. L. (1990). *Science*, 250: 1411.

Morton, R. C., Mangroo, D., and Gerber, G. E. (1988). *Can. J. Chem.*, 66: 1701.

Nanni, R. G., Ding, J., Jacobo-Molina, A., Hughes, S. H., and Arnold, E. (1993). *Perspect. Drug Discovery Des.*, 1: 129.

Nunberg, J. H., Schleif, W. A., Boots, E. J., O'Brien, J. A., Quintero, J. C., Hoffman, J. M., Emini, E. A., and Goldman, M. E., (1991). *J. Virol.*, 65: 4887.

Pauwels, R., Andries, K., Debyser, Z., Kukla, M. J., Schols, D., Breslin, H. J., Woestenborghs, R., Desmyter, J., Janssen, M. A. C., De Clercq, E., and Janssen, P. A. J. (1994). *Antimicrob. Agents Chemother.*, 38: 2863.

Pierce, M. E., Parsons, R. L., Radesca, L. A., Lo, Y. S., Silverman, S., Moore, J. R., Islam, Q., Choudhury, A., Fortunak, J. M. D., Nguyen, D., Luo, C., Morgan, S. J., Davis, W. P., Confalone, P. N., Chen, C., Tillyer, R. D., Frey, L., Tan, L., Xu, F., Zhao, D., Thompson, A. S., Corley, E. G., Grabowski, E. J. J., Reamer, R., and Reider, P. J. (1998). *J. Org. Chem.*, 63: 8536.

Prasad, V. R. and Goff, S. P. (1990). *Ann. N.Y. Acad. Sci. USA*, 616: 11.

Richman, D., Shih, C.-K., Lowy, I., Rose, J., Prodanovich, P., Goff, S., and Griffin, J. (1991). *Proc. Natl. Acad. Sci USA.*, 88: 11241.

Richman, D. D., Havlir, D., Carbeil, J., Looney, D., Inacio, C., Spector, S. A., Sullivan, J., Cheeseman, S., Barringer, K., Pauletti, D., Shih, C.-K., Myers, M., and Griffin, J. (1994). *J. Virol.*, 68: 1660.

Romero, D. L. (1994). *Drugs of the Future*, 19: 238.

Romero, D. L., Morge, R. A., Genin, M. J., Biles, C., Busso, M., Resnick, L., Althaus, I. W., Reusser, F., Thomas, R.C., and Tarpley, W. G. (1993). *J. Med. Chem.*, 36: 1505.

Romero, D. L., Morge, R. A., Biles, C., Berrios-Pena, N., May, P. D., Palmer, J. R., Johnson, P. D., Smith, H. W., Busso, M., Tan, C.-K., Voorman, R. L., Reusser, F., Althaus, I. W., Downey, K. M., So, A. G., Resnick, L., Tarpley, W. G., and Aristoff, P. A. (1994). *J. Med. Chem.*, 37: 999.

Salahuddin, S. Z. (1983). *Virology*, 129: 51.

Sardana, V. V., Emini, E. A., Gotlieb, L., Graham, D. J., Lineberger, D. W., Long, W. J., Schlabach, A. J., Wolfgang, J. A., and Condra, J. H. (1992). *J. Biol. Chem.*, 267: 17526.

Sheehan, J. C., Cruickshank, P. A., and Boshart, G. I. (1961). *J. Org. Chem.*, 26: 2525.

Smerdon, S. J., Jager, J., Wang, J., Kohlstaedt, L. A., Friedman, J. M., Rice, P. A., and Steitz, T. A. (1994). *Proc. Natl. Acad. Sci. USA*, 91: 3911.

Tantillo, C., Ding, J., Jacobo-Molina, A., Nanni, R. G., Boyer, P. L., Hughes, S. H., Pauwels, R., Andries, K., Janssen, P. A. J., and Arnold, E. (1994). *J. Mol. Biol.*, 243: 369.

Thompson, A. S., Corley, E. G., Grabowski, E. J. J., and Yasuda, (1996). N. WO 96/37457.

Thompson, A. S., Corley, E. G., Huntington, M. F., and Grabowski, E. J. J. (1995). *Tetrahedron Lett.*, 36: 8937.

Thompson, A., Corley, E. G., Hungington, M. F., Grabowski, E. J. J., and Collum, D. B. (1998). *J. Am. Chem. Soc.*, 120: 2028.

White, E. L., Buckheit, R. W., Jr., Ross, L. J., Germany, J. M., Andries, K., Pauwels, R., Janssen, P. A. J., Shannon, W. M., and Chirigos, M. A. (1991). *Antiviral Res.*, 16: 257.

Young, S. D., Britcher, S. F., Payne, L. S., Tran, L. O., and Lumma (1996). W. C. US 5519021.

Young, D. S., Britcher, S. F., Tran, L. O., Payne, L. S., Lumma, W. C., Jr., Lyle, T. A., Huff, J. R., Anderson, P. S., and Olsen, D. B. (1995). *Antimicrob. Agents Chemother.*, 39: 2602.

Yu, C. and Levy, G. C. (1984). *J. Am. Chem. Soc.*, 106: 6533.

7

NEUROMINIDASE INHIBITORS FOR INFLUENZA: OSELTAMIVIR PHOSPHATE (TAMIFLU®) AND ZANAMIVIR (RELENZA®)

Douglas S. Johnson and Jie Jack Li

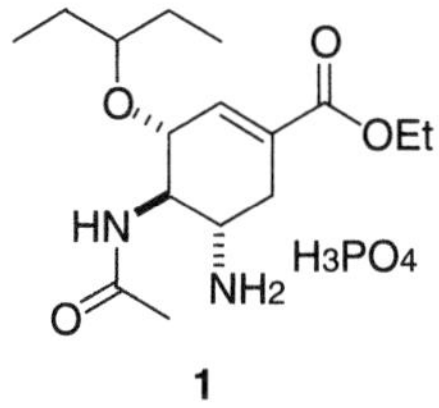

1

USAN: Oseltamivir phosphate
Trade name: Tamiflu®
Company: Gilead, Roche
Launched: 1999
M.W. 312.4 (parent)

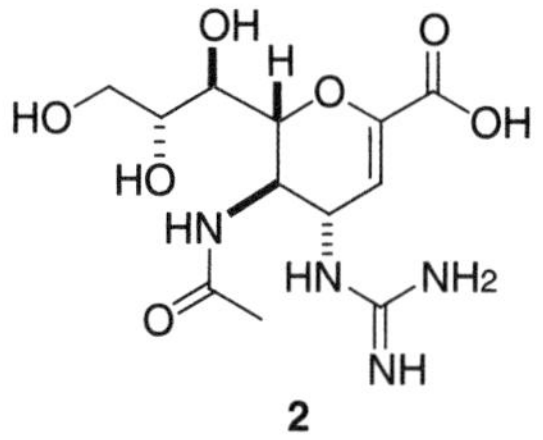

2

USAN: Zanamivir
Trade name: Relenza®
Company: Biota Holdings, GlaxoSmithKline
Launched: 1999
M.W. 332.3

7.1 INTRODUCTION (Chand, 2005; Couch, 2000; Gubareva et al., 2000; Liu et al., 2005; Moscona, 2005)

Influenza develops in approximately 20% of the global population each year. The 1918 influenza killed 50 million people worldwide, when an avian flu was passed on to humans. Chemistry genius Robert Burns Woodward lost his father to the 1918 Spanish influenza pandemic when he was only two years old. The Asian influenza occurred in

The Art of Drug Synthesis. Edited by Douglas S. Johnson and Jie Jack Li

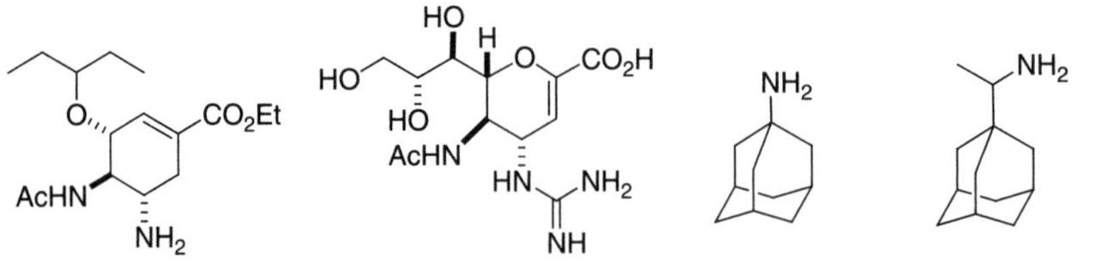

Oseltamivir (**1**) Zanamivir (**2**) Amantadine (**3**) Rimantadine (**4**)

Figure 7.1. The four drugs available for treatment of influenza infections.

1957 and the Hong Kong influenza took place in 1968, respectively, claiming 70,000 and 34,000 lives in the United States alone. In 2004–2005 around 70 people in Asia died of the H5N1 strain of avian flu, although the virus has not yet passed easily among humans. The fear of hundreds of millions of people dying in a pandemic influenza outbreak has pushed antiviral drugs into the spotlight, worldwide. Overnight, oseltamivir (**1**, Tamiflu®) and zanamivir (**2**, Relenza®) have been transformed from ugly ducklings into beautiful swans on the drug market.

Currently, there are four drugs available for the treatment or prophylaxis of influenza infections (Fig. 7.1): the adamantanes, including amantadine (**3**) and rimantadine (**4**), and the neuraminidase inhibitors, including oseltamivir (**1**, Tamiflu®) and zanamivir (**2**, Relenza®). The adamantanes, as M_2 ion channel inhibitors, interfere with viral uncoating inside the cell. They are effective only against influenza A and are associated with several toxic effects as well as drug resistance. The neuraminidase inhibitors are newer drugs, which have shown greater efficacy for both influenza A and B and are associated with fewer side-effects in comparison with the adamantanes.

Oseltamivir (**1**), the first orally active neuraminidase inhibitor, was discovered by Choung U. Kim (of "Corey–Kim oxidation" fame) and co-workers at Gilead Sciences in 1995. Gilead and Roche began co-developing it in 1995, and the FDA approved it in 1999. Zanamivir (**2**), however, was discovered by Biota Holdings, a small Austrian biotechnology concern. In the United States, Biota established an alliance with GlaxoSmithKline for development and marketing of zanamivir (**2**).

The influenza neuraminidase is one of two major glycoproteins located on the influenza virus membrane envelope (the other one is haemagglutinin, HA). As the name suggests,

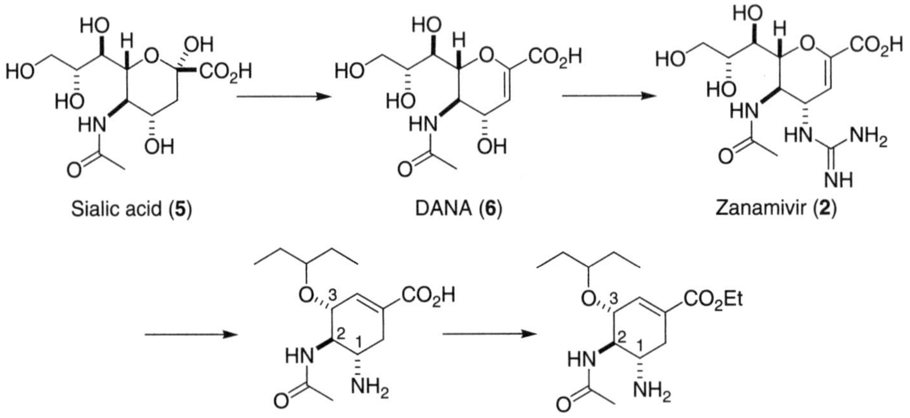

Sialic acid (**5**) DANA (**6**) Zanamivir (**2**) GS–4071 (**7**) Oseltamivir (**1**)

Figure 7.2. Progression of zanamivir and oseltamivir from sialic acid.

neuraminidase is an enzyme that is responsible for the cleavage of terminal sialic acid (**5**) residues from glucoconjugates, and is essential for virus replication and infectivity (Fig. 7.2). As a consequence, neuraminidase is sometimes also known as sialidase. Theoretically, sialic acid itself would be a neuraminidase inhibitor. However, it is a sugar-like molecule, and is thus rapidly cleared in the stomach. Therefore, efforts have been focused on sialic acid analogs, which possess better bioavailability, thus allowing oral administration. These efforts have been very fruitful so far thanks to the availability of a high-resolution crystal structure of influenza neuraminidase and its complex with sialic acid, which has tremendously aided rational drug design.

7.1.1 Relenza (Cheer and Wagstaff, 2002a, 2002b; Colman, 2005; Fromtling and Castaner, 1996; Waghorn and Goa, 1998)

Although oseltamivir (**1**, Tamiflu) has risen to be a superstar in the wake of the avian flu, zanamivir (**2**, Relenza) seems to be underappreciated by the world medical society, possibly because zanamivir (**2**) is a mouth spray and may cause problems for patients with breathing problems such as asthma. In reality, zanamivir (**2**) has a great tolerability, similar to that of placebo, for otherwise healthy adults and higher-risk patients such as the elderly and children. Recommended doses (10 mg twice daily for 5 days) of zanamivir (**2**) did not adversely affect pulmonary function in patients with respiratory disorders in a well-controlled trial, although there have been reports of bronchospasm and/or decline in respiratory function.

The framework of zanamivir (**2**) is the transition state analog of neuraminidase, Neu5Ac2en (DANA, **6** in Fig. 7.2). In fact, it is actually the 4-guanadino-derivative of DANA (**6**). The genesis of zanamivir (**2**) is one of the early examples of structure-based drug design (SBDD) in the sense that the protein structure directed the design of the ligand and the protein-bound ligand conformation is close to the designed structure.

Zanamivir (**2**) is a potent competitive inhibitor of viral neuraminidase glycoprotein, which is essential in the infective cycle of both influenza A and B viruses. It inhibits a wide range of influenza A and B types in vitro as well as in vivo. The concentrations of inhibiting in vitro plaque formation of influenza A and B virus by 50% in Madin–Darby canine kidney (MDCK) cells were 0.004–0.014 μmol/L in laboratory-passaged strains, and 0.002–16 μmol/L in assays of clinical isolates. Due to its low bioavailability, it is delivered by inhalation via the Diskhaler®, 10 mg twice daily, or intranasally 2–4 times daily for 5 days. After an intravenous dose of 1–16 mg, the median elimination half-life was $t_{1/2} = 7$ h, the volume of distribution at steady state was $V_{dss} = 16$ L, and 90% of the dose was excreted unchanged in the urine. After intranasal and inhaled (dry powder) administration, maximum serum concentrations occurred within 2 h and the terminal phase half-lives were 3.4 and 2.9 h, respectively. The bioavailabilities were 10 and 25%, respectively, and 20% after inhalation of zanamivir (**2**) by nebulizer.

7.1.2 Tamiflu (Doucette and Aoki, 2001; Lew et al., 2000; McClellan and Perry, 2001)

Oseltamivir (**1**) is a prodrug of GS-4071 (**7**, Fig. 7.2). It is also a potent competitive inhibitor of viral neuraminidase glycoprotein for both influenza A and B types. Similar to zanamivir (**2**), oseltamivir (**1**) is also the fruit of SBDD in the sense that the protein structure directed the design of the ligand (sialic acid) and the protein-bound ligand conformation is close to the designed structure. In particular, the neuraminidase active site contains several well-defined pockets. All residues that make direct contact with the substrate are strictly conserved among both influenza A and B neuraminidases. In an attempt to achieve high-affinity inhibitors, Gilead scientists took out the oxygen atom of the sialic-acid framework, resulting in the cyclohexene core structure. They discovered that the length, substitution, and geometry of the C-3 (see **7** and **1** in Fig. 7.2) alkyl side-chain profoundly influence neuraminidase inhibitory activity. At the end, the optimal alkyl substitution was the 3-pentyl group, which is active for both type A and B inhibition. The outcome of their structure–activity relationship (SAR) optimization provided GS-4071 (**7**), a very potent neuraminidase inhibitor for both type A and B viruses.

Unfortunately, although GS-4071 (**7**) was designed with the intent of developing an orally active influenza neuraminidase inhibitor, pharmacokinetic experiments (Table 7.1) demonstrated that the oral bioavailability of GS-4071 (**7**) was only ~5%. Gratifyingly, conversion of GS-4071 (**7**) into its corresponding ethyl ester resulted in the highly orally bioavailable drug oseltamivir (**1**). The oral bioavailability in rats following oral administration of oseltamivir (**1**) was found to be more than five-fold higher that that of the parent compound **7**. In addition, a high concentration of GS-4071 (**7**) in plasma following oral administration of oseltamivir (**1**) was observed in mice (30%), dog (70%), and humans (80%). After oral administration, oseltamivir (**1**) is rapidly absorbed from the GI tract with a bioavailability of 79% in human. It is then, not surprisingly, extensively metabolized into its corresponding acid, GS-4071 (**7**), predominantly by hepatic esterases. GS-4071 is then rapidly distributed to the primary site of influenza virus replication and to the middle ear and sinuses. It is eliminated by a first-order process, primarily by glomerular filtration and renal tubular secretion, with a terminal elimination half-life of 6–10 h. Its mean renal clearance is 21.7 L/h in healthy adult volunteers after administration of 20–1000 mg. As Tamiflu (**1**) does not bind to hepatic cytochrome P450 enzymes, there have been no significant drug–drug interactions.

Tamiflu and Relenza did not sell well at first for common flu, because doctors and the public choose the flu vaccine for prevention. Therefore, antivirals as treatment were not profitable, so much so that Roche scaled back their marketing efforts for Tamiflu. Gilead accused Roche of a "consistent record of inactivity and neglect" and wanted to take back its right in 2005. The issue was settled after Roche increased their compensation to Gilead.

Interestingly, the third neuraminidase inhibitor, peramivir, saw an incarnation after the avian flu scare. Peramivir was developed by BioCryst Pharmaceuticals and licensed to Johnson & Johnson (Chand et al., 2005). However, because the bioavailability of peramivir was low and the prospective market small, Johnson & Johnson pulled out of the alliance. In 2005, peramivir was regaining its life as a possible drug delivered intravenously.

TABLE 7.1. The Pharmacokinetic Profile of Oseltamivir (1) and Its Active Metabolite GS-4071 (7)

	Oseltamivir		
Parameters	Plasma	BALF	GS-4071
C_{max} (mg/L)	7.8	—	—
T_{max} (h)	1.0	—	—
$T_{1/2}$ (h)	0.7	1.32	1.32
$AUC_{0-\infty}$ (mg h/L)	19.0	28.7	52.6

BALF, bronchoalveolar lining fluid.

7.2 SYNTHESIS OF OSELTAMIVIR PHOSPHATE (TAMIFLU®) (Abrecht et al., 2001, 2004; Bischofberger et al., 1999; Federspiel et al., 1999; Graul et al., 1999; Harrington et al., 2004; Iding et al., 2001; Karpf and Trussardi, 2001; Kent et al., 1998, 1999; Kim, et al., 1997, 1998; McGowan and Berchtold, 1981; Rohloff et al., 1998)

Oseltamivir phosphate (GS-4104, **1**) was discovered at Gilead Sciences and co-developed with Hoffmann-La Roche. The initial Gilead discovery synthesis of GS-4071 (**7**) started from (−)-shikimic acid (**8**), with the goal of transforming it to a key intermediate that could be easily derivatized to form carbocyclic analogs of sialic acid (Abrecht et al., 2004; Chand et al., 2005; Kim et al., 1997, 1998). To this end, (−)-shikimic acid (**8**) was treated with diethyl azodicarboxylate (DEAD) in the presence of triphenylphosphine to give the *syn*-epoxide **9** (Scheme 7.1) (Kim et al., 1998). After protection of the hydroxyl group in **9** with MOMCl, the epoxide was opened with sodium azide in the presence of ammonium chloride to give the azido alcohol **10**. The alcohol was converted to the mesylate and the azide was reduced with triphenylphosphine to the amine, which then displaced the mesylate to provide aziridine **11**. Ring opening of aziridine **11** with sodium azide in the presence of ammonium chloride followed by deprotection of the MOM-ether gave amino alcohol **12**. The aziridine **13** was formed in a one-pot process by protection of the amino group in **12** with trityl chloride followed by mesylation of the hydroxyl in the presence of triethylamine. *N*-Tritylaziridine **13** was opened with 3-pentanol in the presence of $BF_3{\cdot}Et_2O$, and the resulting crude amino ether was acetylated with acetic anhydride to afford **14** as a single regio- and stereoisomer. Finally, reduction of the azide with triphenylphosphine followed by saponification of the methyl ester provided **7**.

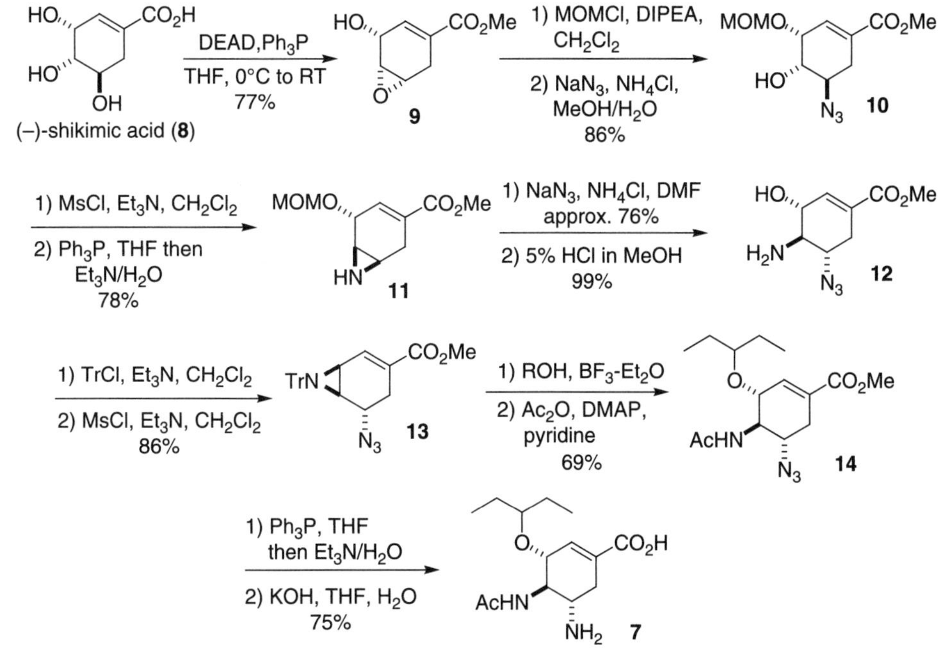

Scheme 7.1. Gilead synthesis of GS-4071 (**7**) from (−)-shikimic acid (**8**).

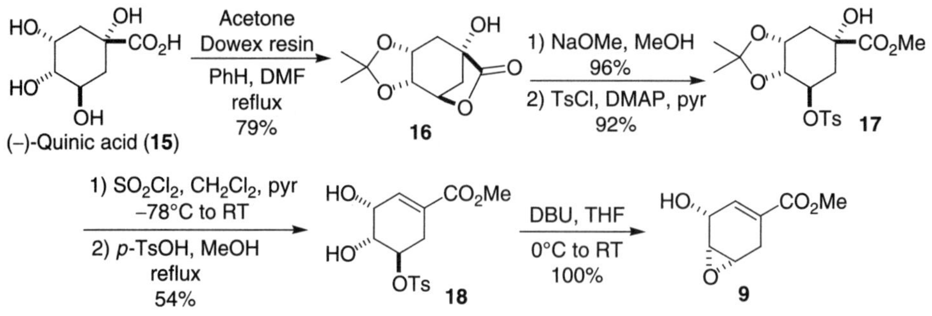

Scheme 7.2. Synthesis of epoxide **9** from (−)-quinic acid (**15**).

Due to the high cost and low availability of (−)-shikimic acid, extracted from the Chinese spice star anise, on scale, the use of (−)-quinic acid (**15**) as a chiral starting material for the synthesis of GS-4071 (**7**) was explored (Abrecht et al., 2004). (−)-Quinic acid is much cheaper and is commercially abundant. The transformation of **15** to the key epoxide intermediate **9** is shown in Scheme 7.2. (−)-Quinic acid was converted to the acetonide with concomitant lactonization to give **16**. The lactone was opened with sodium methoxide to afford the methyl ester and the resulting secondary alcohol was tosylated selectively to provide **17**. Compound **17** was treated with sulfuryl chloride in pyridine to effect dehydration of the tertiary alcohol and the acetonide was removed using *p*-toluenesulfonic acid in refluxing methanol to give cyclohexene **18**, which crystallized out of the reaction mixture in 54% yield. The other olefinic regioisomer aromatized under the reaction conditions and was easily separated by crystallization. Finally, **18** was treated with DBU in tetrahydrofuran to produce the epoxide **9** in quantitative yield.

The first scalable synthesis of oseltamivir phosphate (**1**) was described by the Gilead process chemistry group (Scheme 7.3) (Bischofberger et al., 1999; Graul et al., 1999; Kent et al., 1998, 1999; McGowan and Berchtold, 1981; Rohloff et al., 1998). The synthesis started with the conversion of (−)-quinic acid (**15**) to compound **19** in a similar fashion as that described for compound **17** in Scheme 7.2. Dehydration of **17** was accomplished using sulfuryl chloride and pyridine to give a 4 : 1 mixture of 1,2- and 1,6-olefin regioisomers **18** and **19** in 60% yield. It was not possible to separate **18** and **19** by fractional crystallization so the crude mixture was treated with pyrrolidine and catalytic $Pd(Ph_3P)_4$, which led to the selective conversion of **19** to **20**. Compound **20** was then removed by aqueous sulfuric acid extraction and the pure 1,2-olefin isomer **18** was isolated by crystallization in 42% yield from **17**. Transketalization of **18** with 3-pentanone in the presence of catalytic perchloric acid gave **21**. The 3,4-pentylidene ketal was not incorporated at the beginning of the synthesis because the corresponding intermediates were not crystalline. Reductive opening of the 3,4-pentylidene ketal **21** was accomplished using trimethylsilyl triflate and borane dimethyl sulfide complex to give a 10 : 1 : 1 mixture of **22**, the isomeric pentyl ether and the diol. This crude mixture was treated with potassium bicarbonate in aqueous ethanol followed by heptane extraction to give the crystalline epoxide **23** in 60% yield from **21**. Epoxide **23** was heated with sodium azide and ammonium chloride in aqueous ethanol to give a 10 : 1 mixture of azido alcohol **24** and its corresponding regioisomer. Reductive cyclization of **24** with trimethylphosphine afforded aziridine **25**. Ring-opening of aziridine **25** with sodium azide in the presence of ammonium chloride provided the azidoamine, which was directly acylated with

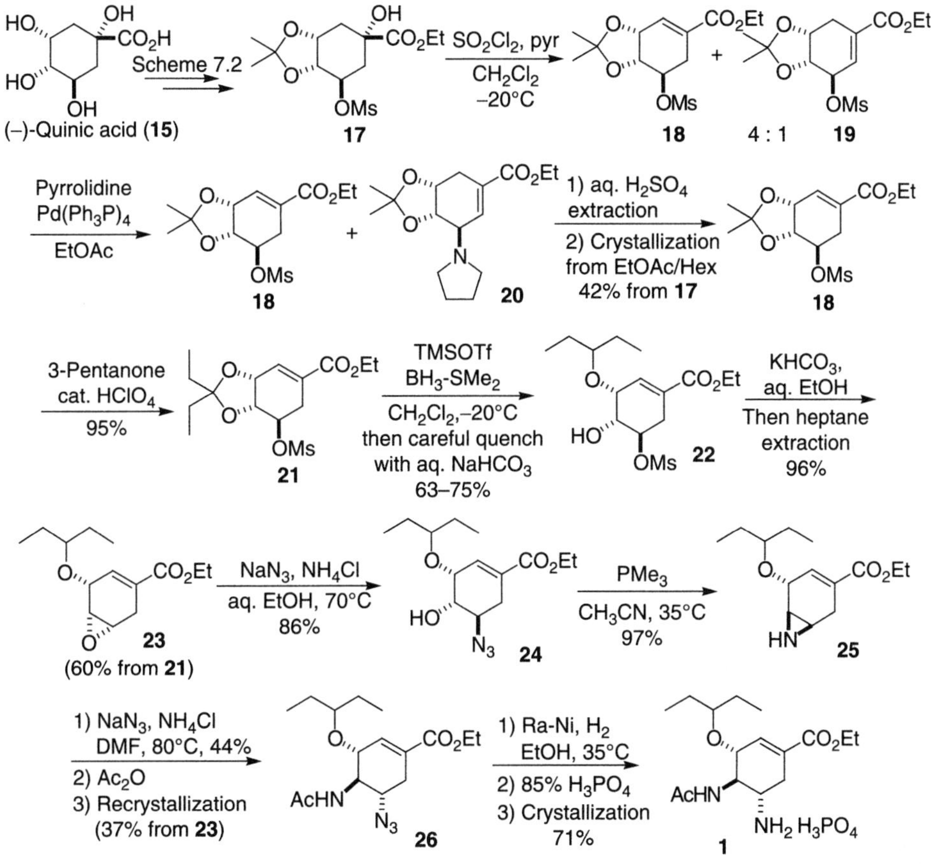

Scheme 7.3. Gilead's first process route to oseltamivir phosphate (**1**).

acetic anhydride to afford azidoacetamide **26** in 37% yield (from **23**) after recrystallization. The azide in **26** was reduced using catalytic hydrogenation with Raney nickel in ethanol. Following the removal of the catalyst, 85% phosphoric acid (1 equiv.) was added and oseltamivir (**1**) crystallized as the phosphoric acid salt, which was isolated in 71% yield from **26**. This route was amenable to kilogram-scale synthesis of **1**, as it required only three isolated crystalline intermediates, with no chromatography, and was used to provide the drug for phase I clinical trials.

Additional process development by Gilead Sciences, Roche, and various third parties led to further optimization of the synthesis, allowing for the production of tons of **1** for launch (Scheme 7.4) (Abrecht et al., 2004; Federspiel et al., 1999). (−)-Shikimic acid was now available in ton quantities, either by extraction of Chinese star anise or ginkgo leaves, or by fermentation using a genetically engineered *E. coli* strain. Thus, (−)-shikimic acid (**8**) was esterified under acidic conditions using $SOCl_2$ in refluxing ethanol to provide ester **27**. The solvent was exchanged from ethanol to ethyl acetate and **27** was treated with 2,2-dimethoxypropane in the presence of *p*-toluenesulfonic acid to afford acetonide **28**. The alcohol in **28** was mesylated and the product was crystallized from methanol to give **18**. Transketalization of **18** to **21** was accomplished using catalytic trifluoromethylsulfonic acid instead of the potentially hazardous perchloric acid (used in Scheme 7.3). Next it was necessary to optimize the conditions for the reductive

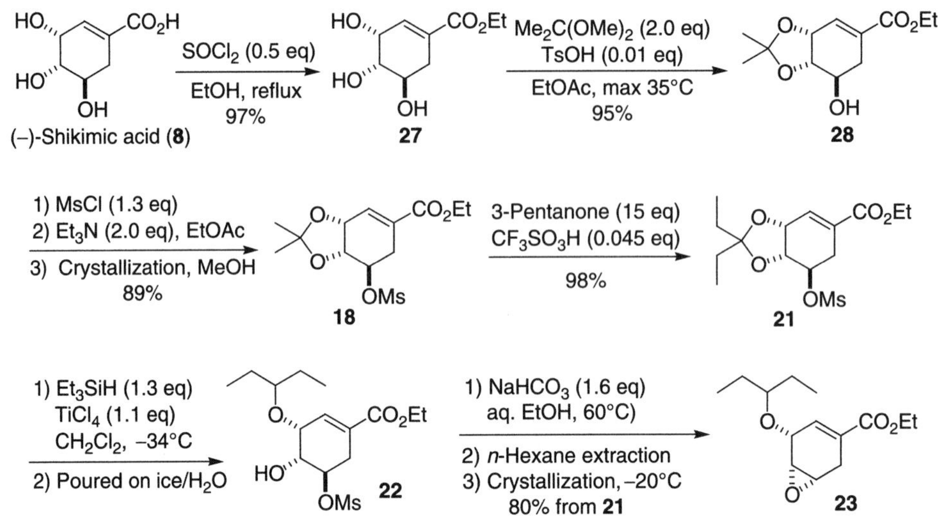

Scheme 7.4. Roche synthesis of epoxide **23**.

opening of the 3,4-pentylidene ketal **21**, as the trimethylsilyl triflate and borane dimethyl sulfide complex used in Scheme 7.3 were not ideally suitable for large-scale production. After screening many reducing agents and Lewis acids, it was found that the reagent combination of triethylsilane and titanium tetrachloride afforded a very selective reduction to give a 32 : 1 mixture of **22** and the isomeric pentyl ether along with 2–4% of the diol. The reaction was preferentially carried out in methylene chloride at −32 to −36°C, because at higher temperatures a substantial amount of the diol was formed. Crude **22** was treated with sodium bicarbonate in aqueous ethanol at 60°C to give **23**. Extraction with hexane at 35°C and then cooling of this solution to −20°C resulted in crystallization of epoxide **23**, which was isolated as a white, crystalline solid in 80% yield. Epoxide **23** was converted to oseltamivir phosphate (**1**) using azide chemistry similar to that shown in Scheme 7.3 (Abrecht et al., 2004).

The cost of the starting material, (−)-shikimic acid, and the azide chemistry was still an issue; therefore additional routes were also evaluated with the goal to use inexpensive and abundant starting materials and azide-free transformations. The first azide-free synthesis was reported by Karpf and Trussardi (Scheme 7.5) (Karpf and Trussardi, 2001). Initial attempts to open epoxide **23** with various amines or related nitrogen nucleophiles resulted in incomplete conversion along with aromatization of the highly functionalized cyclohexene. It was discovered that ytterbium trifluoromethanesulfonate [$Yb(OTf)_3$] and magnesium bromide etherate catalyzed the addition of benzylamine or allylamine to give the amino diol adduct **29** with high regioselectivity (10 : 1 to 13 : 1). Optimization of the epoxide ring-opening conditions led to the selection of allylamine and magnesium bromide etherate (0.2 equiv.) as the reagents of choice in a 9 : 1 mixture of *t*-butylmethyl ether and acetonitrile at 55°C. The magnesium salts were removed using a simple work-up of stirring the reaction mixture with 1 M aqueous ammonium sulfate, yielding **29** (87% of mixture) with the corresponding regioisomer (7% of mixture), which was taken directly into the next reaction.

Deallylation of **29** was achieved using 10% Pd/C in the presence of ethanolamine in refluxing ethanol to afford **30** in 77% yield (with 4% of the regioisomer). The reaction

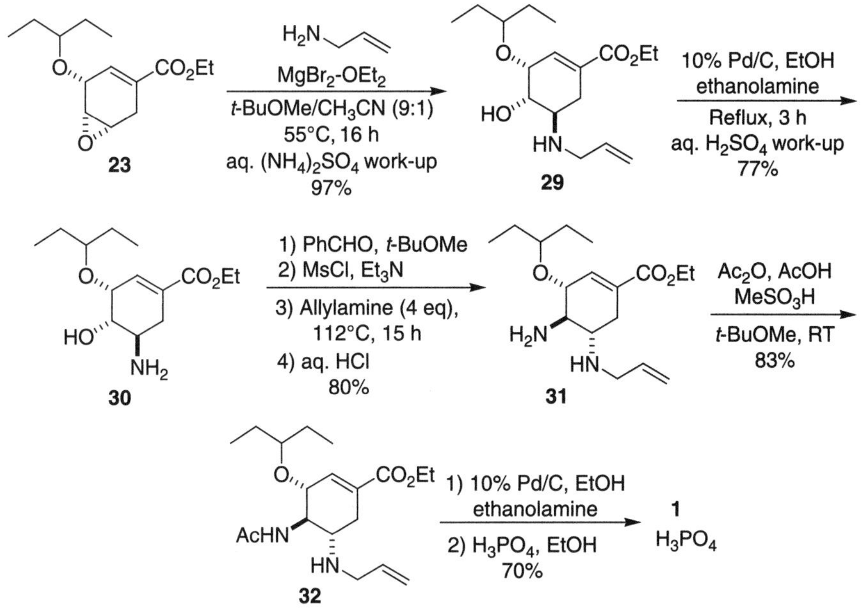

Scheme 7.5. Roche-Basel route to oseltamivir phosphate (**1**) from epoxide **23**.

works by heterogenous palladium(0)-catalyzed allyl isomerization followed by enamine hydrolysis. The ethanolamine helped promote the deallylation process and also reduced the amount of the *N*-propylamino by-product to a level of about 1–2%.

The domino sequence leading to the conversion of **30** to **31** was performed without isolation of the intermediates. The sequence is thought to proceed as shown in Scheme 7.6. Treatment of **30** with benzaldehyde with azeotropic removal of water led to the protection of the amine as the benzaldehyde imine **33**, which was directly mesylated to give **34**. The solution of **34** was treated with allylamine (4 equiv.) and autoclaved at 112°C for 15 h to provide **31** in 80% yield after acidic hydrolysis. It is assumed that the benzaldehyde imine of **33** is exchanged with allyl amine to form the aminomesylate **35**, which undergoes fast ring closure to the aziridine **36**. Then allylamine adds to the

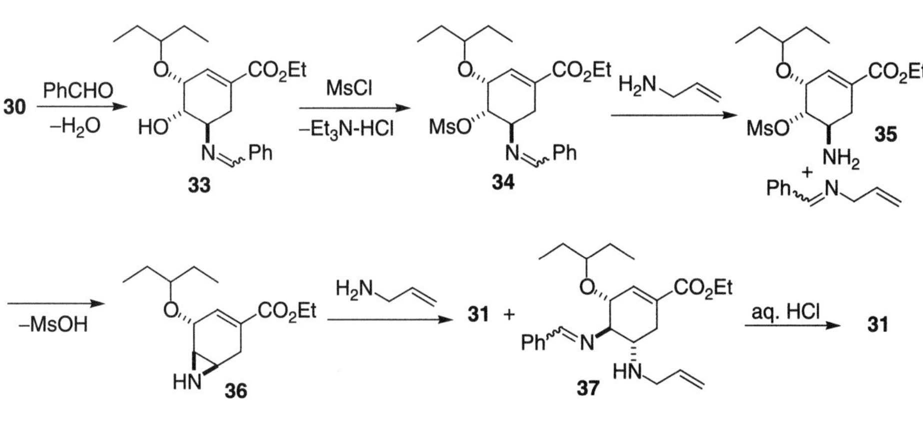

Scheme 7.6. Domino sequence transforming **30** to **31**.

aziridine, potentially catalyzed by the methanesulfonic acid that was liberated during the formation of **36**, leading to the diamine **31** and imine **37**, which forms from **31** under the reaction conditions. Finally, acidic hydrolysis leads to the isolation of **31** in 80% yield.

The 4-amino group of **31** was selectively acetylated under acidic conditions using acetic anhydride (1 equiv.) and methanesulfonic acid (1 equiv.) in acetic acid and *t*-butylmethyl ether to give **32**. Deallylation of **32** using 10% Pd/C in the presence of ethanolamine in refluxing ethanol proceeded as before to afford **1**, which was converted to the phosphate salt in 70% yield with high purity (99.7%). The overall yield of **1** from epoxide **23** was 35–38%.

A second-generation azide-free approach using *t*-butylamine and diallylamine was developed by the Roche group in Colorado (Scheme 7.7) (Harrington et al., 2004). Although the magnesium bromide diethyl etherate catalyzed ring opening of epoxide **23** with *t*-butyl amine worked to provide **38**, catalysis with magnesium chloride was examined in an effort to reduce catalyst cost and eliminate the diethyl ether by-product. It was found that the order of addition was important, with the preferred order being magnesium chloride, followed by *t*-butyl amine, and then epoxide **23**. Therefore, epoxide **23** was added to a preformed complex of magnesium chloride and *t*-butylamine in toluene. The resulting suspension was heated at 50°C followed by work-up with aqueous citric acid to give **38** in 96% yield. The alcohol of amino alcohol **38** was selectively mesylated and aqueous potassium carbonate was added to effect cyclization to aziridine **39**. Benzenesulfonic-acid-promoted ring opening of aziridine **39** with diallylamine (1.35 equiv.) proceeded at 120°C to give **40** in 93% yield.

Acetylation of the sterically hindered amine in **40** required treatment with acetic anhydride and sodium acetate at 116°C. The resulting *t*-butyl acetamide was treated with one equivalent of hydrogen chloride in ethanol followed by filtration of the precipitate to give clean hydrochloride salt **41** in 87% yield. This was the only purification step in the entire

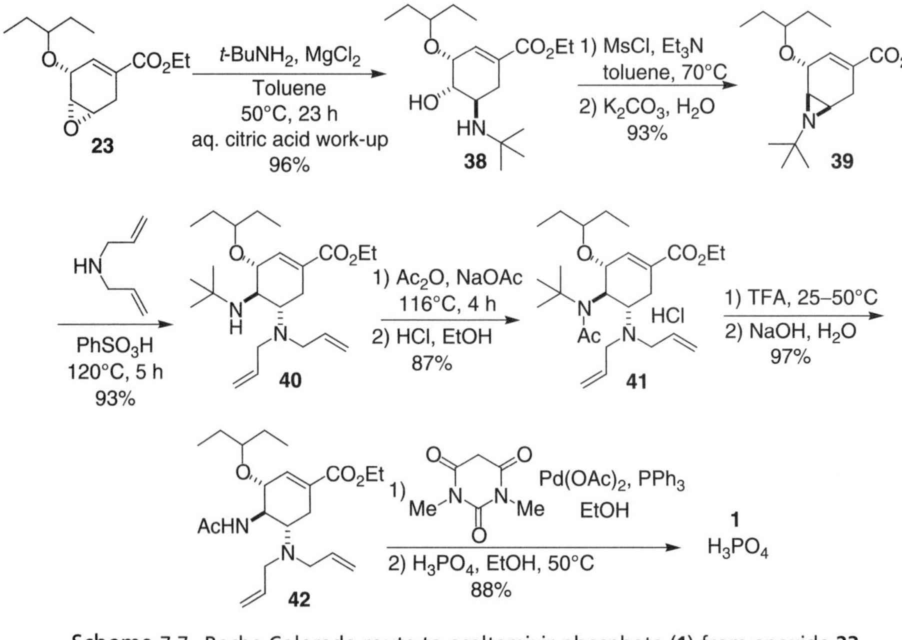

Scheme 7.7. Roche-Colorado route to oseltamivir phosphate (**1**) from epoxide **23**.

sequence. The *tert*-butyl group of salt **41** was cleaved with trifluoroacetic acid at 25–50°C to give **42**. The allyl protecting groups were removed via a palladium-catalyzed allyl transfer to 1,3-dimethylbarbituric acid in ethanol. This reaction mixture was then added to a solution of 85% phosphoric acid in ethanol at 50°C. Seed crystals of **1** were added to initiate crystallization and **1** was collected as a colorless solid in 88% yield. The overall yield of **1** from epoxide **23** was increased to 61% with this route.

Approaches to oseltamivir phosphate (**1**) that were independent of (−)-shikimic acid as the raw material were also evaluated. The furan-ethyl acrylate Diels–Alder approach is shown in Scheme 7.8 (Abrecht et al., 2001, 2004). The zinc-catalyzed Diels–Alder reaction between furan and ethyl acrylate was heated at 50°C for 72 h to provide a 9 : 1 mixture favoring *exo*-isomer ***rac*-43** over the *endo*-isomer. The *endo*-isomer was kinetically preferred, but with increased reaction times an equilibrium ratio of 9 : 1 was achieved favoring the thermodynamically preferred *exo*-isomer ***rac*-43**. The optical resolution of ***rac*-43** was achieved via enantioselective ester hydrolysis using Chirazyme L-2 to give (−)-**43** in 97%

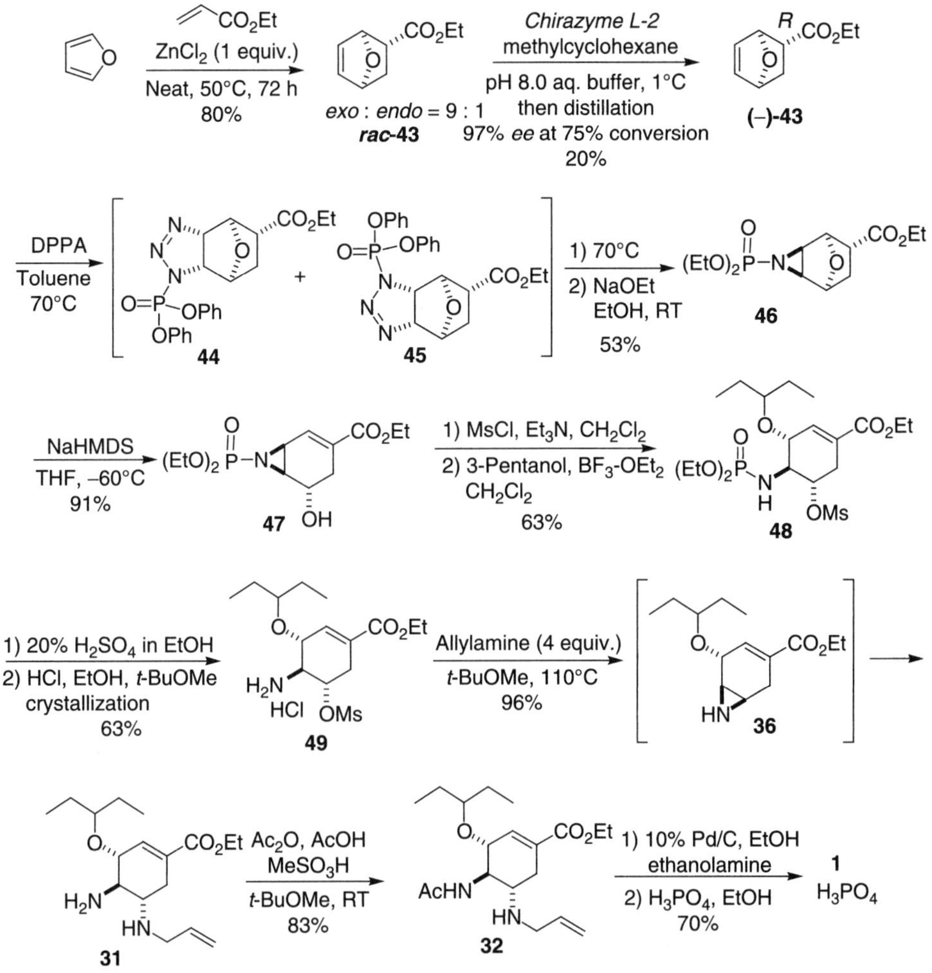

Scheme 7.8. The Diels–Alder approach to oseltamivir phosphate (**1**).

ee after intensive reaction optimization. Another key step in this synthesis was the conversion of (−)-**43** to aziridine **46**. Thus, [3 + 2]-cycloaddition of (−)-**43** with diphenylphosphoryl azide provided a mixture of the *exo*-triazoles **44** and **45**, which with continued heating at 70°C resulted in the thermal extrusion of nitrogen to give the *endo*-*N*-diphenylphosphoryl aziridine. A potential explanation for this "inversion" could be that the [3 + 2] cycloaddition reaction produces small equilibrium amounts of the *endo*-triazoles, which may undergo nitrogen extrusion at a much faster rate due to the higher steric strain. Transesterification with sodium ethoxide in the same pot then led to the *N*-diethylphosphoryl aziridine **46**. Treatment of *endo*-aziridine **46** with sodium hexamethyldisilazane led to smooth ring opening to give cyclohexene **47**. After O-mesylation of **47**, the *N*-diethylphosphoryl aziridine was opened with 3-pentanol in the presence of $BF_3{\cdot}EtO_2$ as in Scheme 7.1 to afford **48**. The phosphoryl amide of **48** was hydrolyzed with 20% sulfuric acid in ethanol and **49** was isolated as the hydrochloride salt. **49** was reacted with four equivalents of allylamine in *t*-butylmethyl ether at 110°C to give **31** via the aziridine intermediate **36**. The 4-amino group of **31** was selectively acetylated using acetic anhydride (1 equiv.) and methanesulfonic acid (1 equiv.) in acetic acid and *t*-butylmethyl ether to give **32**. Deallylation of **32** using 10% Pd/C in the presence of ethanolamine in refluxing ethanol proceeded as in Scheme 7.5 to afford **1**, which was converted to the phosphate salt.

Another approach to oseltamivir phosphate (**1**) that was independent of (−)-shikimic acid as the raw material employed an enzymatic hydrolytic desymmetrization of a *cis*-*meso*-diester on a penta-substituted cyclohexane (Scheme 7.9) (Abrecht et al., 2004; Iding et al., 2001). The synthesis started with the alkylation of 2,6-dimethoxy phenol with the mesylate of 3-pentanol using potassium *t*-butoxide as the base, followed by bromination with *N*-bromosuccinimide to provide the dibromide **51**. Palladium-catalyzed double ethoxycarbonylation of dibromide **51** furnished the diester **52**. Hydrogenation of **52** over Ru/Al_2O_3 at 100 bar afforded the all-*cis*-pentasubstituted cyclohexane **53** in 82% yield. The methyl ethers of **53** were cleaved using in situ trimethylsilyl iodide generated from trimethylsilyl chloride and sodium iodide to give *meso*-diol-diester **54**. The hydrolytic desymmetrization of *cis*-*meso*-diester **54** was carried out with pig liver esterase (PLE) to afford the mono-acid **55** in high yield and enantioselectivity. The 5-amino group was introduced by treating the acid **55** with diphenylphosphoryl azide to give an acyl azide intermediate that underwent Curtius rearrangement to the isocyanate, which was trapped by the adjacent alcohol to furnish oxazolidinone **56**. Oxazolidinone **56** was treated with Boc-anhydride and a catalytic amount of 4-dimethylaminopyridine (DMAP) to give the *N*-Boc-protected oxazolidinone, which was treated with a catalytic amount of sodium hydride to trigger the cleavage of the oxazolidinone by the intramolecular attack of the alcohol at the "2-position" with formation of a 1,3-diaxial cyclic carbonate. The stereoelectronics of this ring system were set up for efficient base-promoted decarboxylative elimination of the cyclic carbonate to the cyclohexenol, which was converted to triflate **57** with triflic anhydride and pyridine. Azide displacement of the triflate in **57** with inversion of configuration afforded azide **58**. In a one-pot sequence, the azide was hydrogenated with a Raney-cobalt catalyst to give the amine, which was *N*-acetylated with acetic anhydride in the presence of triethylamine. The *N*-Boc group was removed using hydrogen bromide in acetic acid and, after work-up, a solution of phosphoric acid in ethanol was added and **1** was isolated as the crystalline phosphoric acid salt.

In May 2006, E. J. Corey at Harvard University and two of his co-workers published a short enantioselective pathway for the synthesis of oseltamivir. Corey came up with the idea after reading a New York Times article on the shortage of Tamiflu supply, after

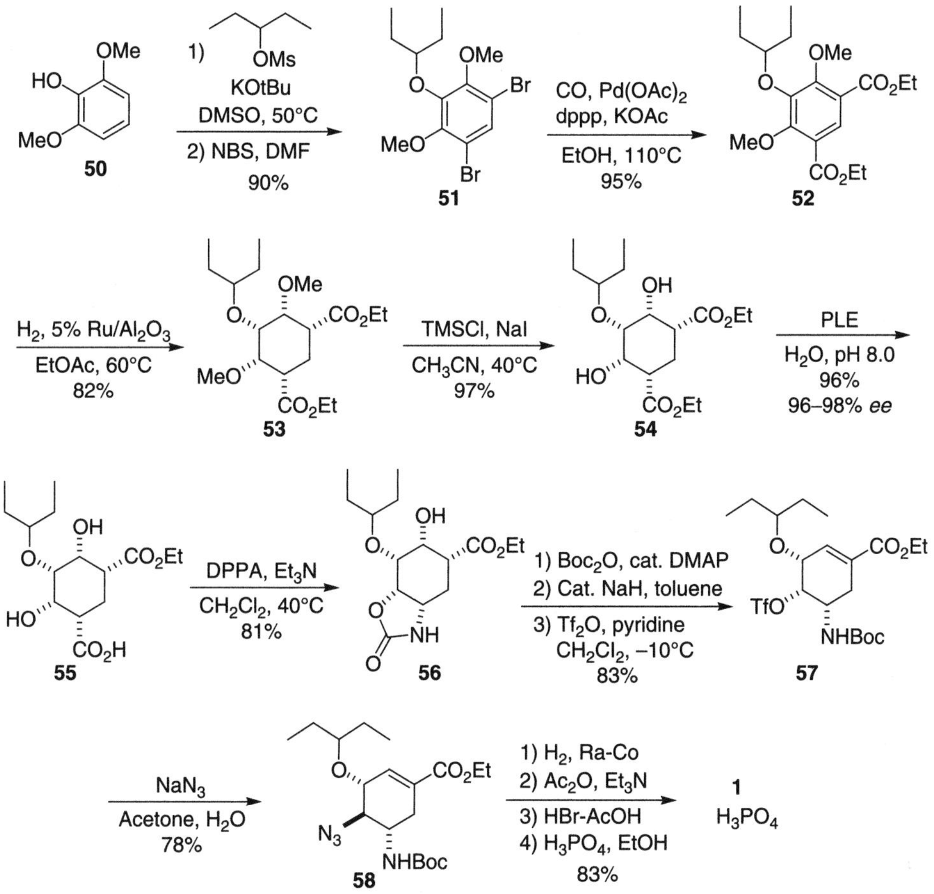

Scheme 7.9. Synthesis of oseltamivir phosphate (**1**) employing an enzymatic monohydrolysis desymmetrization.

which, his two associates completed the synthesis in a mere two months. The synthesis provides a number of advantages:

1. Use of inexpensive and abundant starting materials;
2. Complete enantio-, regio-, and diastereocontrol;
3. Avoidance of explosive, azide-type intermediates; and
4. Good overall yield (~30%, still not completely optimized).

Corey did not patent his synthesis, so anybody could use his process. A BBC correspondent interviewed Corey for his spectacular synthesis—a testiment to the importance of his contributions.

The Corey synthesis began with an asymmetric Diels–Alder reaction between butadiene and 2,2,2-trifluoroethyl acrylate in the presence of the *S*-proline-derived catalyst *ent*-**59** to form the adduct *ent*-**60** in excellent yield (97%) and with >97% *ee* (Scheme 7.10). Ammonolysis of **60** produced amide **61** quantitatively, which underwent iodolactamization using the Knapp protocol to generate lactam **62**. *N*-Acylation of **62** with

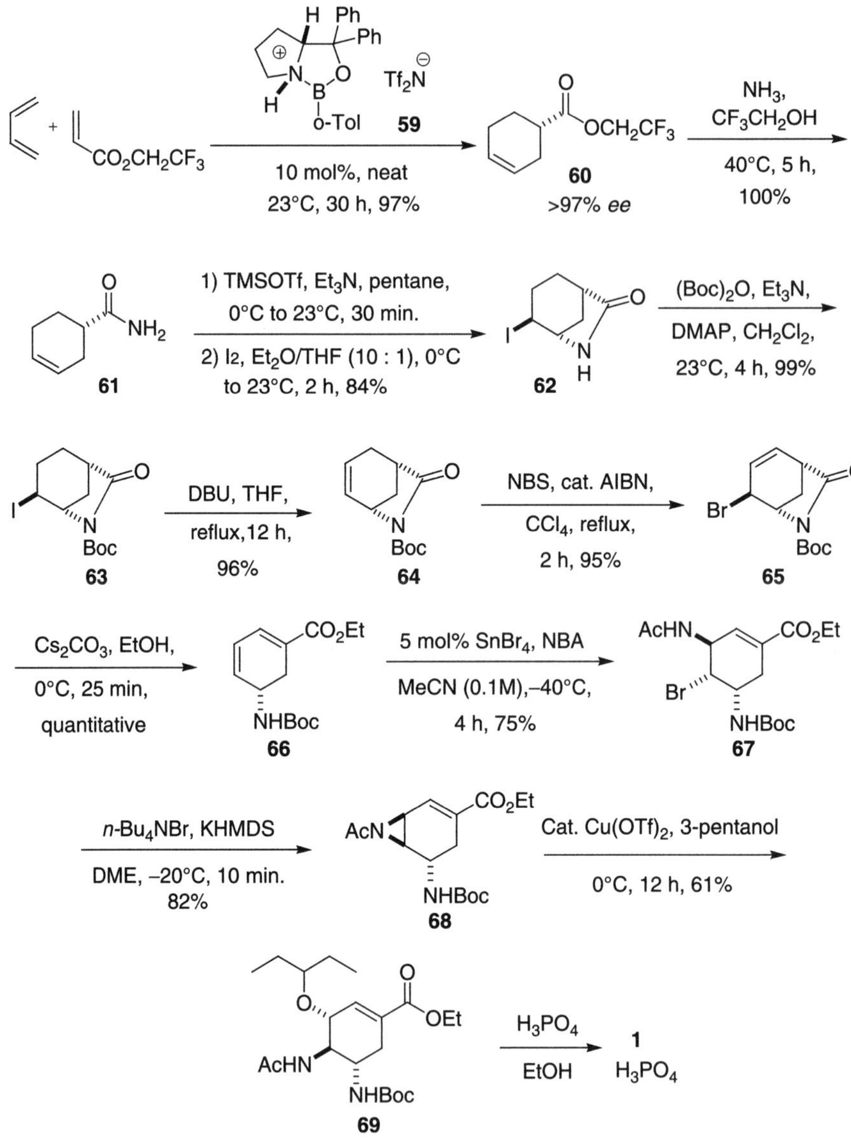

Scheme 7.10. Corey synthesis of oseltamivir phosphate (**1**).

di-*tert*-butyl dicarbonate provided the *tert*-butoxycarbonyl (Boc) derivative **63** in 99% yield. Dehydroiodination of **63** occurred cleanly with 1,8-diazabicyclo[5.4.0]undec-7-ene (DBU) to give **64**, which was allylically brominated using *N*-bromosuccinimide to generate **65** in excellent yield. Treatment of **65** with cesium carbonate in ethanol afforded the diene ethyl ester **66** quantitatively. The diene **66** was converted to the bromodiamide **67** using a novel $SnBr_4$-catalyzed bromoacetamidation with *N*-bromoacetamide (NBA) in acetonitrile at −40°C. The reaction was completely regio- and stereoselective and the structure was verified by single-crystal X-ray diffraction analysis (of the racemic methyl ester). Cyclization of **67** to the *N*-acetylaziridine **68** was rapid and efficient using in situ generated tetra-*n*-butylammonium hexamethyldisilazane. Reaction of **68** with 3-pentanol in the presence of a catalytic amount of cupric triflate at 0°C occurred regioselectively to generate ether **69**. Finally, removal of the Boc group and salt formation with phosphoric acid in ethanol afforded **1**.

The Shibasaki group has also recently completed a synthesis of oseltamivir phosphate (**1**) utilizing a catalytic asymmetric ring-opening of a *meso*-aziridine as detailed in Scheme 7.11 (Fukuta et al., 2006). Catalytic desymmetrization of *meso*-aziridine **70** with trimethylsilyl azide using the rare earth alkoxide $Y(O^iPr)_3$ and ligand **71** provided **72** in 96% yield and 91% *ee*. Recrystallization from isopropanol enhanced the *ee* to 99%. The amide **72** was Boc-protected and the *N*-3,5-dinitrobenzoyl group was hydrolyzed with sodium hydroxide to afford **73**. The azide was reduced with triphenylphosphine and the resulting amine was Boc-protected to give the optically pure C_2 symmetric 1,2-diamine **74**. Allylic oxidation of **74** with selenium dioxide in the presence of Dess–Martin periodinane produced a mixture of enone **75** and the corresponding allylic alcohol, which was oxidized with Dess–Martin periodinane to give enone **75** in 68% yield. Trimethylsilyl cyanide was added to **75** in the presence of $Ni(COD)_2$ to give the 1,4-adduct, which was brominated with *N*-bromosuccinimide followed by elimination

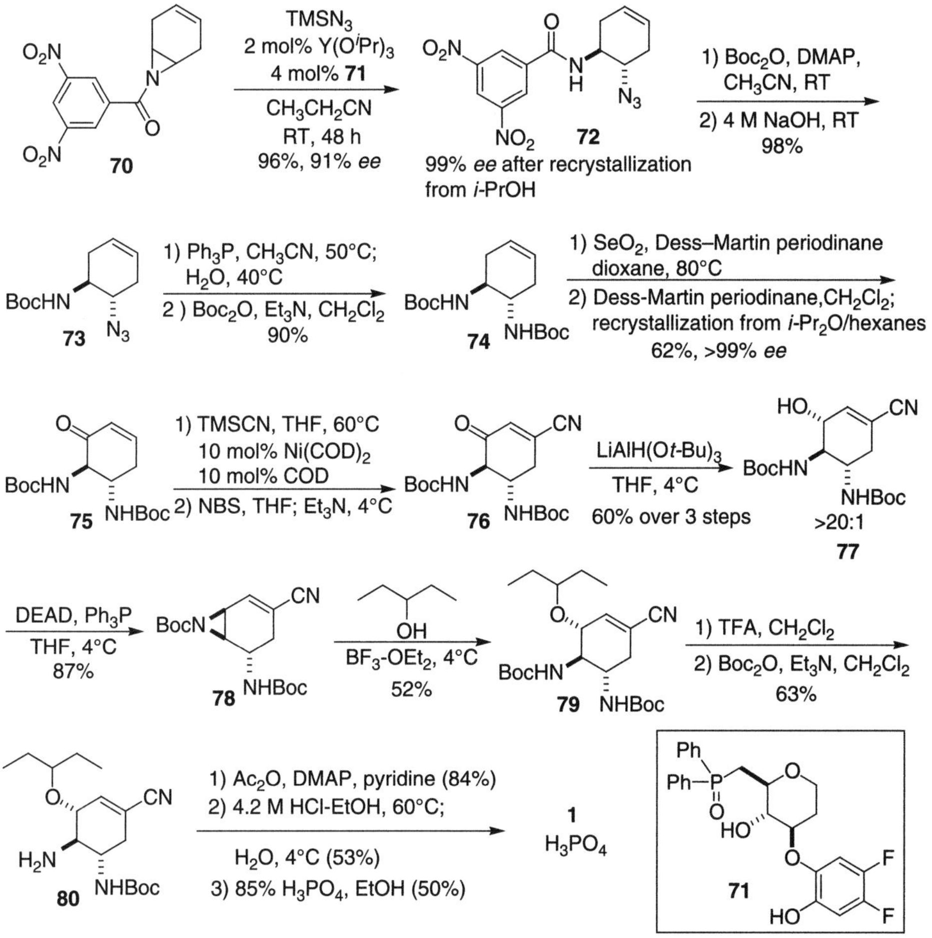

Scheme 7.11. Shibasaki synthesis of oseltamivir phosphate (**1**).

of the bromide with triethylamine to afford γ-keto nitrile **76**. Diastereoselective reduction of the ketone was accomplished using the bulky aluminum reagent $LiAlH(O\textit{t}\text{-}Bu)_3$ to give alcohol **77** with >20 : 1 diastereoselectivity. The aziridine **78** was formed under Mitsunobu conditions and then opened with 3-pentanol in the presence of $BF_3{\cdot}Et_2O$ in a similar fashion as that described in Scheme 7.1 to afford **79**. The Boc protecting groups were removed by treatment of **79** with trifluoroacetic acid and a Boc protecting group was reintroduced on the sterically less hindered amine to give **80**.

The unprotected amine was acetylated with acetic anhydride and the nitrile was converted to the ethyl ester in acidic ethanol with concomitant removal of the Boc group to provide oseltamivir. Finally, salt formation with 85% phosphoric acid in ethanol afforded oseltamivir phosphate (**1**).

7.3 SYNTHESIS OF ZANAMIVIR (RELENZA®) (Chandler and Weir, 1993; Chandler et al., 1995; Fromtling and Castaner, 1996; Patel, 1994; von Itzstein et al., 1991, 1993, 1994; Weir et al., 1994; Westerberg et al., 1996)

Zanamivir (GG167, **2**) was discovered at Biota Holdings and licensed to Glaxo (now GlaxoSmithKline) for clinical and commercial development. The initial synthesis from the von Itzstein group is described in Scheme 7.12 (von Itzstein et al., 1991). Methyl 5-acetamido-4,7,8,9-tetra-*O*-acetyl-2,3,5-trideoxy-*D*-glycero-*D*-galacto-non-2-enopryano sonate (**81**) was treated with $BF_3{\cdot}Et_2O$ and methanol followed by acetic acid and water to affect a selective deprotection of the C-4 *O*-acetyl on the sialic acid template to give **82**.

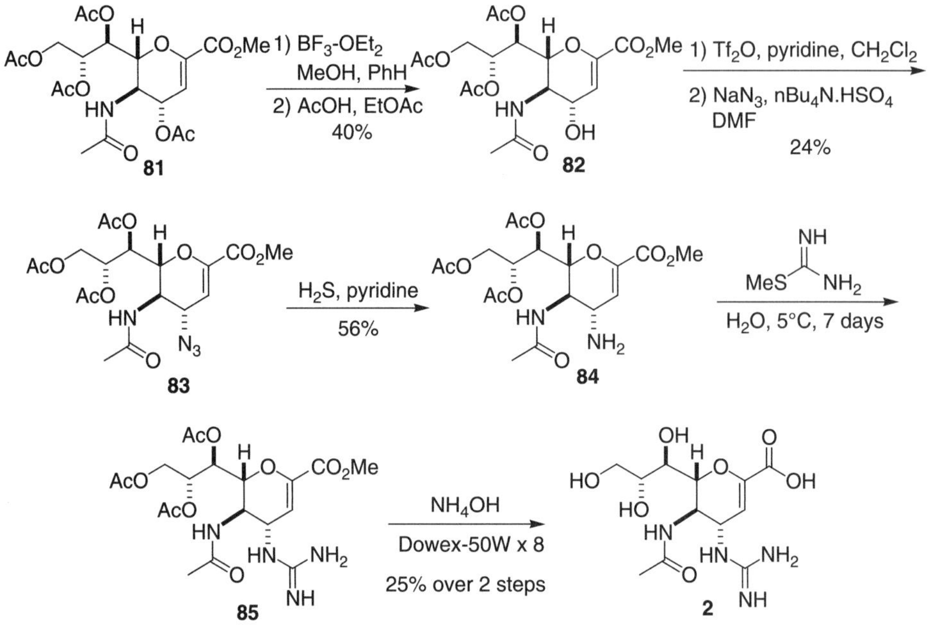

Scheme 7.12. The Biota synthesis of zanamivir (**2**) from the patent literature.

The alcohol was converted to the triflate and then displaced with sodium azide to afford **83**. Azide **83** was reduced with hydrogen sulfide in pyridine to give the corresponding amine **84**. The guanidine was introduced by reacting **84** with *S*-methylisothiourea in water. Finally, **85** was saponified by passing it through a column of Dowex 50W × 8 eluting with 1.5 M ammonium hydroxide to give zanamivir (**2**) in low yield.

The route in Scheme 7.12 was further optimized resulting in improved yields and a more efficient introduction of the C4-amino group (Scheme 7.13) (von Itzstein et al., 1993, 1994). Compound **81** was treated with $BF_3 \cdot Et_2O$ to give the oxazoline intermediate **86** in high yield. Addition of trimethylsilyl azide to the activated allylic oxazoline group led to the stereoselective introduction of azide at the C4 position on the sialic acid template to provide **83** in good yield. Hydrogenation of **83** in the presence of 10% Pd/C in toluene, methanol, and acetic acid afforded amine **84** in 72% yield. These particular reaction conditions minimized unwanted by-products resulting from acetate migration or over-reduction. Hydrolysis of the ester and the acetate protecting groups led to **87**, followed by introduction of the guanidine by treatment with aminoiminomethanesulfonic acid in the presence of aqueous potassium carbonate to afford zanamivir (**2**).

The Glaxo synthesis of zanamivir (**2**) started with the esterification of commercially available *N*-acetyl-neuraminic acid (**88**) with methanolic HCl to give the methyl ester as shown in Scheme 7.14 (Chandler and Weir, 1993; Chandler et al., 1995; Patel, 1994; Weir et al., 1994). Global acetylation of all the hydroxyl groups with acetic anhydride in pyridine with catalysis by 4-(dimethylamino)pyridine (DMAP) led to the penta-acetoxy compound **89**. Treatment of **89** with trimethylsilyl triflate in ethyl acetate at 52°C introduced the oxazoline as well as the 2,3-double bond to provide **86**. Addition of trimethysilyl azide to the activated allylic oxazoline group led to the stereoselective introduction of azide at the C-4 position to afford **83** as in Scheme 7.13.

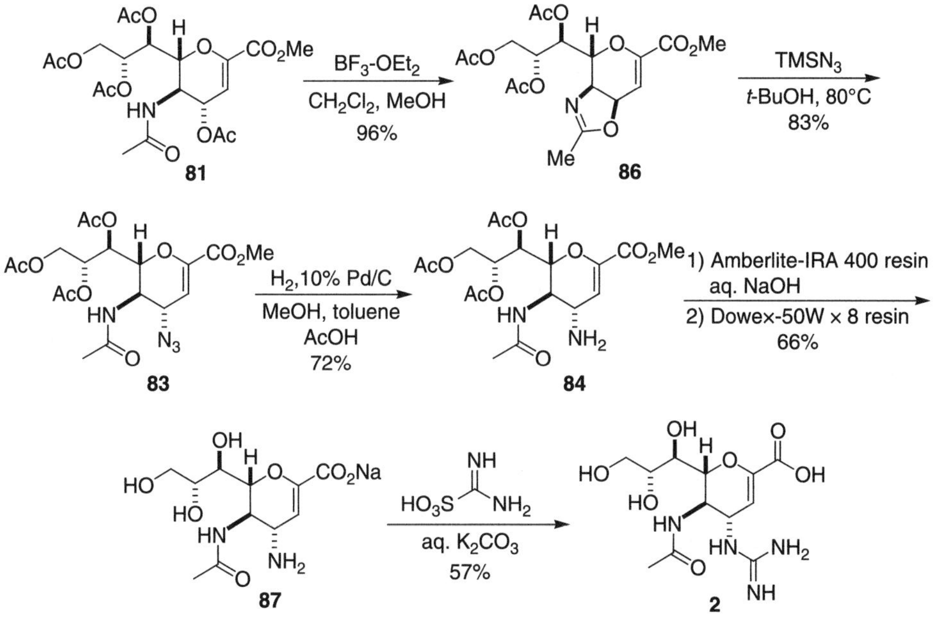

81 → 86: BF_3-OEt_2, CH_2Cl_2, MeOH, 96%
86 → 83: $TMSN_3$, *t*-BuOH, 80°C, 83%
83 → 84: H_2, 10% Pd/C, MeOH, toluene, AcOH, 72%
84 → 87: 1) Amberlite-IRA 400 resin, aq. NaOH; 2) Dowex-50W × 8 resin, 66%
87 → 2: HO_3SC(=NH)NH_2, aq. K_2CO_3, 57%

Scheme 7.13. Improved Biota synthesis of zanamivir (**2**).

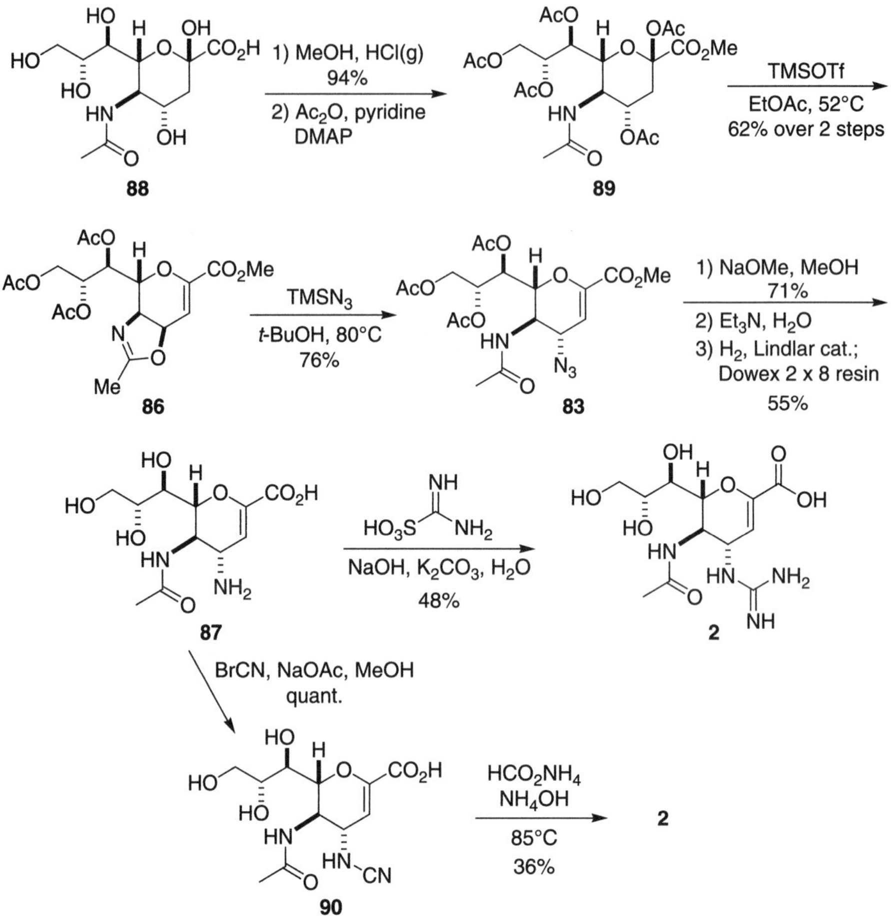

Scheme 7.14. Glaxo synthesis of zanamivir (**2**).

The acetate protecting groups were removed with catalytic sodium methoxide in methanol, resulting in a compound with improved water solubility. The methyl ester was then hydrolyzed using triethylamine in water and the azide was hydrogenated in the presence of Lindlar's catalyst to afford the triethylamine salt of the free amine, which was desalted to give amino acid **87**. Finally, treatment of **87** with three equivalents of aminoiminomethanesulfonic acid in the presence of aqueous potassium carbonate introduced the guanidine, and afforded crystalline zanamivir (**2**) after ion-exchange chromatography. Alternatively, the guanidine could be introduced by treating **87** with cyanogen bromide in the presence of sodium acetate in methanol to give cyanamide **90**, which was treated with ammonium hydroxide and ammonium formate followed by ion-exchange chromatography and crystallization to give zanamivir (**2**).

[^{11}C]-Zanamivir (**2**) was also synthesized with a ^{11}C-label on the guanidino group for use in PET (positron emission tomography) studies by treating **87** with [^{11}C]-cyanogen bromide followed by treatment of the intermediate cyanamide with a solution of aqueous ammonium hydroxide and ammonium chloride (Westerberg et al., 1996).

REFERENCES

Abrecht, S., Harrington, P., Iding, H., Karpf, M., Trussardi, R., Wirz, B., and Zutter, U. (2004). *Chimia*, 58: 621–629.

Abrecht, S., Karpf, M., Trussardi, R., and Wirz, B. (2001). EP 1127872.

Bischofberger, N. W., Dahl, T. C., Hitchcock, M. J. M., Kim, C. U., Lew, W., Liu, H., Mills, R. G., and Williams, M. A. (1999). WO 1999/014185.

Chand, P. (2005). "Recent advances in the discovery of neuraminidase inhibitors," *Expert Opin. Ther. Patents*, 15: 1009–1025.

Chand, P., Bantia, S., Kotian, P. L., El-Kattan, Y., Lin, T-H., and Babu, Y. S. (2005). *Bioorg. Med. Chem.*, 13: 4071–4077.

Chandler, M. and Weir, N. G. (1993). WO 1993/12105 (Glaxo).

Chandler, M., Bamford, M. J., Conroy, R., Lamont, B., Patel, B., Patel, V. K., Steeples, I. P., Storer, R., Weir, N. G., Wright, M., and Williamson, C. (1995). *J. Chem. Soc. Perkin Trans., 1*, 1173–1180.

Cheer, S. M. and Wagstaff, A. J. (2002). "Spotlight on zanamivir in influenza," *American Journal of Respiratory Medicine*, 1: 147–152.

Cheer, S. M. and Wagstaff, A. J. (2002). "Zanamivir: An update of its use in influenza," *Drugs*, 62: 71–106.

Colman, P. M. (2005). "Zanamivir: an influenza virus neuraminidase inhibitor," *Expert Review of Anti-Infective Therapy*, 3: 191–199.

Couch, R. B. (2000). "Prevention and treatment of influenza," *New Engl. J. Med.*, 343: 1778–1787.

Doucette, K. E. and Aoki, F. Y. (2001). "Oseltamivir, a clinical and pharmacological perspective," *Expert Opinions of Pharmacotherapy*, 2: 1671–1683.

Federspiel, M., Fischer, R., Hennig, M., Mair, H.-J., Oberhauser, T., Rimmler, G., Albiez, T., Bruhin, J., Estermann, H., Gandert, C., Gockel, V., Gotzo, S., Hoffmann, U., Huber, G., Janatsch, G., Lauper, S., Rocker-Stabler, O., Trussardi, R., and Zwahlen, A. G. (1999). *Org. Process Res. Dev.*, 3: 266–274.

Fromtling, R. A. and Castaner, J. (1996). *Drugs of the Future*, 21: 375–382.

Fukuta, Y., Mita, T., Fukuda, N., Kanai, M., and Shibasaki, M. (2006). *J. Am. Chem. Soc.*, 128: 6312–6313.

Graul, A., Leeson, P. A., and Castaner, J. (1999). *Drugs of the Future*, 24: 1189–1202.

Gubareva, L.V., Kaiser, L., and Hayden, F. G. (2000). "Influenza virus neuraminidase inhibitors," *Lancet*, 355: 827–835.

Harrington, P. J., Brown, J. D., Foderaro, T., Hughes, R. C. (2004). *Org. Process Res. Dev.*, 8: 86–91.

Iding, H., Wirz, B., and Zutter, U. (2001). EP 1146036.

Karpf, M. and Trussardi, R. (2001). *J. Org. Chem.*, 66: 2044–2051.

Kent, K. M., Kim, C. U., McGee, L. R., Munger, J. D., Prisbe, E. J., Postich, M. J., Rohloff, J. C., Kelly, D. E., Williams, M. A., and Zhang, L. (1999). US 5886213 (Gilead Sciences, Inc.).

Kent, K. M., Kim, C. U., McGee, L. R., Munger, J. D., Prisbe, E. J., Postich, M. J., Rohloff, J. C., St. John, D. E., Williams, M. A., and Zhang, L. (1998). WO 1998/007685.

Kim, C. U., Lew, W., Williams, M. A., Liu, H., Zhang, L., Swaminathan, S., Bischofberger, N., Chen, M. S., Mendel, D. B., Tai, C. Y., Laver, W. G., and Stevens, R. C. (1997). *J. Am. Chem. Soc.*, 119: 681–690.

Kim, C. U., Lew, W., Williams, M. A., Wu, H., Zhang, L., Chen, X., Escarpe, P. A., Mendel, D. B., Laver, W. G., and Stevens, R. C. (1998). *J. Med. Chem.*, 41: 2451–2460.

Lew, W., Chen, X., and Kim, C. U. (2000). "Discovery and development of GS 4104 (oseltamivir): an orally active influenza neuraminidase inhibitor," *Current Medicinal Chemistry*, 7: 663–672.

Liu, A-L., Wang, Yi-Tao., and Du, G.-H. (2005). "Neuraminidase as a target for drugs for the treatment of influenza," *Drugs Fut.*, 30: 799–806.

McClellan, K. and Perry, C. M. (2001). "Oseltamivir, a review of its use in influenza," *Drugs*, 61: 263–283.

McGowan, D. A. and Berchtold, G. A. (1981). *J. Org. Chem.*, 46: 2381–2383.

Moscona, A. (2005). "Neuraminidase inhibitors for influenza," *New Engl. J. Med.*, 353: 1363–1373.

Patel, V. (1994). WO 1994/07886 (Glaxo).

Rohloff, J. C., Kent, K. M., Postich, M. J., Becker, M. W., Chapman, H. H., Kelly, D. E., Lew, W., Louie, M. S., McGee, L. R., Prisbe, E. J., Schultze, L. M., Yu, R. H., and Zhang, L. (1998). *J. Org. Chem.*, 63: 4545–4550.

von Itzstein, M., Jin, B., Wu, W.-Y., and Chandler, M. (1993). *Carbohydr. Res.*, 244: 181–185.

von Itzstein, M., Wu, W.-Y., and Jin, B. (1994). *Carbohydr. Res.*, 259: 301–305.

von Itzstein, M., Wu, W.-Y., Phan, T., Danylec, B., and Jin, B. (1991). WO 1991/16320 (Biota).

Waghorn, S. L. and Goa, K. L. (1998). "Zanamivir," *Drugs*, 55: 721–725.

Weir, N. G., Chandler, M., and Bamford, M. J. (1994). WO 1994/07885 (Glaxo).

Westerberg, G., Bamford, M., Daniel, M. J., Langstrom, B., and Sutherland, D. R. (1996). *J. Label. Compd. Radiopharm.*, 38: 585–589.

Yeung, Y-Y., Hong, S., and Corey, E. J. (2006). *J. Am. Chem. Soc.*, 128: 6310–6311.

II

CARDIOVASCULAR AND METABOLIC DISEASES

8

PEROXISOME PROLIFERATOR-ACTIVATED RECEPTOR (PPAR) AGONISTS FOR TYPE 2 DIABETES

Jin Li

USAN: Rosiglitazone maleate
Trade name: Avandia®
GlaxoSmithKline
Launched: 1999
M.W. 357.11

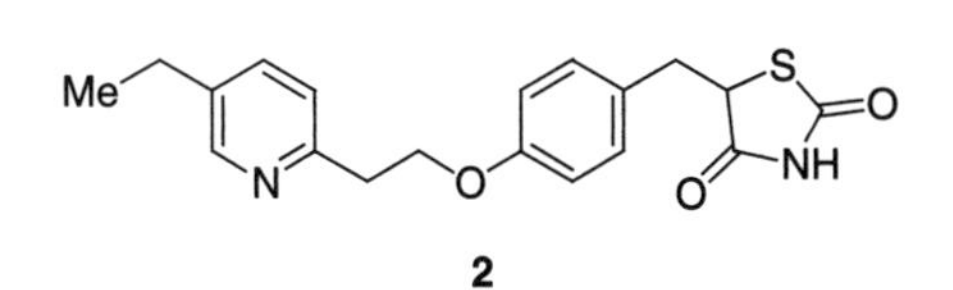

USAN: Pioglitazone hydrochloride
Trade name: Actos®
Takeda
Launched: 1999
M.W. 356.12

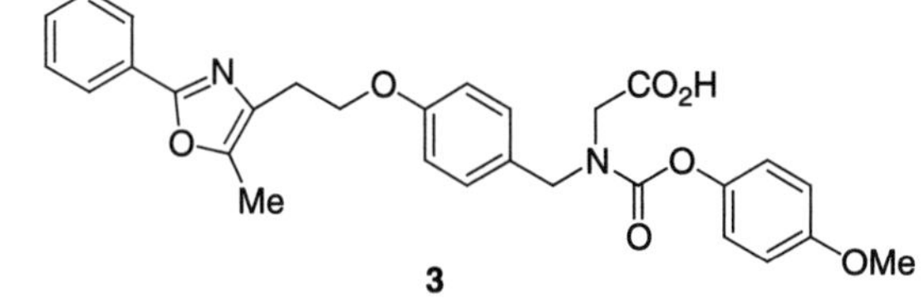

USAN: Muraglitazar
Trade name: Pargluva®
Bristol-Myers Squibb
Withdrawn
M.W. 516.54

8.1 INTRODUCTION

Diabetes is a disorder of metabolism—the way our bodies use digested food for growth and energy. Most of the food we eat is broken down into glucose, the form of sugar in the blood. Glucose is the main source of fuel for the body. After digestion, glucose

The Art of Drug Synthesis. Edited by Douglas S. Johnson and Jie Jack Li

passes into the bloodstream, where it is used by cells for growth and energy. For glucose to get into cells, insulin must be present. In people with diabetes, however, the pancreas either produces little or no insulin, or the cells do not respond appropriately to the insulin that is produced (Poretsky, 2002). Glucose builds up in the blood, overflows into the urine, and is excreted out of the body. Thus, the body loses its main source of fuel, even though the blood contains large amounts of glucose.

Type 1 diabetes is an autoimmune disease (Sperling, 2003). An autoimmune disease results when the body's system for fighting infection (the immune system) turns against a part of the body. In diabetes, the immune system attacks the insulin-producing β cells in the pancreas and destroys them. The pancreas then produces little or no insulin. At present, scientists do not know exactly what causes the body's immune system to attack the β cells, but they believe that autoimmune, genetic, and environmental factors, possibly viruses, are involved. Type 1 diabetes accounts for about 5 to 10% of diagnosed diabetes in the United States. It develops most often in children and young adults, but can appear at any age. Symptoms of type 1 diabetes usually develop over a short period, although β cell destruction can begin years earlier. Symptoms include increased thirst and urination, constant hunger, weight loss, blurred vision, and extreme fatigue. If not diagnosed and treated with insulin, a person with type 1 diabetes can lapse into a life-threatening diabetic coma, also known as diabetic ketoacidosis.

The most common form of diabetes is type 2 diabetes. About 90 to 95% of people with diabetes have type 2. It is associated with older age, obesity, and family history of diabetes, previous history of gestational diabetes, physical inactivity, and ethnicity. About 80% of people with type 2 diabetes are overweight (Laakso, 1999). Type 2 diabetes is increasingly being diagnosed in children and adolescents. In type 2 diabetes, the pancreas is usually producing enough insulin, but the body cannot use the insulin effectively, a condition called insulin resistance. After several years, insulin production decreases. The result is the same as for type 1 diabetes—glucose builds up in the blood and the body cannot make efficient use of its main source of fuel. The symptoms of type 2 diabetes develop gradually. Their onset is not as sudden as in type 1 diabetes. Symptoms may include fatigue or nausea, frequent urination, unusual thirst, weight loss, blurred vision, frequent infections, and slow healing of wounds or sores. The current arsenal for the treatment of diabetes is outlined in the following sections (Fig. 8.1).

8.1.1 Insulin

Everyone with type 1 diabetes and some people with type 2 diabetes must take insulin every day to replace what their pancreases are unable to produce. Unfortunately, insulin cannot be taken in pill form because enzymes in the stomach break it down so that it becomes ineffective. Therefore many people inject themselves with insulin using a syringe. Others may use an insulin pump, which provides a continuous supply of insulin, eliminating the need for daily shots. The most widely used form of insulin is synthetic human insulin, which is chemically identical to human insulin but manufactured in a laboratory. Unfortunately, synthetic human insulin is not perfect. One of its chief failings is that it does not mimic the way natural insulin is secreted. However, newer types of insulin, known as insulin analogs, more closely resemble the way natural insulin acts in the body. Among these are lispro (Humalog) (Czupryniak et al., 2005), insulin aspart (NovoLog) (Umpierrez et al., 2003), and glargine (Lantus) (Owens and Griffiths, 2002).

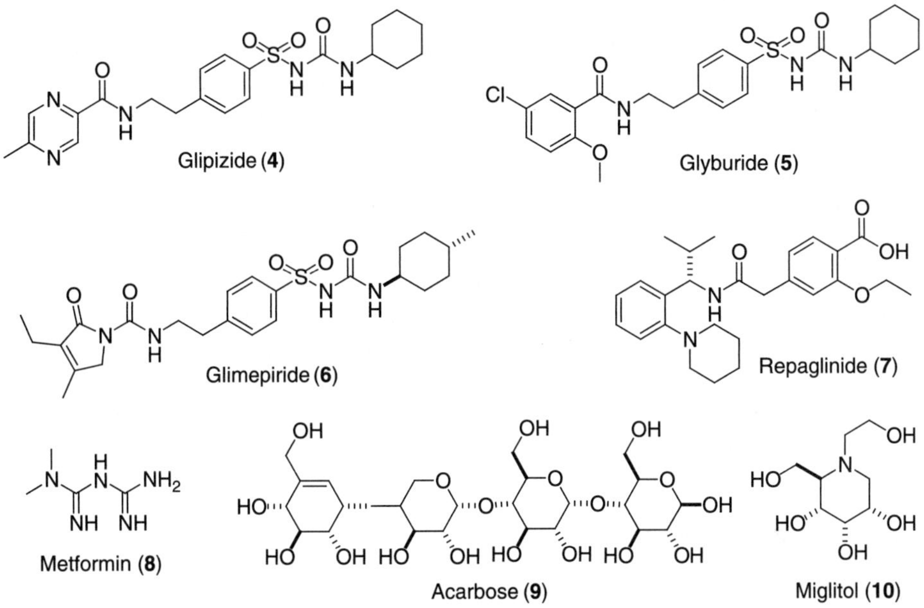

Glipizide (**4**) Glyburide (**5**)

Glimepiride (**6**) Repaglinide (**7**)

Metformin (**8**) Acarbose (**9**) Miglitol (**10**)

Figure 8.1. Selected anti-diabetes drugs.

8.1.2 Sulfonylurea Drugs

These drugs stimulate the pancreas to produce and release more insulin. For them to be effective, the pancreas must produce some insulin on its own. Second-generation sulfonylureas such as glipizide (**4**) (Glucotrol, Glucotrol XL) (Feinglos et al., 2003; Madsbad et al., 2001), glyburide (**5**) (DiaBeta, Glynase PresTab, Micronase) (St. John Sutton et al., 2002), and glimepiride (**6**) (Amaryl) (Prous and Castaner, 1992) are prescribed most often. The most common side effect of sulfonylureas is low blood sugar, especially during the first four months of therapy.

8.1.3 Meglitinides

These medications, such as repaglinide (**7**) (Prandin) (Rachman and Turner, 1995), have effects similar to sulfonylureas, but with fewer side effects in developing low blood sugar. Meglitinides work quickly, and the results fade rapidly as well.

8.1.4 Biguanides

Metformin (**8**) (Glucophage, Glucophage XR) (DeFronzo and Goodman, 1995; Saulie et al., 1995) is the only drug in this class available in the United States. It works by inhibiting the production and release of glucose from the liver, which means you need less insulin to transport blood sugar into the cells. One advantage of metformin (**8**) is that it tends to cause less weight gain than do other diabetes medications. Possible side effects include a metallic taste in your mouth, loss of appetite, nausea or vomiting, abdominal bloating, or pain, gas and diarrhea. These effects usually decrease over time and are less likely to occur if people take the medication with food. A rare but serious side effect is lactic acidosis, which results when lactic acid builds up in the body. Symptoms include tiredness, weakness, muscle aches, dizziness, and drowsiness. Lactic acidosis is especially likely to occur if people mix this medication with alcohol or have impaired kidney function.

8.1.5 Alpha-Glucosidase Inhibitors

These drugs block the action of enzymes in the digestive tract that break down carbohydrates. This means that sugar is absorbed into the bloodstream more slowly, which helps prevent the rapid rise in blood sugar that usually occurs right after a meal. Drugs in this class include acarbose (**9**) (Precose) (Balfour and McTavish, 1993; Wolever et al., 1993) and miglitol (**10**) (Glyset) (Campbell et al., 2000; Kingma et al., 1992). Although safe and effective, alpha-glucosidase inhibitors can cause abdominal bloating, gas, and diarrhea. If taken in high doses, they may also cause reversible liver damage.

8.1.6 Thiazolidinediones

The thiazolidinediones (TZDs) or "glitazones" are a new class of oral anti-diabetic drugs that improve metabolic control in patients with type 2 diabetes through the improvement of insulin sensitivity. The TZDs exert their anti-diabetic effects through a mechanism that involves activation of the gamma isoform of the peroxisome-proliferators-activated receptor (PPAR-γ) (Kota et al., 2005), a nuclear receptor (Fig. 8.2). The TZD-induced activation of PPAR-γ alters the transcription of several genes involved in glucose and lipid metabolism and energy balance, including those that code for lipoprotein lipase, fatty acid transporter protein, adipocyte fatty acid binding protein, fatty acyl-CoA synthase, malic enzyme, glucokinase, and the GLUT4 glucose transporter (Smith, 2004). The TZDs reduce insulin resistance in adipose tissue, muscle and the liver. However, PPAR-γ is predominantly expressed in adipose tissue. It is possible that the effect of TZDs on insulin resistance in muscle and liver is promoted via endocrine signaling from adipocytes. Potential signaling factors include free fatty acids (well-known mediators of insulin resistance linked to obesity) or adipocyte-derived tumor necrosis factor-α (TNF-α), which is over-expressed in obesity and insulin resistance. Although there are still many unknown facts about the mechanism of action of TZDs in type 2

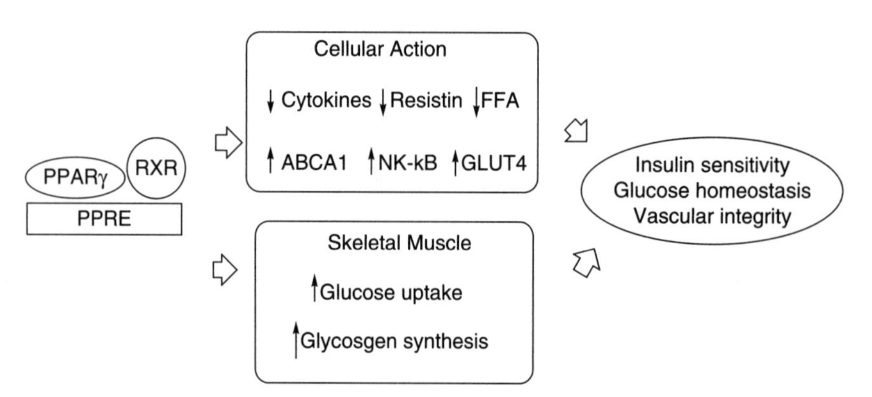

Figure 8.2. PPAR-γ gene transcription mechanism and its biologic effects.

diabetes, it is clear that these agents have the potential to benefit the full "insulin resistance syndrome" associated with the disease. Therefore, TZDs may also have potential benefits on the secondary complications of type 2 diabetes, such as cardiovascular diseases.

The process of transcription begins with the binding of ligands (endogenous or exogenous) to the PAAR-γ receptor. Ligand-bound PPAR heterodimerizes with retinoid X receptor (RXR) (IJpenberg et al., 1997) and this heterodimer binds to the promoter region of peroxisome proliferator response elements (PPREs) with the recruitment of co-activators (Fig. 8.2). This results in the increase in transcription activities of various genes involved in diverse biological process (Kota et al., 2005).

8.2 SYNTHESIS OF ROSIGLITAZONE

Rosiglitazone (**1**) is a potent and selective PPAR-γ agonist with a K_d of approximately 40 nM. In vitro studies have shown the total tissue GLUT4 glucose transporter number to increase in response to oral administration of rosiglitazone (**1**). PPAR-γ activation leads to the expression of adipose uncoupling protein and the differentiation of white adipose to brown adipose tissue. It has also been shown that in C2C12N myoblasts and mouse muscle satellite cells, rosiglitazone induces adipogenesis (Mais et al., 1997; Teboul et al., 1995), probably by its effect on PPAR-γ. The in vitro high affinity of rosiglitazone for the PPAR-γ receptor has been confirmed in studies with the pluripotent C3H10T/2 cell line, which demonstrated differentiation to adipocytes in response to exposure to rosiglitazone (Lehman et al., 1995). In animal models, rosiglitazone's antidiabetic activity was shown to be mediated by increased sensitivity to insulin's action in the liver, muscle, and adipose tissues. The expression of the insulin regulated glucose transporter GLUT-4 was increased in adipose tissue. Rosiglitazone did not induce hypoglycemia in animal models of type 2 diabetes and/or impaired glucose tolerance.

In terms of pharmacokinetics (Bolton et al., 1996; Freed et al., 1998; www.fda.gov), the maximum plasma concentration (C_{max}) and the area under the curve (AUC) of rosiglitazone increase in a dose-proportional manner over the therapeutic dose range. The elimination half-life is 3–4 h and is independent of dose. The absolute bioavailability of rosiglitazone is 99%. Peak plasma concentrations are observed about 1 h after dosing. Administration of rosiglitazone with food resulted in no change in overall exposure (AUC), but there was an approximately 28% decrease in C_{max} and a delay in T_{max} (1.75 h). The mean (CV%) oral volume of distribution (V_{ss}/F) of rosiglitazone is approximately 17.6 L, based on a population pharmacokinetic analysis. Rosiglitazone is approximately 99.8% bound to plasma proteins, primarily albumin. Rosiglitazone is extensively metabolized with no unchanged drug excreted in the urine. The major routes of metabolism were *N*-demethylation and hydroxylation, followed by conjugation with sulfate and glucuronic acid. All the circulating metabolites are considerably less potent than parent and, therefore, are not expected to contribute to the insulin-sensitizing activity of rosiglitazone. In vitro data demonstrate that rosiglitazone is predominantly metabolized by cytochrome P450 (CYP) isoenzyme 2C8, with CYP2C9 contributing as a minor pathway. Following oral or intravenous administration of [^{14}C]-rosiglitazone maleate, approximately 64% and 23% of the dose was eliminated in the urine and in the feces, respectively. The plasma half-life of [^{14}C]-related material ranged from 103 to 158 h.

The synthesis of rosiglitazone is quite straightforward (Scheme 8.1). Reaction of 2-chloropyridine (**11**) with 2-(*N*-methylamino)ethanol (**12**) by heating at 150°C gave 2-[*N*-methyl-*N*-(2-pyridyl)]aminoethanol (**13**, 95% yield), which was then reacted with

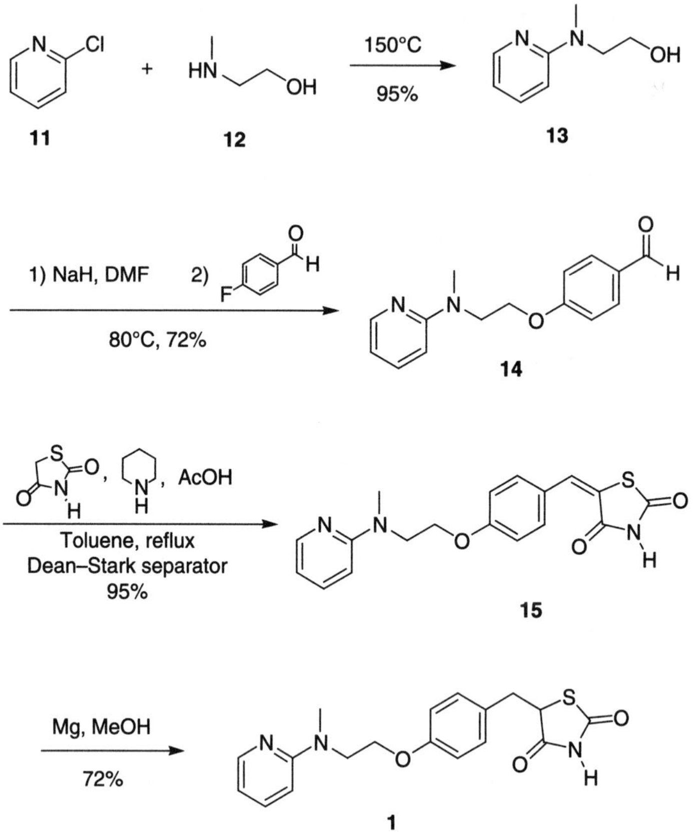

Scheme 8.1. Synthesis of rosiglitazone (1).

4-fluorobenzaldehyde by means of NaH in DMF, to yield 4-[2-[*N*-methyl-*N*-(2-pyridylamino)]-ethoxy]benzaldehyde (**14**) (Cantello et al., 1994a, b; Hindly). The reaction of aldehyde **14** with thiazolidine-2,4-dione by means of piperidinium acetate in refluxing toluene afforded amino pyridine **15**. Finally, **15** was reduced with Mg in methanol to provide rosiglitazone (**1**) in 95% yield.

8.3 SYNTHESIS OF PIOGLITAZONE

Pioglitazone (**2**) is a thiazolidinedione anti-diabetic agent that depends on the presence of insulin for its mechanism of action (DelloRussou et al., 2003). Pioglitazone decreases insulin resistance in the periphery and in the liver, resulting in increased insulin-dependent glucose disposal and decreased hepatic glucose output. Pioglitazone is a potent and highly selective agonist for PPARγ. In animal models of diabetes, pioglitazone reduces the hyperglycemia, hyperinsulinemia, and hypertriglyceridemia characteristic of insulin-resistant states such as type 2 diabetes. The metabolic changes produced by pioglitazone result in increased responsiveness of insulin-dependent tissues and are observed in many animal models of insulin resistance.

In terms of pharmacokinetics (www.fda.gov), serum concentrations of total pioglitazone (pioglitazone plus active metabolites) remain elevated 24 h after once-daily dosing.

Following oral administration, in the fasting state, pioglitazone is first measurable in serum within 30 min, with peak concentrations observed within 2 h. Food slightly delays the time to peak serum concentration to 3–4 h, but does not alter the extent of absorption. The mean apparent volume of distribution (V_d/F) of pioglitazone following single-dose administration is 0.63 ± 0.41 (mean $\pm$ SD) L/kg of body weight. Pioglitazone is extensively protein-bound (>99%) in human serum, principally to serum albumin. Pioglitazone also binds to other serum proteins, but with lower affinity. Pioglitazone is extensively metabolized by hydroxylation and oxidation; the metabolites also partly convert to glucuronide or sulfate conjugates. In vitro data demonstrate that multiple CYP isoforms are involved in the metabolism of pioglitazone. The cytochrome P450 isoforms involved are CYP2C8 and, to a lesser degree, CYP3A4 with additional contributions from a variety of other isoforms, including the mainly extra hepatic CYP1A1. Following oral administration, approximately 15 to 30% of the pioglitazone dose is recovered in the urine. Renal elimination of pioglitazone is negligible, and the drug is excreted primarily as metabolites and their conjugates. It is presumed that most of the oral dose is excreted into the bile either unchanged or as metabolites and eliminated in the feces. The mean serum half-lives of pioglitazone and total pioglitazone ranges from 3 to 7 h and 16 to 24 h, respectively. Pioglitazone has an apparent clearance, CL/F, calculated to be 5–7 L/h.

A number of syntheses of pioglitazone have been disclosed (Arita and Mizuno, 1992; Fischer et al., 2005; Les et al., 2004; Meguro and Fujita, 1986, 1987; Momose et al., 1991; Prous and Castaner, 1990; Saito et al., 1998). Two related syntheses (Fischer et al., 2005; Les et al., 2004) of pioglitazone hydrochloride are described in Scheme 8.2. The tosylate of 2-(5-ethylpyridin-2-yl)ethanol (**16**), formed in situ with tosyl chloride, was displaced by 4-hydroxybenzaldehyde (**17**) by means of benzyltributylammonium chloride and NaOH to give 4-[2-(5-ethylpyridin-2-yl)ethoxy]benzaldehyde (**20**). Condensation of **20** with thiazolidine-2,4-dione in basic medium afforded 5-[4-[2-(5-ethylpyridin-2-yl)ethoxy]benzylidene]thiazolidine-2,4-dione (**21**). Finally, this compound was hydrogenated to provide pioglitazone (**2**). Alternatively, a nucleophilic aromatic substitution reaction

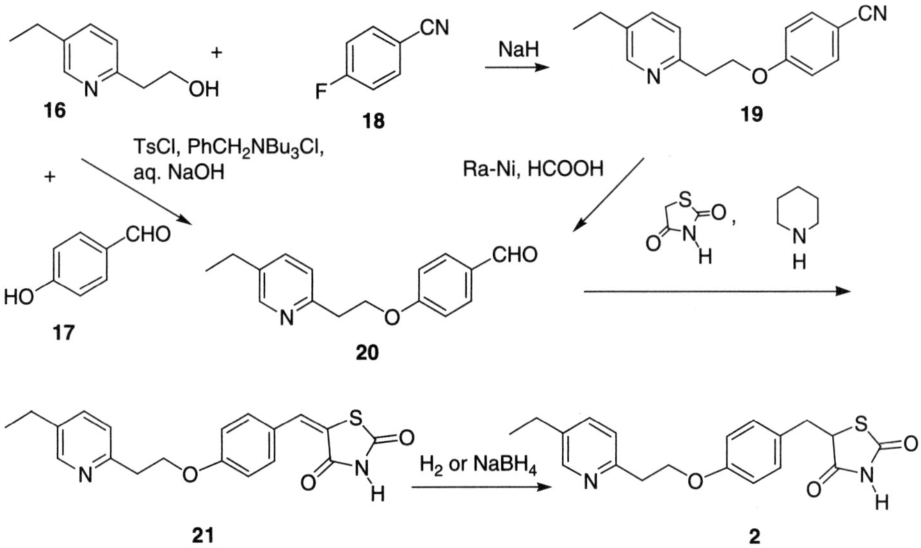

16 + 18 —NaH→ 19

16 + 17 —TsCl, $PhCH_2NBu_3Cl$, aq. NaOH→ 20

19 —Ra-Ni, HCOOH→ 20

20 —thiazolidine-2,4-dione, piperidine→ 21 —H_2 or $NaBH_4$→ 2

Scheme 8.2. Synthesis of pioglitazone (**2**).

between alcohol **16** and 4-fluorobenzonitrile (**18**) using NaH as the base provided 4-[2-(5-ethylpyridin-2-yl)ethoxy]benzonitrile (**19**), which was reduced with Raney nickel and formic acid to aldehyde **20**.

8.4 SYNTHESIS OF MURAGLITAZAR

Muraglitazar (**3**) is the first PPAR dual agonist (γ/α) being reviewed by the FDA for marketing approval in the United States (Devasthale et al., 2005). The rationale for the development of PPAR dual agonists is to simultaneously treat hyperglycemia and dyslipidemia via effects mediated by PPARγ and α activation, respectively (McIntyre and Castaner, 2004). PPARα agonists decrease triglycerides and increase HDL cholesterol with minimal effects on LDL cholesterol. PPARα-dependent stimulation of lipoprotein lipase and apolipoprotein A-V expression are thought to increase the metabolism of triglycerides, and PPARα-mediated induction of apolipoproteins A-1 and A-II are thought to mediate the increase in plasma HDL. Preclinical studies demonstrate that muraglitazar exhibits potent binding to recombinant human PPAR-α and PPAR-γ receptors in vitro (DelloRussou et al., 2003). Muraglitazar also induces efficient PPAR-α- and PPAR-γ-dependent transactivation of reporter gene expression in cell-based functional assays. Thus, muraglitazar possesses binding and activation properties characteristic of both PPAR-α and PPAR-γ, the intended molecular targets. Specificity is also apparent as muraglitazar does not bind to, or transactivate, a number of related nuclear hormone receptors, including the PPAR-δ isoform, the PPAR heterodimeric binding partner retinoid X receptor (RXR), estrogen receptor (ER), or progesterone receptor (PR). In vivo, muraglitazar exhibits potent antihyperglycemic and lipid-lowering activity in several animal models of diabetes, hyperglycemia, obesity, and hyperlipidemia. In a rodent model of diabetes (*db/db* mouse), muraglitazar improves insulin sensitivity and lipid metabolism. This effect is demonstrated by the correction of plasma glucose, insulin, triglycerides, and free fatty acids to levels normally observed in the nondiabetic mouse model. Muraglitazar also maintains glycemic control, preserves pancreatic islet β cell function, and increases pancreatic islet insulin content in a prediabetic rodent model (young *db/db* mouse).

Muraglitazar has a consistent and predictable PK profile, which has been characterized in healthy subjects and in subjects with type 2 diabetes. Muraglitazar has a low potential for interactions with food or concomitantly administered drugs. After oral administration, muraglitazar is rapidly absorbed, with plasma concentrations detectable within 30 min and peak plasma concentrations observed between 0.5 and 6 h for the 5-mg dose. Administration of muraglitazar with a high fat meal or with an agent that increases gastric pH (famotidine) did not affect overall muraglitazar exposure (AUC). Plasma concentrations of muraglitazar increased in a dose-proportional manner over and beyond the therapeutic dose range (up to 20 mg). Muraglitazar's elimination half-life is approximately 24 h (range, 19–31 h). Muraglitazar is extensively metabolized through both phase 1 and phase 2 pathways. The major biotransformation pathways are glucuronidation, hydroxylation, *O*-demethylation, and isoxazole ring opening. The parent compound contributes the majority (76%) of the drug-related plasma exposure. There are no known major circulating metabolites of muraglitazar with pharmacologic activity. Muraglitazar is primarily eliminated via the biliary pathway as unchanged muraglitazar, metabolites, and conjugates. It has negligible renal elimination. It is a substrate for multiple CYP450 isozymes in humans, including CYP3A4, 2C19, 2C9, 2C8, and 2D6, reducing the probability of significant drug–drug interactions. Clinical studies with inhibitors of CYP3A4 (ketoconazole) and 2C8 (gemfibrozil) found only small increases in muraglitazar

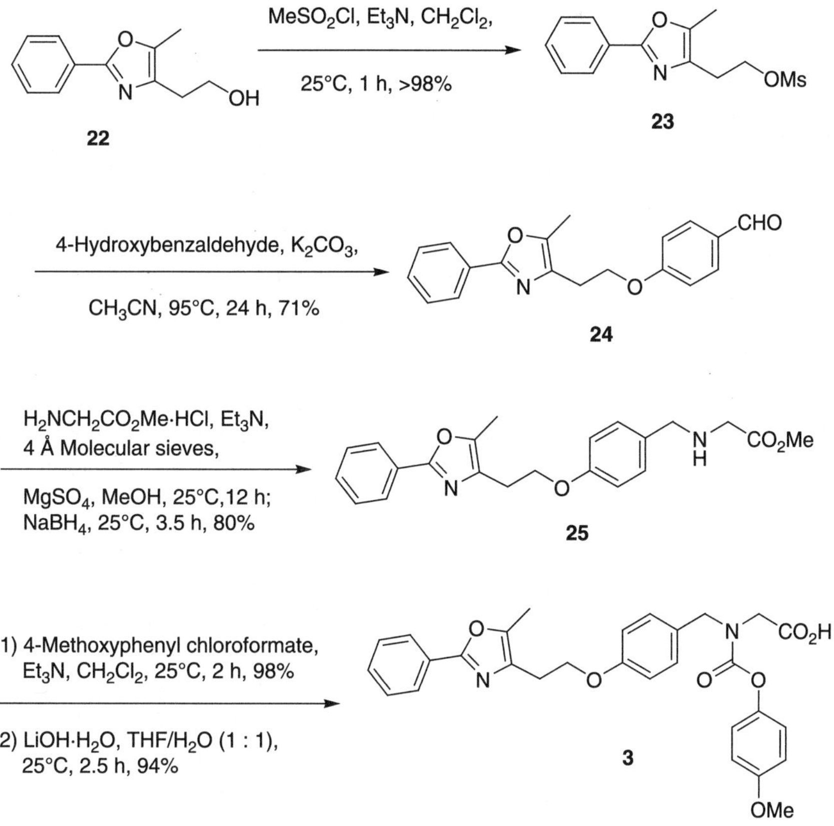

$MeSO_2Cl$, Et_3N, CH_2Cl_2, 25°C, 1 h, >98%

22 → 23

4-Hydroxybenzaldehyde, K_2CO_3, CH_3CN, 95°C, 24 h, 71%

24

$H_2NCH_2CO_2Me{\cdot}HCl$, Et_3N, 4 Å Molecular sieves, $MgSO_4$, MeOH, 25°C,12 h; $NaBH_4$, 25°C, 3.5 h, 80%

25

1) 4-Methoxyphenyl chloroformate, Et_3N, CH_2Cl_2, 25°C, 2 h, 98%
2) $LiOH{\cdot}H_2O$, THF/H_2O (1 : 1), 25°C, 2.5 h, 94%

3

Scheme 8.3. Synthesis of muraglitazar (**3**).

maximal drug concentration [C_{max} (23–27%) or AUC (23–46%)], which are not considered to be clinically relevant. Muraglitazar did not affect the pharmacokinetic profile of a number of drugs, including substrates of CYP2C9 (*S*-warfarin), 2C19 (*R*-warfarin), and 3A4 (atorvastatin or simvastatin) or frequently used antidiabetic or lipid-lowering agents such as metformin, pravastatin, and fenofibrate.

Several syntheses of muraglitazar (**3**) have appeared in patents (Cheng et al.) and journals (Devasthale et al., 2005) (Scheme 8.3). Alkylation of 4-hydroxybenzaldehyde with the phenyloxazolemesylate **23**, prepared readily from commercially available alcohol **22**, yielded aldehyde **24**, which was treated with glycine methyl ester under reductive amination conditions to provide secondary amine **25** in excellent yield. Reaction of amine **25** with 4-methoxyphenyl chloroformate followed by hydrolysis of the methyl ester afforded **3** in 94% yield.

Unfortunately, in May 2006, Bristol–Myers Squibb withdrew from developing muraglitazar (**3**) due to safety concerns, one year after Merck pulled out of the collaboration to co-develop the drug.

REFERENCES

Arita, M. and Mizuno, Y. (1992). EP0506273.

Balfour, J. A. and McTavish, D. (1993). *Drugs*, 46: 1025.

Bolton, G. C., Keogh, J. P., East, P. D., Hollis, F. J., and Shore, A. D. (1996). *Xenobiotica*, 26: 627.

Campbell, L. K., Baker, D. E., and Campbell, R. K. (2000). *Ann Pharmacother.*, 34: 1291.

Cantello, B. C. C., Cawthorne, M. A., Haigh, D., Hindley, R. M., Smith, S. A., and Thurlby, P. L. (1994a). *Bioorg. Med. Chem. Lett.*, 4: 1181.

Cantello, B. C. C., Cawthorne, M. A., Cottam, G. P., Duff, P. T., Haigh, D., Hindley, R. M., Lister, C. A., Smith, S. A., and Thurlby, P. L. (1994b). *J. Med. Chem.*, 37: 3977.

Cheng, P. T., Devasthale, P., Jeon, Y., Chen, S., and Zhang, H. JP 2003509503, US 6414002, WO 0221602.

Czupryniak, L., Saryuz-Wolska, M., Pawlowski, M., and Loba, J. (2005). *Am. J. Med.*, 118: 96.

DeFronzo, R. A. and Goodman, A. M. (1995). *New Engl. J. Med.*, 333: 541.

DelloRussou, C., Gavrilyuk, V., Weinberg, G., Almeida, A., Bolanos, J. P., Palmer, J., Pelligrino, D., Galea, E., and Feinstein, D. L. (2003). *J. Biol. Chem.*, 278: 5828.

Devasthale, P. V., Chen, S., Jeon, Y., Qu, F., Shao, C., Wang, W., Zhang, H., Farrelly, D., Golla, R., Grover, G., Harrity, T., Ma, Z., Moore, L., Ren, J., Seethala, R., Cheng, L., Sleph, P., Sun, W., Tieman, A., Wetterau, J. R., Doweyko, A., Chandrasena, G., Chang, S. Y., Humphreys, W. G., Sasseville, V. G., Biller, S. A., Ryono, D. E., Selan, F., Hariharan, N., and Cheng, P. T. W. (2005). *J. Med. Chem.*, 48: 2248.

Feinglos, M., Gibson, E., Chaiken, R., and Kourides, I. (2003). *Diabetes*, 52 (Suppl. 1): Abs 509.

Fischer, J., Fodor, T., Levai, S., and Petenyi, E. (2005). WO2005058827.

Freed, M. I., Miller, A., Jorkasky, D. K., and Dicicco, R. A. (1998). *Diabetes,* 47 (Suppl. 1): Abs 1365.

Hindly, R. M. Patents: AU8821738, EP306228, EP842925, JP8913169, JP97183726, JP97183771, JP97183772, JP98194970, JP98194971, US5002953, US5194443, US5232925, US5646169.

IJpenberg, A., Jeannin, E., Wahli, W., and Desvergne, B. (1997). *J. Biol. Chem.*, 272: 20108.

Kingma, P. J., Menheere, P. P. C. A., Sels, J. P., and Nieuwenhuijzen-Kruseman, A. (1992). *C. Diabetes Care*, 15: 478.

Kota, B. P., Huang, T. H., and Roufogalis, B. D. (2005). *Pharmacol. Res.*, 51: 85.

Laakso, M. (1999). *Diabetes*, 48: 937.

Lehmann, J. M., Moore, L. B., Smith, O., Wilkison, W. O., Willson, T. M., and Kliewer, S. A. (1995). *J. Biol. Chem.*, 270: 12953.

Les, A., Pucko, W., and Szelejewski, W. (2004). *Org. Process Res. Dev.*, 8: 157.

Madsbad, S., Kilhovd, B., Lager, I., Mustajoki, P., and Dejgaard, A. (2001). *Diabet. Med.*, 18: 395.

Mais, D. E., Hamann, L. G., Klausing, K. U., Paterniti, J. R., and Mukherjee, R. (1997). *Med. Chem. Res.*, 7: 324.

McIntyre, J. A. and Castaner, J. (2004). *Drugs Future*, 29: 1084.

Meguro, K. and Fujita, T. (1986). EP0193256.

Meguro, K. and Fujita, T. (1987). US4687777.

Momose, Y., Meguro, K., Ikeda, H., Hatanaka, C., Oi, S., and Sohda, T. (1991). *Chem. Pharm. Bull.*, 39: 1440.

Owens, D. R. and Griffiths, S. (2002). *Int. J. Clin. Pract.*, 460.

Poretsky, L. (ed.) (2002) *Principles of Diabetes Mellitus*, Kluwer Academic, Boston.

Prous, J. and Castaner, J. (1990). *Drugs Future*, 15: 1080.

Prous, J. and Castaner, J. (1992). *Drugs Future*, 17: 774.

Rachman, J. and Turner, R. C. (1995). *Diabet. Med.*, 12: 467.

Saito, Y., Mizufune, H., and Yamashita, M. (1998). EP0816340.

Saulie, B., White, J., and Campbell, R. K. (1995). *Diabetes Educ.*, 21: 441.

Smith, S. I. (2004). *Mol. Cel. Biochem.*, 263: 189.

Sperling, M. A. (2003). *Type 1 Diabetes: Etiology and Treatment*, Humana Press Inc., Totowa.

St. John Sutton, M., Rendell, M., Dandona, P., Dole, J. F., Murphy, K., Patwardhan, R., Patel, J., and Freed, M. (2002). *Diabetes Care*, 25: 2058.

Teboul, L., Gaillard, D., Staccini, L., Inadera, H., Amri, E. Z., and Grimaldi, P. A. (1995). *J. Bio. Chem.*, 270: 28183.

Umpierrez, G., Latif, K., Cuervo, R., Karabell, A., Freire, A., and Kitabchi, A. (2003). *Diabetes*, 52 (Suppl.1); Abst 584.

Wolever, T. M. S., Singer, W., Chiasson, J.-L., and Palmason, C. (1993). *Clin. Res.*, 41: 360A.

9

ANGIOTENSIN AT_1 ANTAGONISTS FOR HYPERTENSION

Larry Yet

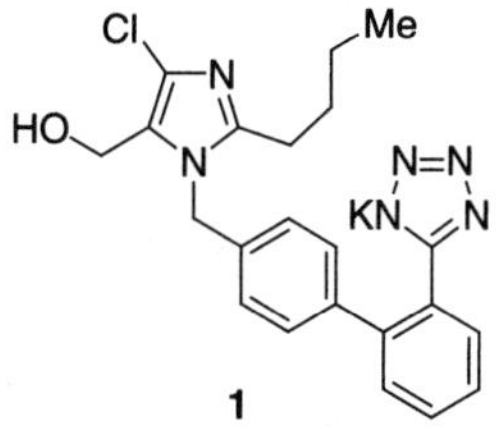

1

USAN: Losartan potassium
Trade name: Cozaar®
Merck
Launched: 1995
M.W. 461.01 (parent: 422.91)

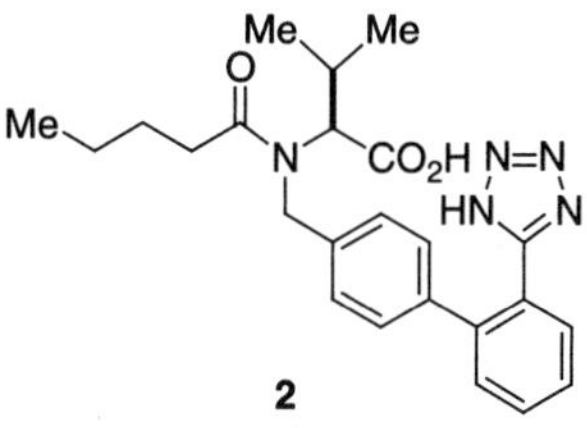

2

USAN: Valsartan
Trade name: Diovan®
Novartis/Ciba-Geigy
Launched: 1997
M.W. 425.52

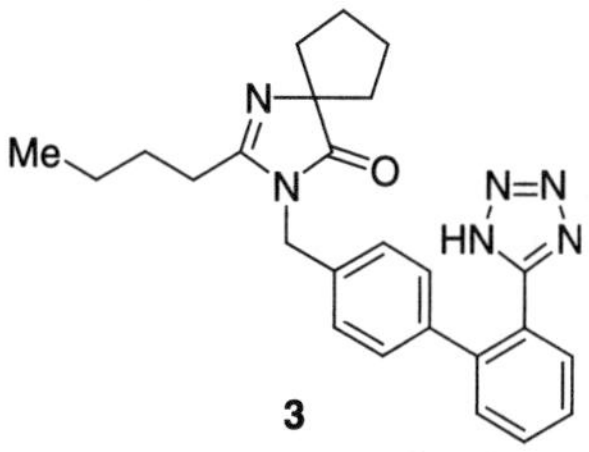

3

USAN: Irbesartan
Trade name: Avapro®
Sanofi
Launched: 1998
M.W. 428.53

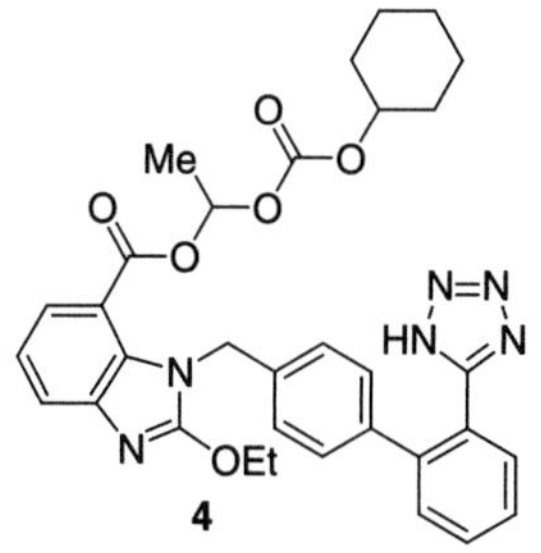

4

USAN: Candesartan cilexetil
Trade name: Atacand®
Takeda
Launched: 1998
M.W. 610.66 (active drug: 440.45)

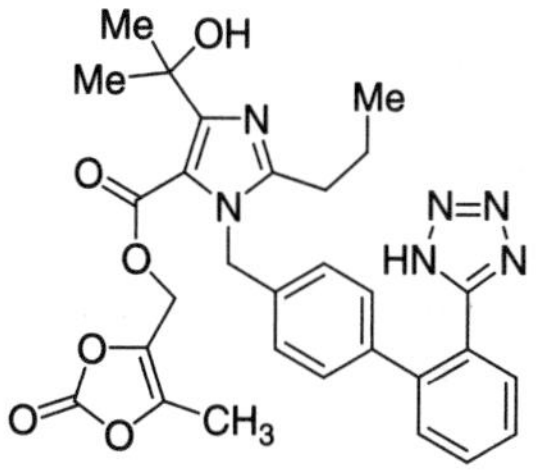

5

USAN: Olmesartan medoxomil
Trade name: Benicar®
Sankyo
Launched: 2002
M.W. 558.22 (active drug: 446.50)

The Art of Drug Synthesis. Edited by Douglas S. Johnson and Jie Jack Li

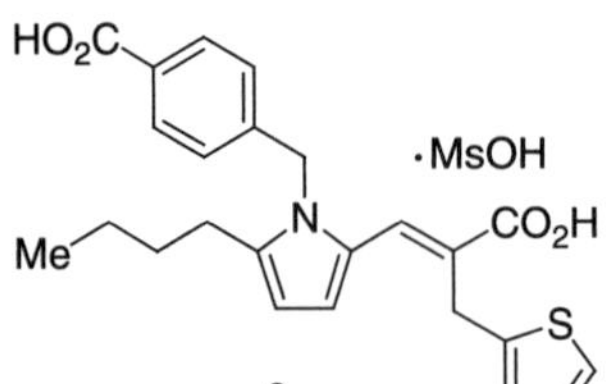

6

USAN: Eprosartan mesylate
Trade name: Teveten®
SmithKline Beecham
Launched: 1997
M.W. 423.5 (parent)

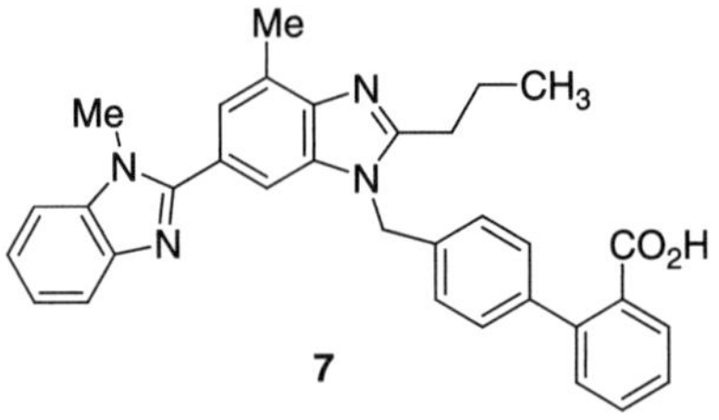

7

USAN: Telmisartan
Trade name: Micardis®
Boehringer Ingelheim
Launched: 1999
M.W. 514.63

9.1 INTRODUCTION

Coronary heart disease is one of the leading causes of death in the industrialized world (Packer, 1992; Remme and Swedberg, 2001). Hypertension is a risk factor for cardiovascular disease and is associated with an increased incidence of stroke and coronary heart disease. Other risk factors for cardiovascular disease include also high cholesterol, diabetes, and obesity. Although there have been many advances in treatment over the past several decades, less than a quarter of all hypertensive patients have their blood pressure adequately controlled with available therapies.

The renin angiotensin system (RAS) is one of the most powerful regulators of blood pressure. Renin, a proteinase enzyme, is secreted by the kidney in response to a reduction in renal blood flow, blood pressure, or sodium concentration. Renin then converts angiotensinogen, which is secreted by the liver, to angiotensin I, an inactive decapeptide (Scheme 9.1). Angiotensin I can also be generated by nonrenin enzymes such as tonin or cathepsin. Angiotensin I is then cleaved by angiotensin-converting enzyme (ACE) to angiotensin II, an octapeptide. Angiotensin I can also be converted to angiotensin II by enzymes such as trypsin or chymase. The final step of the renin angiotensin cascade is the activation of angiotensin II receptors AT_1 and AT_2 by angiotensin II. AT_1 and AT_2 belong to the superfamily of G-protein-coupled receptors that contain seven transmembrane regions (Bergsma et al., 1992; Mukoyama et al., 1993). AT_1 receptors are located in the kidney, heart, vascular smooth-muscle cells, brain, adrenal glands, platelets, adipocytes, and in the placenta. AT_1 receptors promote aldosterone secretion and sodium retention by stimulation of angiotension receptors present on the adrenal cortex, resulting in elevated blood pressure due to vasoconstriction. AT_2 receptors are present only in low levels, mainly in the uterus, adrenal, central nervous system, heart, and kidney, but are not associated with cardiovascular homeostasis.

Thus, research efforts over the last several decades have been directed toward developing drugs capable of suppressing the renin angiotensin system by inhibition of renin release, by blocking the formation of angiotensin II by inhibition of ACE, and by antagonism of angiotensin II at its physiologic receptors AT_1. Until 1995, ACE inhibitors were the only drugs capable of the RAS blockade, which produced undesirable side effects such as a dry cough. In recent years, renin and angiotensin II AT_1 receptor antagonists have been targeted for development as more specific inhibitors of the RAS (Burnier and Brunner, 2000; Kirch et al., 2001; Schmidt and Schieffer, 2003).

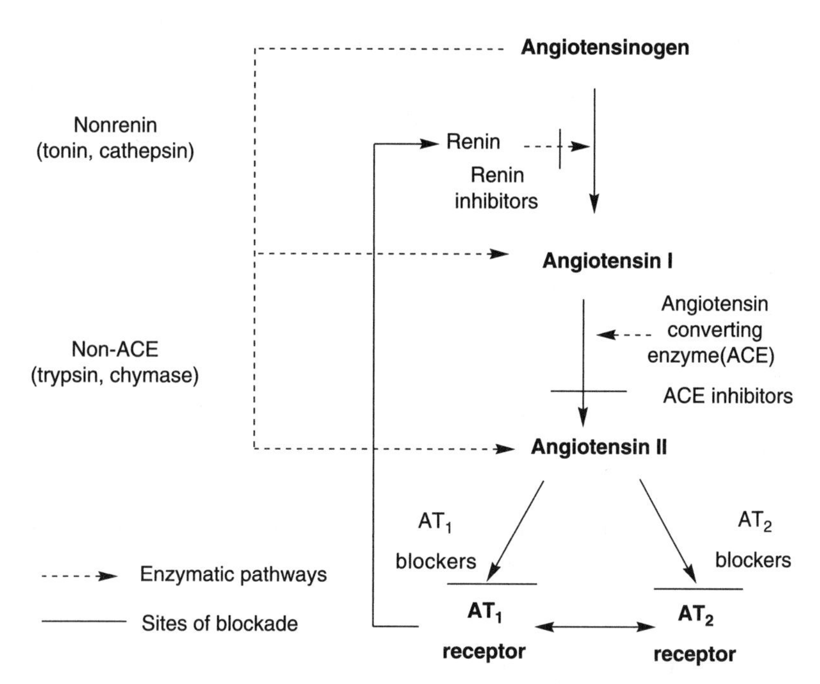

Scheme 9.1. Renin-angiotensin system with sites of blockade by inhibitors or antagonists.

Highly selective nonpeptidic angiotensin II AT_1 receptor antagonists have been designed and developed as competitive antagonists with virtually no agonistic effect at the receptor level (Burnier and Brunner, 2000; Kirch et al., 2001; Schmidt and Schieffer, 2003). Losartan potassium (**1**), which appeared in 1995, was described as the first nonpeptide AT_1 receptor antagonist, and the coined group name was sartans. Since then, six orally active AT_1 receptor antagonists, valsartan (**2**), irbesartan (**3**), candesartan cilexetil (**4**), olmesartan medoxomil (**5**), eprosartan mesylate (**6**), and telmisartan (**7**) have been accepted by the FDA and launched in the United States and various European countries for the treatment of hypertension. Many of these share the same biphenyltetrazole unit, taken from the losartan lead.

The pharmacokinetic properties of each of the sartans are listed in Table 9.1 (Burnier and Brunner, 2000; Kirch et al., 2001; Schmidt and Schieffer, 2003). All seven drugs have very potent IC_{50} antagonist values in the low nanomolar range, with bioavailability varying considerably from 15% for eprosartan mesylate (**6**) to as high as 70% for irbesartan (**3**). Losartan potassium (**1**), candesartan cilexetil (**4**), and olmesartan medoxomil (**5**) have metabolites (see bracketed elements in Table 9.1) that are far more active than their parent molecules; these will be described in their respective sections. Telmisartan (**7**) is minimally transformed, but the remaining AT_1 receptor antagonists act unchanged at their receptors without biotransformations. The elimination half-lives of these drugs average about 9 h, somewhat less for losartan (2 h), and higher for irbesartan and telmisartan (15 h and 24 h, respectively). All the AT_1 receptor antagonists are lipophilic and plasma protein binding to albumin is high ($>90\%$). All of the AT_1 receptor antagonists are excreted by the bile and kidney as feces and urine, respectively, in roughly 60/40 to 98/2 ratios, depending on the drugs.

TABLE 9.1. Pharmacokinetic Properties of the Angiotensin II AT_1 Antagonist Drugs

Drug	AT_1 Receptor IC_{50} (nM)	Oral Bioavailability (%)	$t_{1/2}$ (h)	Approved Daily Doses (mg)
Losartan potassium (**1**)				
(**EXP 3174**)	20 (1.3)	(33)	2 (6–9)	50–100
Valsartan (**2**)	1.6	25	6	80–320
Irbesartan (**3**)	1.3	70	15	150–300
Candesartan cilexetil (**4**)				
(**CV11974**)	(0.1)	(15)	>9	4–16
Olmesartan medoxomil (**5**)				
(**RNH-6270**)	(0.3)	(26)	14	20–40
Eprosartan mesylate (**6**)	1.5	15	5–7	400–800
Telmisartan (**7**)	1.0	43	24	40–80

Other methods for reducing hypertension include ACE inhibitors and calcium channel blockers; these are discussed in subsequent chapters.

9.2 LOSARTAN POTASSIUM

9.2.1 Introduction to Losartan Potassium

Losartan potassium (**1**), the prototype highly selective AT_1 receptor antagonist (Goa and Wagstaff, 1996), was derived from the Takeda series of 1-benzylimidazole-5-acetic acid derivatives, which were themselves weak angiotensin II antagonists. Losartan potassium

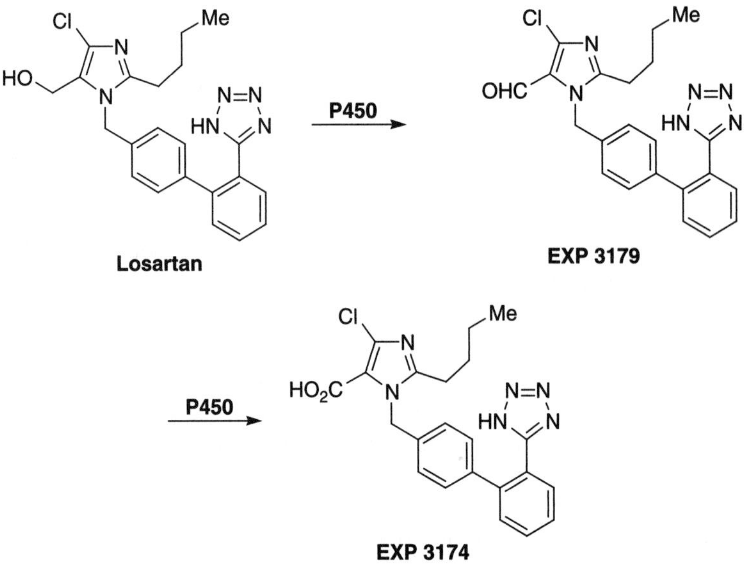

Scheme 9.2. Conversion of losartan to its active metabolite **EXP 3174** by cytochrome P450 oxidation.

(**1**) is converted in the liver by cytochrome P450 CYP2C9 and CYP3A4 oxidation to its active carboxylic acid metabolite **EXP 3174** via **EXP 3179** (Scheme 9.2). **EXP 3174** is 20 times more potent than **1** and has a longer duration of action (see Table 9.1). Because the oral bioavailability of the active metabolite is very low, the drug on the market is losartan potassium (**1**). The metabolite is eliminated from the bile and kidneys in a 60 : 40 ratio.

9.2.2 Synthesis of Losartan Potassium

The synthesis of losartan potassium (**1**) by the process research chemists at Merck is outlined in the following (Griffiths et al., 1999; Larsen et al., 1994). Phenyltetrazole (**8**) is protected as the trityl phenyltetrazole **9** (Scheme 9.3). *Ortho*-lithiation of **9** followed by quenching with triisopropyl borate afforded boronic acid **10** after treatment with aqueous ammonium chloride. Reaction of glycine (**11**) with methyl pentanimidate (**12**) in a methanol/water mixture yielded (pentanimidoylamino) acetic acid (**13**), which underwent a Vilsmeier reaction with phosphorous oxychloride in DMF followed by hydrolysis to give imidazole-4-carbaldehyde **14** in moderate yield.

Imidazole-4-carbaldehyde **14** is alkylated with 4-bromobenzyl bromide (**15**) in the presence of potassium carbonate to unstable intermediate **16** and reduced to the imidazole alcohol **17** with sodium borohydride (Scheme 9.4). Imidazole alcohol **17** participated in a Suzuki cross-coupling reaction with boronic acid **10** followed by deprotection of the triphenylmethyl group with dilute sulfuric acid and conversion to losartan potassium (**1**) with potassium *tert*-butoxide. Potassium *tert*-butoxide in warm methanol is an improved process over the potassium hydroxide method previously employed in this step (Kumar et al., 2004).

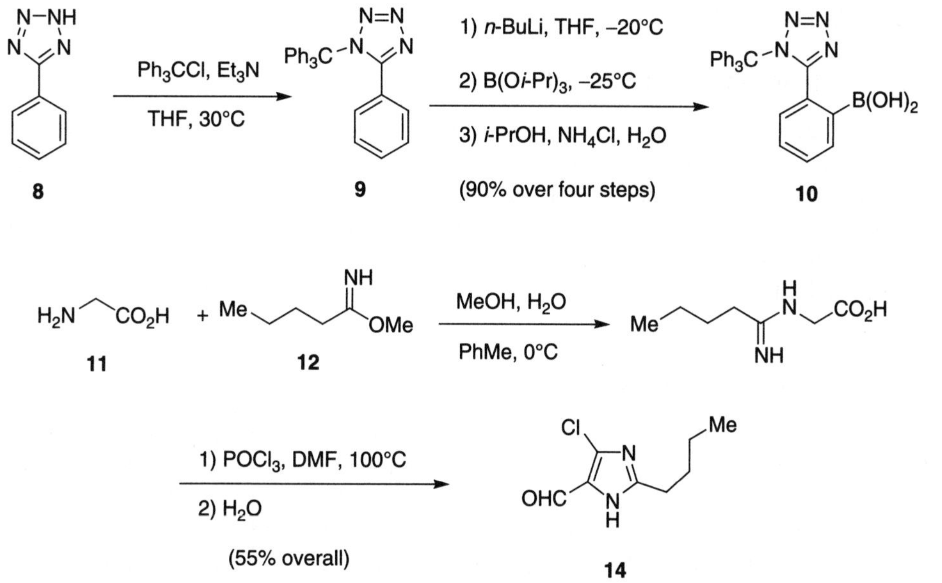

Scheme 9.3. Synthesis of boronic acid **10** and imidazole-4-carbaldehyde **14**.

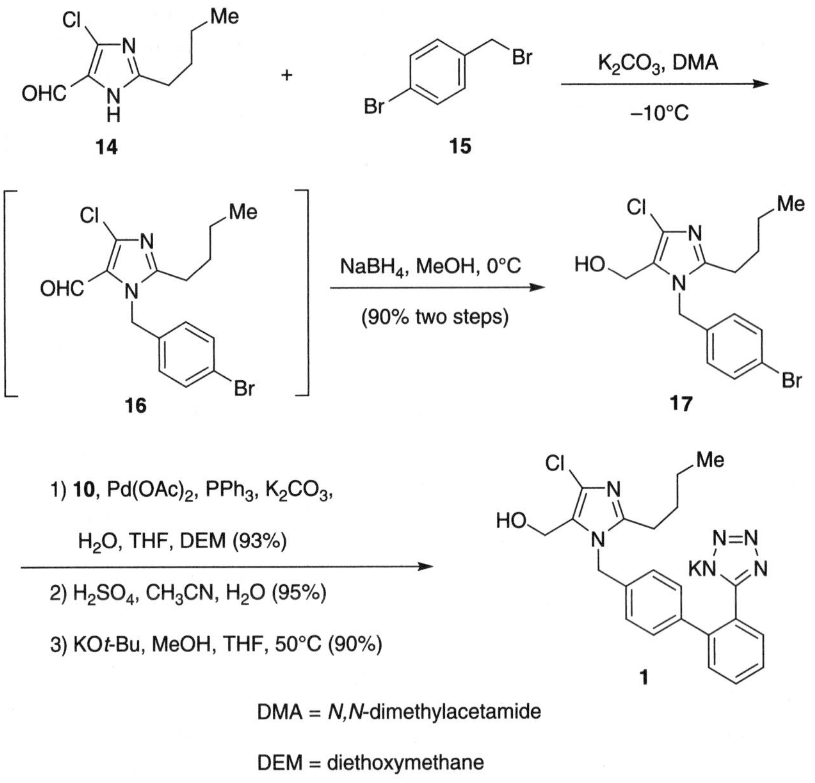

K_2CO_3, DMA, −10°C

$NaBH_4$, MeOH, 0°C (90% two steps)

1) **10**, $Pd(OAc)_2$, PPh_3, K_2CO_3, H_2O, THF, DEM (93%)

2) H_2SO_4, CH_3CN, H_2O (95%)

3) KO*t*-Bu, MeOH, THF, 50°C (90%)

DMA = *N,N*-dimethylacetamide

DEM = diethoxymethane

Scheme 9.4. The Merck process synthesis of losartan potassium (**1**).

9.3 VALSARTAN

9.3.1 Introduction to Valsartan

Valsartan (**2**) is a nonheterocyclic antagonist in which the imidazole of losartan has been replaced with an acylated amino acid. It is a very potent AT_1 antagonist ($IC_{50} = 1.6$ nM). There is only one metabolite, valeryl 4-hydroxy valsartan, and it is inactive. The enzymes responsible for valsartan metabolism have not been identified, but do not seem to be P450 CYP isozymes. Food decreases the absorption by 40%. Valsartan (**2**) is excreted in the bile (70%) and by the kidneys (30%). [See Chiolero and Burnier (1998).]

9.3.2 Synthesis of Valsartan

The synthesis of valsartan (**2**) by Novartis/Ciba-Geigy chemists is highlighted in Scheme 9.5. Biphenylbenzyl bromide **18** is converted to biphenyl acetate **19** in the presence of sodium acetate in acetic acid. Hydrolysis of **19** followed by Swern oxidation delivered the biphenyl aldehyde **20**, which underwent reductive amination with (*L*)-valine methyl ester (**21**) to give biphenyl amino acid **22**. Acylation of **22** with pentanoyl chloride (**23**) afforded biphenyl nitrile **24**, which is reacted with tributyltin azide to form the tetrazole followed by ester hydrolysis and acidification to provide valsartan (**2**). [See Bühlmayer et al. (1994, 1995).]

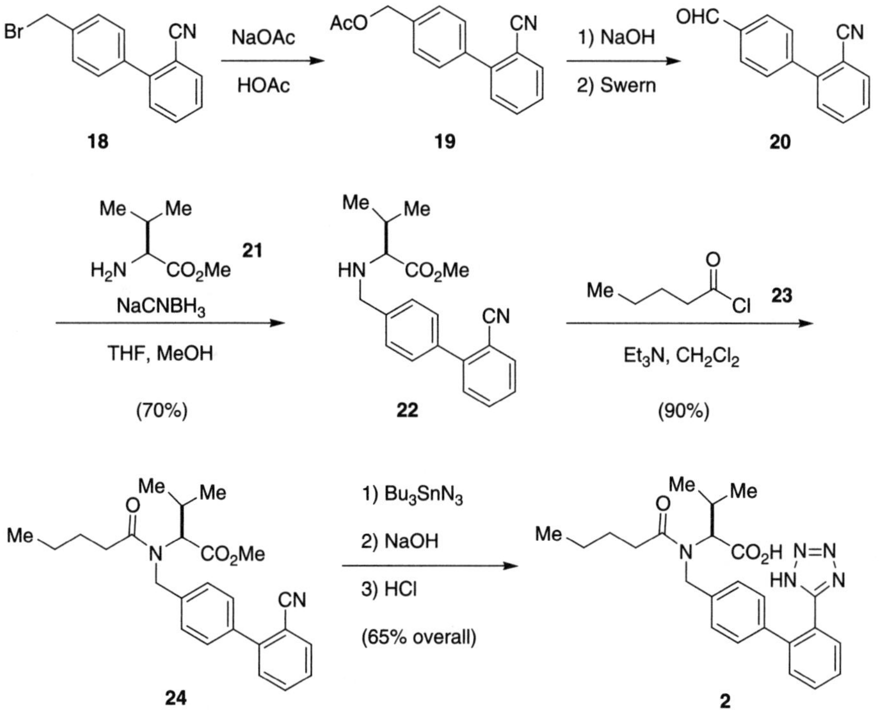

Scheme 9.5. The Novartis/Ciba-Geigy synthesis of valsartan (**2**).

9.4 IRBESARTAN

9.4.1 Introduction to Irbesartan

Irbesartan (**3**) is longer acting than losartan potassium (**1**) and valsartan (**2**). Irbesartan (**3**) has the highest oral bioavailability of the AT_1 antagonists and is more completely absorbed from the gastrointestinal tract. Peak plasma concentration is reached within 2 h. It is metabolized hepatically to inactive metabolites via cytochrome P450 CYP2C9 glucuronide conjugation and oxidation, and is excreted by both the bile and kidneys in about a 80:20 ratio. It has the second highest elimination half-life of all the angiotensin II AT_1 receptor antagonists. [See Gillis and Markham (1997).]

9.4.2 Synthesis of Irbesartan

The short Sanofi route to irbesartan (**3**) is outlined in Scheme 9.6. Dihydroimidazolone **27**, which is prepared from the reaction of 1-amino-cyclopentanecarboxylic acid ester (**25**) with ethyl pentanimidate (**26**) in the presence of acetic acid in refluxing xylene, is alkylated with biphenylbenzyl bromide **18** in the presence of sodium hydride in DMF to give **28**. Finally, the synthesis of irbesartan (**3**) is completed by the tetrazole formation from reaction of the nitrile group of **28** with tributyltin azide in refluxing xylene. [See Bernhart et al. (1993a, b).]

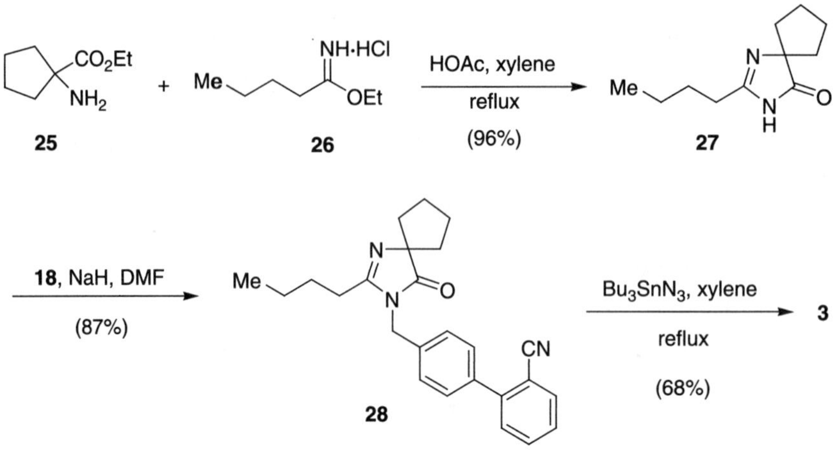

Scheme 9.6. The Sanofi synthesis of irbesartan (**3**).

9.5 CANDESARTAN CILEXETIL

9.5.1 Introduction to Candesartan Cilexetil

Candesartan cilexetil (**4**) is a potent long-acting antagonist. Because of poor oral absorption, ester prodrugs were synthesized and **4** was identified as the compound with the best angiotensin II AT_1 antagonistic activity profile after oral administration. Candesartan cilexetil (**4**) is rapidly and completely bioactivated by ester hydrolysis during absorption from the gastrointestinal tract to candesartan (**CV11974**). The absolute bioavailability of candesartan is estimated to be 15% and peak serum concentration is reached after 3–4 h. Candesartan is excreted unchanged by the bile and kidneys in a 70:30 ratio. [See Sever (1997).]

9.5.2 Synthesis of Candesartan Cilexetil

The synthesis of candesartan cilexetil (**4**) by Takeda Chemical Industries is illustrated in Scheme 9.7. Cyclization of methyl 2,3-diaminobenzoate (**29**) with tetraethoxymethane

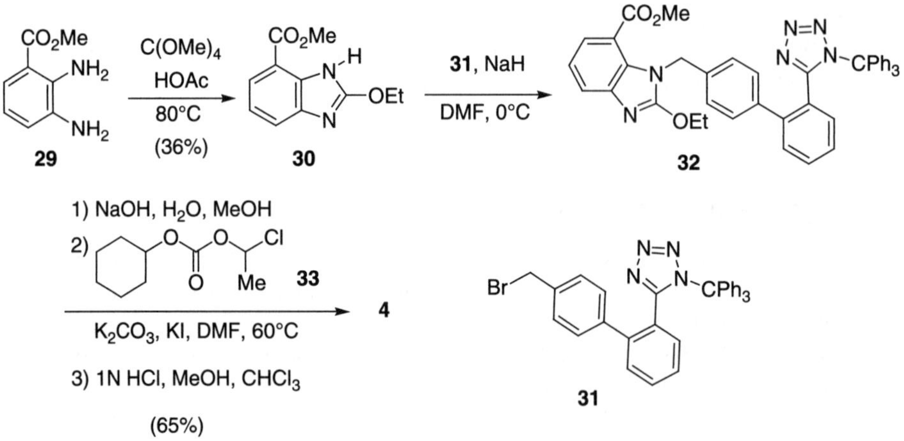

Scheme 9.7. The Takeda synthesis of candesartan cilexetil (**4**).

afforded 2-ethoxybenzimidazole **30**, which was alkylated with biphenyltetrazolyl bromide **31** to give benzimidazole **32**. Ester hydrolysis of **32** followed by esterification with cyclohexyl 1-chloroethyl carbonate (**33**) and deprotection of the triphenylmethyl protecting group using dilute hydrochloric acid provided candesartan cilexetil (**4**). [See Kubo et al. (1993a, b).]

9.6 OLMESARTAN MEDOXOMIL

9.6.1 Introduction to Olmesartan Medoxomil

Olmesartan medoxomil (**5**) is also a prodrug that is rapidly and completely bioactivated by ester hydrolysis to olmesartan (**RNH-6270**) during absorption from the gastrointestinal tract (Warner and Jarvis, 2002). Olmesartan is eliminated in a biphasic manner with an elimination half-life of 14 h. The absolute bioavailability of olmesartan is 26% and peak plasma concentration is reached in 1–2 h. Following the rapid and complete conversion of **5** to olmesartan, there is virtually no further metabolism of olmesartan. Olmesartan is eliminated from the bile and kidneys in about a 60 : 40 ratio.

9.6.2 Synthesis of Olmesartan Medoxomil

The synthesis of olmesartan medoxomil (**5**) by the Sankyo Company is illustrated in Scheme 9.8. Heating of diaminomaleonitrile (**34**) with trimethylorthobutyrate provided imidazole-4,5-dicarbonitrile **35**, which underwent hydrolysis to the dicarboxylic acid followed by esterification to give imidazole diester **36**. Addition of methyl magnesium

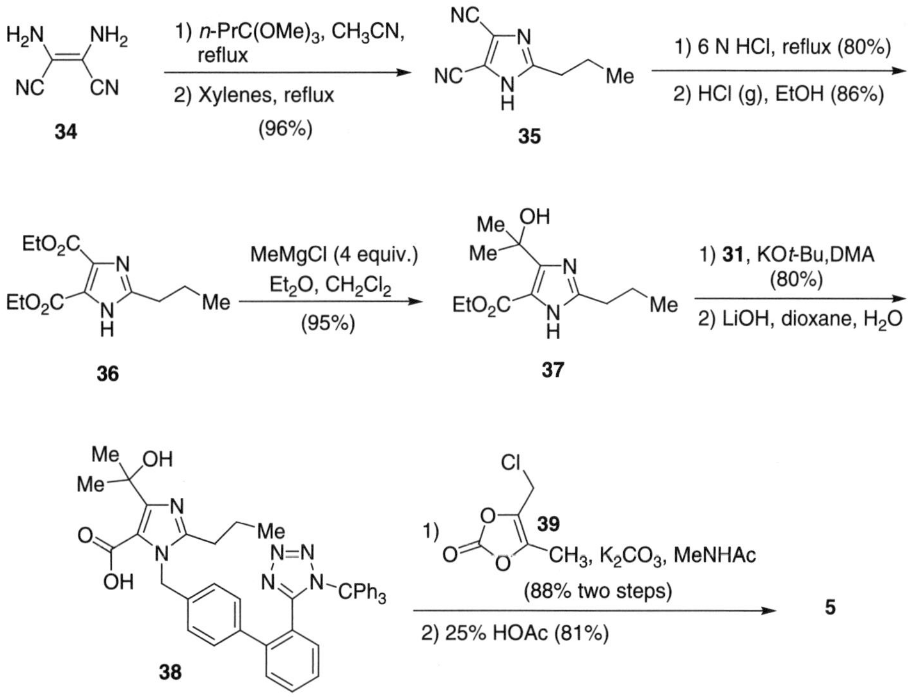

Scheme 9.8. The Sankyo synthesis of olmesartan medoxomil (**5**).

bromide to diester **36** furnished imidazole alcohol **37**. Alkylation of **37** with biphenyltetrazolyl bromide **31**, followed by hydrolysis of the ethyl ester, furnished (biphenylmethyl)-imidazole **38**. Alkylation of acid **38** with 4-chloromethyl-5-methyl-[1,3]dioxol-2-one (**39**) followed by deprotection of the triphenylmethyl protecting group under acidic conditions afforded olmesartan medoxomil (**5**). [See Yanagisawa et al. (1996, 1997).]

9.7 EPROSARTAN MESYLATE

9.7.1 Introduction to Eprosartan Mesylate

Eprosartan mesylate (**6**) was the first non-biphenyltetrazole angiotensin II AT_1 receptor antagonist. The absolute bioavailability of **6** is 15%, and peak plasma concentration is reached 1–2 h after oral administration. Administration of food with **6** delayed absorption by 25%. The elimination half-life varied from 5 to 7 h. Eprosartan mesylate (**6**) is eliminated by the bile and kidneys in a 90 : 10 ratio as unchanged drug. [See McClellan and Balfour (1998).]

9.7.2 Synthesis of Eprosartan Mesylate

The synthesis of eprosartan mesylate (**6**) by SmithKline Beecham Pharmaceuticals is described in Scheme 9.9. The synthesis of the precursors **43** and **45** were required for imidazole formation. Thus, valeronitrile (**40**) was converted to the imidate salt **41** with

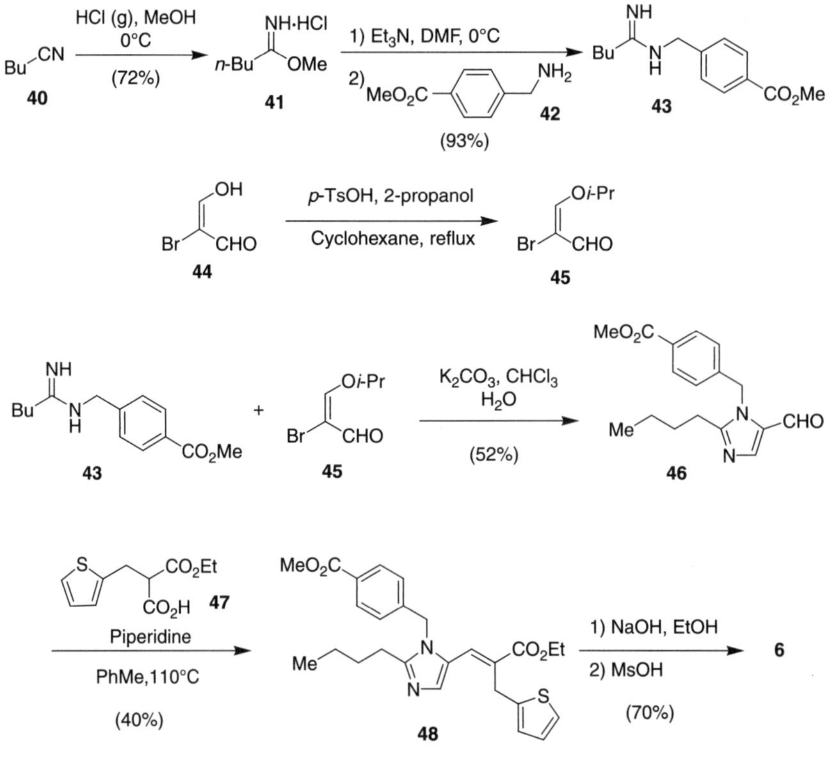

Scheme 9.9. The SmithKline Beecham synthesis of eprosartan mesylate (**6**).

hydrogen chloride gas in methanol. Reaction of **41** with benzylamine **42** provided amidine **43**. 2-Bromomalonaldehyde (**44**) was treated with 2-propanol and *p*-toluenesulfonic acid in refluxing cyclohexane to afford bromo enol ether **45**. Reaction of amidine **43** with bromo enol ether **45** furnished the imidazolecarbaldehyde **46** in moderate yield. Knoevenagel-type condensation of imidazolecarbaldehyde **46** with thiophene acid ester **47** yielded the olefin **48**, in which the ester was hydrolyzed with sodium hydroxide followed by mesylate salt formation to give eprosartan mesylate (**6**). [See Keenan et al. (1993); Shilcrat et al. (1997).]

9.8 TELMISARTAN

9.8.1 Introduction to Telmisartan

Telmisartan (**7**) is another potent non-biphenyltetrazole angiotensin II AT_1 receptor antagonist. Telmisartan (**7**) has an absolute oral bioavailability of 43% and the peak plasma concentration is reached after 0.5–1 h. Administration of food with **7** delayed absorption by 6%. This drug has the longest elimination half-life, at 24 h, of all the AT_1 receptor antagonist drugs. Telmisartan (**7**) is directly active and undergoes minimal biotransformation to its pharmacologically inactive acylglucuronide and is completely eliminated by the bile system (>98%). [See McClellan and Markham (1998).]

9.8.2 Synthesis of Telmisartan

Telmisartan (**7**) was synthesized by Boehringer Ingelheim chemists in eight steps starting with methyl 4-amino-3-methylbenzoate (**49**) as outlined in Scheme 9.10. Acylation of **49**, nitration at the 5-position, followed by reduction and ring closure under acidic conditions afforded benzimidazole **50**. Ester hydrolysis of **50** gave the corresponding benzimidazole carboxylic acid, which was then condensed with *N*-methyl-*o*-phenylenediamine in the presence of polyphosphoric acid (PPA) to give the *bis*-benzimidazole **51**. Alkylation of **51** with biphenylbenzyl bromide **52** followed by ester cleavage furnished telmisartan (**7**). [See Ries et al. (1993).]

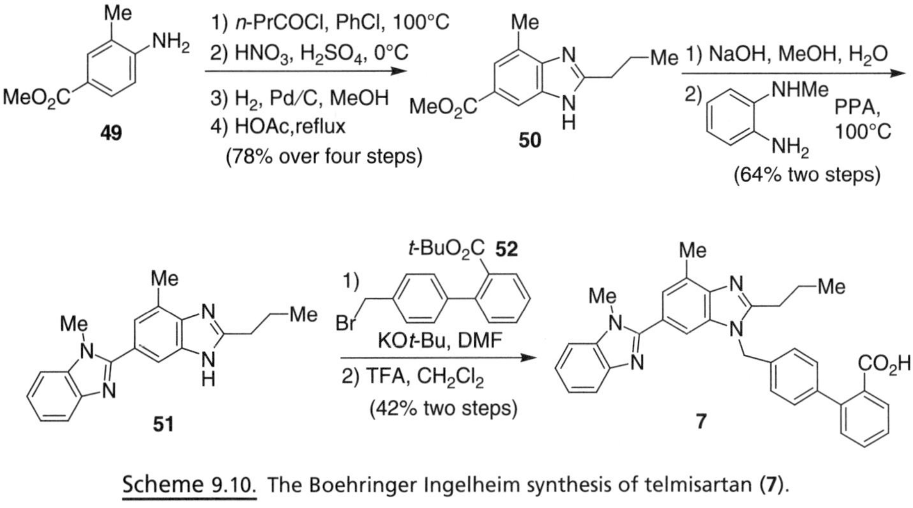

Scheme 9.10. The Boehringer Ingelheim synthesis of telmisartan (**7**).

REFERENCES

Bergsma, D. J., Ellis, C., Kumar, C., Nuthulaganit, P., Kersten, H., Elshourbagy, N., Griffin, E., Stadel, J. M., and Aiyar, N. (1992). *Biochem. Biophys. Res. Commun.*, 183: 989–995.

Bernhart, C., Breliere, J.-C., Clement, J., Nisato, D., Perreault, P., Muneaux, C., and Muneaux, Y. (1993b). US 5270317.

Bernhart, C. A., Perreaut, P. M., Ferrari, B. P., Muneaux, Y. A., Assens, J.-L. A., Clement, J., Haudricourt, F., Muneaux, C. F., Taillades, J. E., Vignal, M.-A., Gougat, J., Guiraudou, P. R., Lacour, C. A., Roccon, A., Cazaubon, C. F., Breliere, J.-C., Le Fur, G., and Nisato, D. (1993a). *J. Med. Chem.*, 36: 3371–3380.

Bühlmayer, P., Furet, P., Criscione, L., de Gasparo, M., Whitebread, S., Schmidlin, T., Lattmann, R., and Wood, J. (1994). *Bioorg. Med. Chem. Lett.*, 4: 29–34.

Bühlmayer, P., Ostermayer, F., and Schmidlin, T. (1995). US 5399578.

Burnier, M. and Brunner, H. R. (2000). *Lancet*, 355: 637–645.

Chiolero, A. and Burnier, M. (1998). *Expert Opin. Investig. Drug*, 7: 1915–1925.

Gillis, J. C. and Markham, A. (1997). *Drugs*, 54: 885–902.

Goa, K. L. and Wagstaff, A. J. (1996). *Drugs*, 51: 820–845.

Griffiths, G. J., Hauck, M. B., Imwinkelried, R., Kohr, J., Roten, C. A., Stucky, G. C., and Gosteli, J. (1999). *J. Org. Chem.*, 64: 8084–8089.

Keenan, R. M., Weinstock, J., Finkelstein, J. A., Franz, R. G., Gaitanopoulos, D. E., Girard, G. R., Hill, D. T., Morgan, T. M., Samanen, J. M., Peishoff, C. E., Tucker, L. M., Aiyar, N., Griffin, E., Ohlstein, E. H., Stack, E. J., Weldley, E. F., and Edwards, R. M. (1993). *J. Med. Chem.*, 36: 1880–1892.

Kirch, W., Horn, B., and Schweizer, J. (2001). *Eur. J. Clin. Invest.*, 31: 698–706.

Kubo, K., Kohara, Y., Imamiya, E., Sugiura, Y., Inada, Y., Furukawa, Y., Nishikawa, K., and Naka, T. (1993a). *J. Med. Chem.*, 36: 2182–2195.

Kubo, K., Kohara, Y., Yoshimura, Y., Inada, Y., Shibouta, Y., Furukawa, Y., Kato, T., Nishikawa, K., and Naka, T. (1993b). *J. Med. Chem.*, 36: 2343–2349.

Kumar, A., Singh, R. K., Panda, N., Upare, A. A., Nimbalkar, M. M., and Soudagar, S. R. (2004). WO 04/087691.

Larsen, R. D., King, A. O., Chen, C. Y., Corley, E. G., Foster, B. S., Roberts, F. E., Yang, C., Lieberman, D. R., Reamer, R. A., Tschaen, D. M., Verhoeven, T. R., Reider, P. J., Lo, Y. S., Rossano, L. T., Brookes, A. S., Meloni, D., Moore, J. R., and Arnett, J. F. (1994). *J. Org. Chem.*, 59: 6391–6394.

McClellan, K. J. and Balfour, J. A. (1998). *Drugs*, 55: 713–718.

McClellan, K. J. and Markham, A. (1998). *Drugs*, 56: 1039–1044.

Mukoyama, M., Nakajima, M., Horiuchi, M., Sasamura, H., Pratt, R. E., and Dzau, V. J. (1993). *J. Biol. Chem.*, 68: 24539–24542.

Packer, M. (1992). *Lancet*, 340: 88–95.

Remme, W. J. and Swedberg, K. (2001). *Eur. Heart J.*, 22: 1529–1560.

Ries, U. J., Mihm, G., Narr, B., Hasselbach, K. M., Wittneben, H., Entzeroth, M., van Meel, J. C. A., Wienen, W., and Hauel, N. H. (1993). *J. Med. Chem.*, 36: 4040–4051.

Schmidt, B. and Schieffer, B. (2003). *J. Med. Chem.*, 46: 2261–2270.

Sever, P. (1997). *J. Hum. Hypertens.*, 11(suppl 2): S91–95.

Shilcrat, S. C., Mokhallalati, M. K., Fortunak, J. M. D., and Pridgen, L. N. (1997). *J. Org. Chem.*, 62: 8449–8454.

Warner, G. T. and Jarvis, B. (2002). *Drugs*, 62: 1345–1353.

Yanagisawa, H., Anemiya, Y., Kanazazki, T., Shimoji, Y., Fujimoto, K., Kitahara, Y., Sada, T., Mizuno, M., Ikeda, M., Miyamoto, S., Furukawa, Y., and Koike, H. (1996). *J. Med. Chem.*, 39: 323–338.

Yanagisawa, H., Fujimoto, K., Amemiya, Y., Shimoji, Y., Kanazaki, T., Koike, H., and Sada, T. (1997). US 5616599.

10

LEADING ACE INHIBITORS FOR HYPERTENSION

Victor J. Cee and Edward J. Olhava

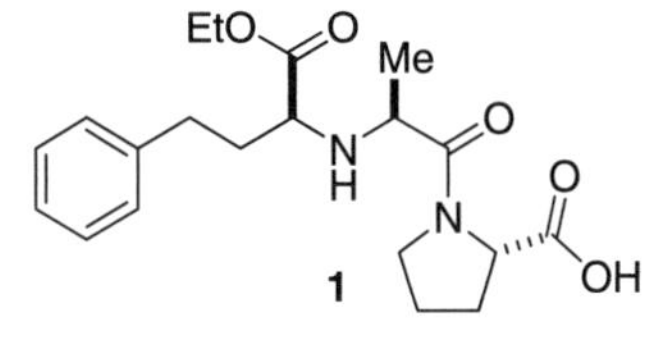

USAN: Enalapril maleate
Trade name: Vasotec®
Merck & Co.
Launched: 1985
M.W. 376.45

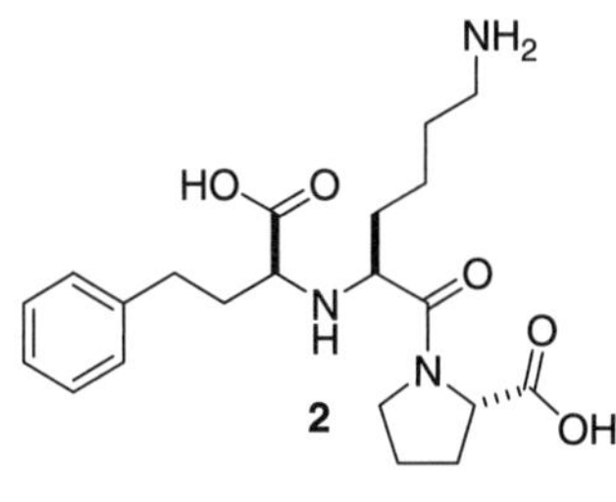

USAN: Lisinopril
Trade name: Zestril®
Merck & Co.
Launched: 1988
M.W. 405.49

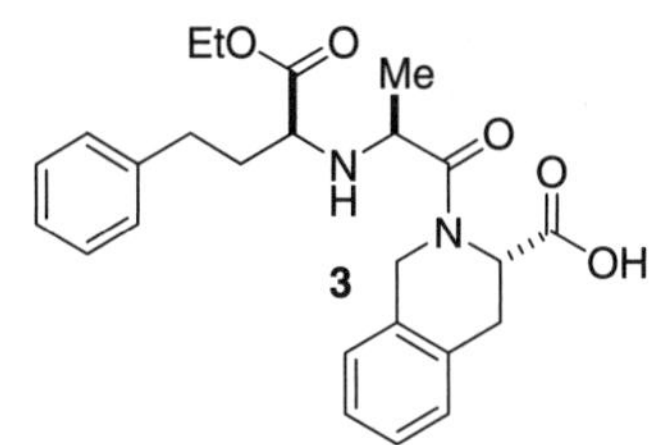

USAN: Quinapril
Trade name: Accupril®
Warner-Lambert
Launched: 1991
M.W. 438.52

The Art of Drug Synthesis. Edited by Douglas S. Johnson and Jie Jack Li

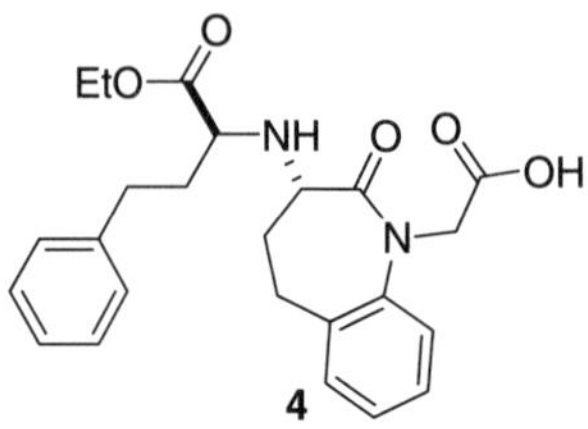

USAN: Benazepril
Trade name: Lotensin®
Ciba-Geigy
Launched: 1991
M.W. 424.49

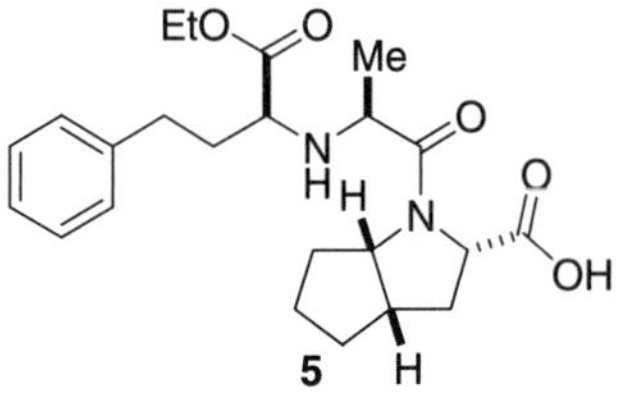

USAN: Ramipril
Trade name: Altace®
Hoechst
Launched: 1991
M.W. 416.51

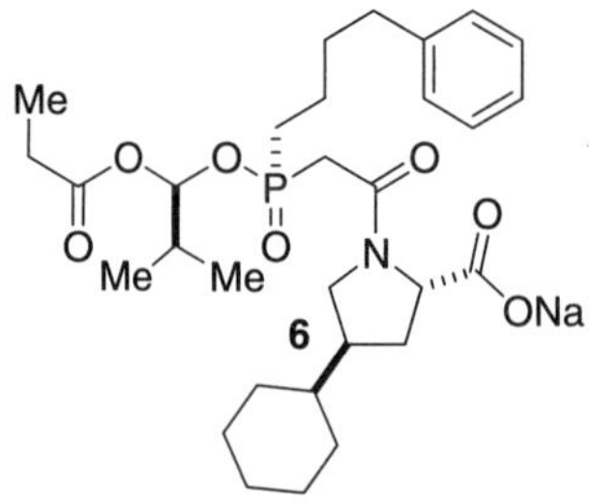

USAN: Fosinopril sodium
Trade name: Monopril®
E.R. Squibb & Sons
Launched: 1991
M.W. 585.64 (parent: 563.66)

10.1 INTRODUCTION

Hypertension is estimated to affect 1 billion individuals worldwide, and is a major risk factor for stroke, coronary heart disease, heart failure, and end-stage renal disease (Frazier et al., 2005). The discovery of the renin-angiotensin system (RAS) as a master regulator of blood pressure and cardiovascular function provided numerous targets for pharmacologic intervention (Zaman et al., 2002). Angiotensin-converting enzyme (ACE) plays a crucial role in the RAS by the production of the vasoconstrictive peptide angiotensin II. In the early 1970s, the first human trial of a polypeptide ACE inhibitor established the clinical relevance of this target (Gavras et al., 1974). This discovery stimulated extensive research into the development of ACE inhibitors, and these agents have become leading treatments for hypertension and hypertension-related target-organ damage. Enalapril maleate (**1**), lisinopril (**2**), quinapril (**3**), benazepril (**4**), ramipril (**5**), and fosinopril sodium (**6**) are discussed herein.

ACE is a chloride-dependent zinc metallopeptidase that is present in both membrane-bound and soluble forms. ACE contains two homologous active sites in which a zinc ion catalyzes the cleavage of C-terminal dipeptides from various proteins, including angiotensin I and bradykinin (Acharya et al., 2003). Cleavage of angiotensin I (Ang I, **7**) by ACE generates the octapeptide angiotensin II (Fig. 10.1). Angiotensin II activates membrane-bound angiotensin receptors, the end result of which is vasoconstriction and sodium retention, leading to a net increase in blood pressure. As vasodilation and salt excretion are among the many processes mediated by bradykinin, the destruction of circulating bradykinin by ACE is also believed to contribute to hypertension (Ram, 2002). Although the

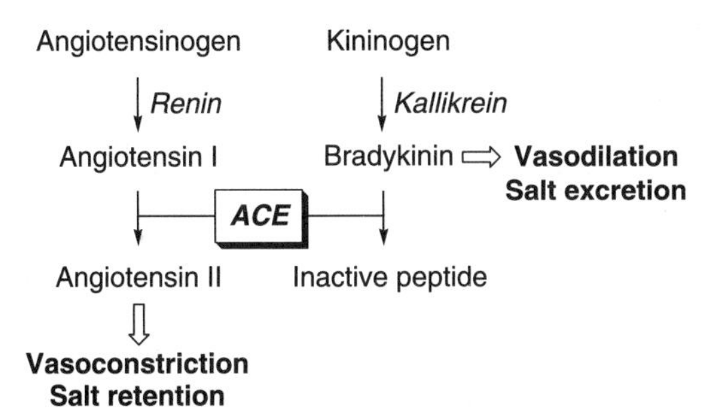

Figure 10.1. ACE in the renin-angiotensin and kallikrein–kinin systems.

accumulation of bradykinin resulting from ACE inhibition may have positive hemodynamic effects, it is believed to be responsible for the adverse effects of ACE inhibitors, including cough and angioedema.

The initial validation of ACE as a therapeutic target was largely due to the discovery that peptides produced by the Brazilian pit viper *Bothrops jararaca* could potentiate the activity of bradykinin (Cushman and Ondetti, 1999; Smith and Vane, 2003). It was further shown that the potentiation of bradykinin activity was due to inhibition of ACE. The most active peptide in vivo, teprotide (**8**), was studied in clinical trials and established ACE as a relevant target for the treatment of hypertension. Teprotide was hypothesized to interact with the active-site zinc ion in a similar fashion to the substrate angiotensin I (**7**) (Fig. 10.2). Studies of teprotide and other peptide ACE inhibitors established proline as the optimal C-terminal inhibitory residue (Ondetti et al., 1971). The first marketed ACE inhibitor, captopril (**9**), contains this C-terminal proline, but utilizes a thiol group as a ligand for the active-site zinc ion. Later generations of ACE inhibitors generally adhere to the captopril prototype, although carboxylate and phosphinic acid zinc-binding elements have replaced the thiol group, the S1 binding pocket is exploited, and the P2′ proline is in some cases replaced by non-natural amino acids.

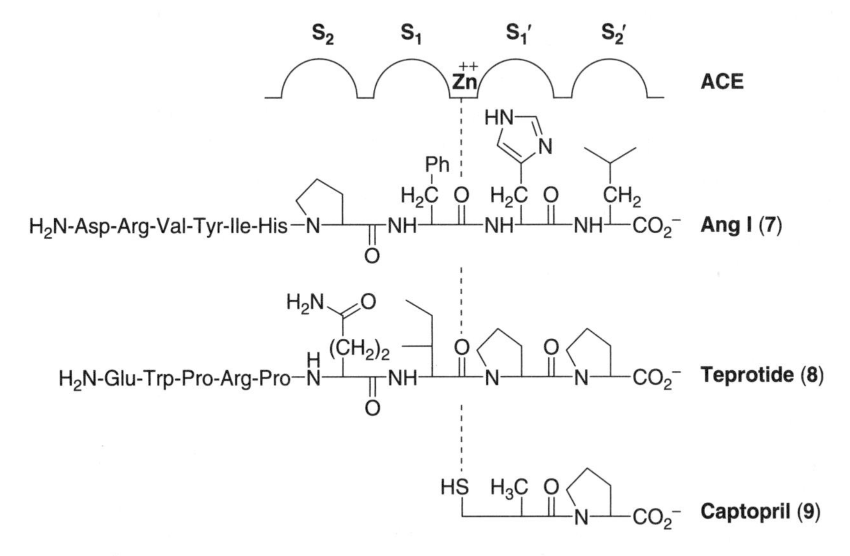

Figure 10.2. Proposed binding of Ang I and early inhibitors to ACE.

TABLE 10.1. Pharmacological Parameters of ACE Inhibitors 1–6

	IC_{50} (nM)[a]	Therapeutic Dose (mg)	Frequency (times/day)	t_{max} (h)[a]	$t_{1/2}$ (h)[a]	F (%)
Enalapril (**1**)	1.2	2.5–40	1–2	2–8	11	60
Lisinopril (**2**)	1.2	5–40	1	6–8	12	25[b]
Quinapril (**3**)	8.3	5–80	1	2	2[c], 25[e]	>60
Benazepril (**4**)	1.7	5–80	1	1–2	10–11	>37
Ramipril (**5**)	NR[f]	1.25–20	1	3	4[c], 9–18[d], >50[e]	50–60
Fosinopril (**6**)	7	10–80	1	3	12	36

[a]Value reported for *bis*-acid; [b]Highly variable; [c]Elimination phase; [d]Apparent elimination phase; [e]Terminal phase, [f]NR = not reported.

The pharmacological and pharmacokinetic parameters of ACE inhibitors discussed herein are summarized in Table 10.1 (Brown and Vaughan, 1998). With the exception of lisinopril, the ACE inhibitors are administered as monoester prodrugs to facilitate absorption. Hydrolysis in vivo by esterases liberates the active *bis*-acid inhibitor. The relatively good absorption of the *bis*-acid lisinopril is attributed to a peptide transporter (Friedman and Amidon, 1989). The active components of **1–6** are reported to inhibit ACE with IC_{50} values in the single-digit nM range. Another common feature of ACE inhibitors **1–6** are the long half-lives of the active drug substances, which supports once-daily dosing.

10.2 SYNTHESIS OF ENALAPRIL MALEATE (VASOTEC®)

Enalapril (**1**, MK-421) was discovered at Merck in the 1970s (Patchett et al., 1980; Patchett, 2002). This molecule was designed in order to move away from the sulfhydryl zinc-binding element of captopril, which was believed to be responsible for captopril's short half-life and unfavorable side-effects. Although Squibb had disclosed carboxyalkanoyl analogs of captopril (**9**) (Cushman et al., 1977), these molecules were several orders of magnitude less potent than the parent sulfhydryl. The breakthrough in Merck's program came from the investigation of *N*-carboxyalkyldipeptides. Although carboxylate **10** was a weak ACE inhibitor, addition of a more lipophilic methyl group to the carboxy terminus (**11**) increased potency by two orders of magnitude (Scheme 10.1). Elaboration to the phenethyl derivative provided the active drug substance enalaprilat (**12**). Due to the poor oral absorption of enalaprilat ($F \leq 10\%$), enalapril was designed as a monoester prodrug. The ethyl ester facilitates absorption, and is hydrolyzed in the liver to enalaprilat, the active drug substance (Ranadive et al., 1992; Vertas et al., 1992). Esterification of the prolinyl carboxylate was also possible, but stability of the drug substance was compromised due to diketopiperazine formation (Patchett, 2002).

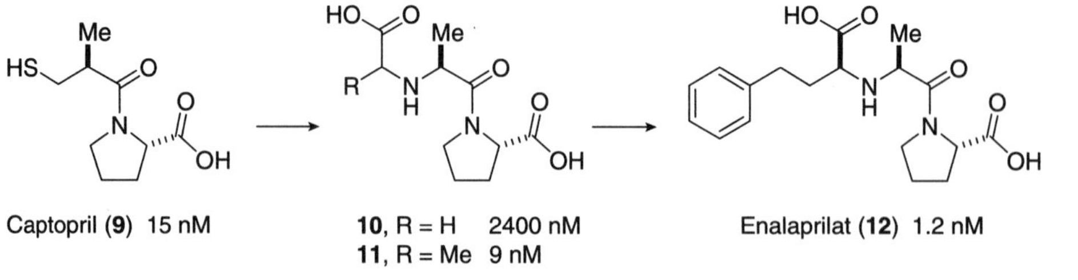

Captopril (**9**) 15 nM

10, R = H 2400 nM
11, R = Me 9 nM

Enalaprilat (**12**) 1.2 nM

Scheme 10.1.

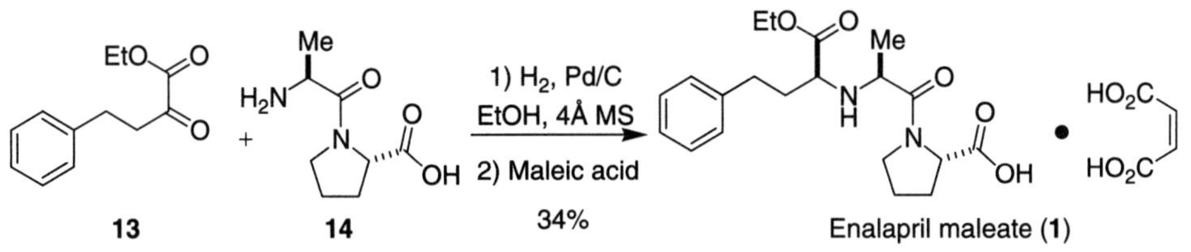

Scheme 10.2.

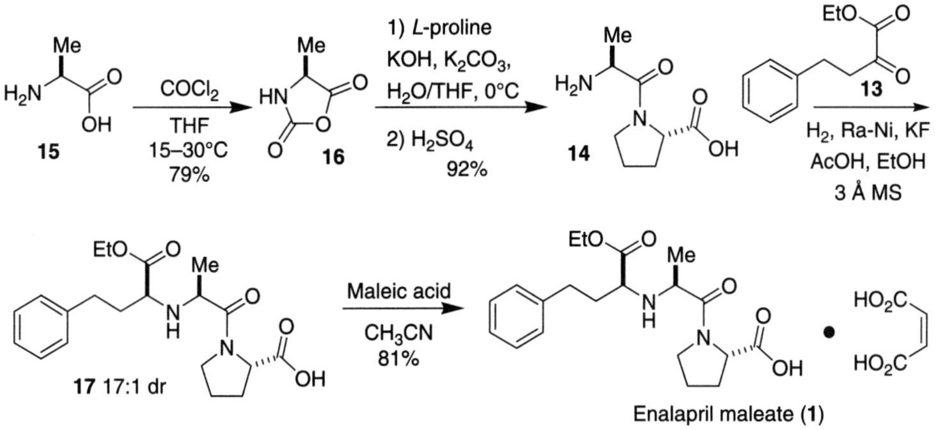

Scheme 10.3.

The initial synthetic route employed by the medicinal chemistry group is straightforward: a moderately diastereoselective reductive amination between ethyl-2-oxo-4-phenylbutyrate (**13**) and alanyl-proline (**14**) generated a 1.6 : 1 mixture of diastereomers favoring the desired (*S*,*S*,*S*)-enantiomer (Scheme 10.2). Fractional recrystallization afforded enalapril as the maleate salt (Harris et al., 1983; Wyvratt et al., 1984). Although requiring only one synthetic step, the diastereoselectivity of the process is poor, reducing the overall yield.

The Merck process group subsequently published a more detailed route amenable towards multikilogram scales (Blacklock et al., 1988). This synthesis begins with treatment of alanine with phosgene to produce *N*-carboxyanhydride (NCA) **16** (Scheme 10.3). Under basic aqueous conditions this anhydride is coupled with proline to produce, upon acidic work-up, the dipeptide alanyl-proline (**14**). Enalapril is then prepared in one synthetic step by a diastereoselective reductive amination between ethyl-2--oxo-4-phenylbutyrate (**13**) and **14**. This reaction was the subject of extensive optimization, and it was found that the highest diastereoselectivity was obtained by hydrogenation over Raney nickel in the presence of acetic acid (25%), KF (4.0 equiv.), and 3 Å molecular sieves (17 : 1 *dr*). Enalapril is then isolated in diastereomerically pure form as its maleate salt (Huffman and Reider, 1999; Huffman et al., 2000).

10.3 SYNTHESIS OF LISINOPRIL

Lisinopril (**2**, MK-521) was discovered at Merck by the same research team who discovered enalapril. It was launched as Prinivil® by Merck in 1987, and since 1988 has been

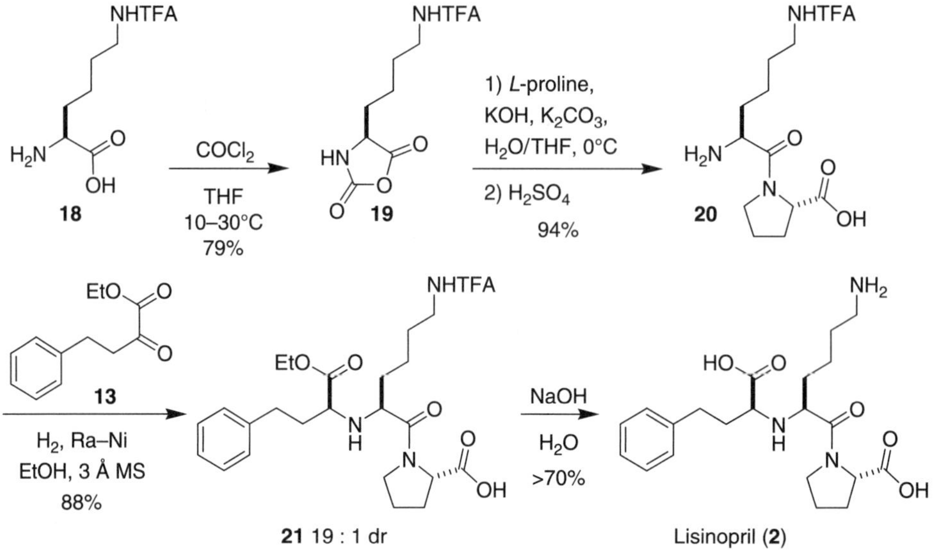

Scheme 10.4.

marketed as Zestril® by AstraZeneca. The Merck group systematically substituted the P1′ position of enalapril with a variety of amino acids (Patchett, 1993). To their surprise, lysine substitution at this position had not only good intrinsic potency, but showed good oral activity. The researchers speculated that an additional beneficial interaction was gained that compensated for the steric bulk of the amino-substituted butyl group. A crystal structure of lisinopril bound to human testis ACE reveals that the lysyl amine forms a hydrogen bond with Glu 162 of the S1′ pocket (Natesh et al., 2003). In contrast with the majority of ACE inhibitors, lisinopril is administered as the diacid. The relatively good, but variable, oral absorption is attributed to active transport involving a peptide carrier (Friedman and Amidon, 1989).

The synthetic development of lisinopril (Scheme 10.4) parallels that of enalapril. The medicinal chemistry synthesis employed a nonselective reductive alkylation for the installation of the carboxyalkyl substituent (Harris et al., 1980; Wu et al., 1985), and the Merck process group later published a more refined route that takes advantage of the stereoselective reductive alkylation developed for enalapril (Blacklock et al., 1988). The *N*-carboxyanhydride **19** was prepared, and coupled with proline under basic conditions to produce dipeptide **20**. Again using Raney nickel, reductive condensation of **13** and **20** proceeds with 19 : 1 selectivity for the desired (*S*,*S*,*S*)-diastereomer **21**. Sodium hydroxide effects the concomitant hydrolysis of the trifluoroacetamide and saponification of the ethyl ester, and after recrystallization, lisinopril (**2**) is obtained in greater than 70% yield as the dihydrate. Of interest, the trifluoroacetamide remains intact in the highly acidic NCA formation and the slightly caustic condensation with proline, but is easily cleaved in this last step.

10.4 SYNTHESIS OF QUINAPRIL

Quinapril (**3**, CI-906), marketed by Pfizer under the trade name Accupril®, was discovered at Warner-Lambert (Klutchko et al., 1986). Like enalapril, quinapril is a prodrug, and is

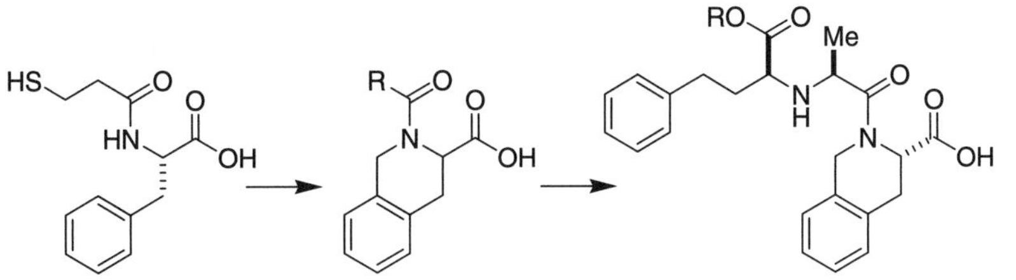

22 430 nM

R = Et: Quinapril (3) 8.3 nM
R = H: Quinaprilat (23) 2.8 nM

Scheme 10.5.

hydrolyzed in vivo to the active drug substance, quinaprilat. An early publication from Squibb described the ACE inhibitory action of sulfhydryl phenylalanine **22** (Cushman et al., 1977). Researchers at Warner-Lambert derived a conformationally restricted analog of **22**, and combined this concept with the highly successful zinc-binding paradigm established by enalapril to give quinapril (Scheme 10.5).

The synthesis of quinapril begins with formation of the *N*-carboxyalkyl alanine intermediate **26** (Scheme 10.6) (Hoefle and Klutchko, 1982; Kaltenbronn et al., 1983; Klutchko et al., 1986). Displacement of ethyl-2-bromo-4-phenylbutanoate (**24**) by (*S*)-alanyl-*tert*-butyl ester provided the secondary amine as a diastereomeric mixture. Deprotection of this mixture allowed for the selective recrystallization of the undesired (*R*,*S*)-isomer, and isolation of the desired (*S*,*S*)-isomer **26** from the mother liquor.

Completion of the synthesis of quinapril involves amide bond formation between **26** and a tetrahydroisoquinoline fragment. Two complementary protected 1,2,3,4-tetrahydro-3-isoquinoline subunits **27** and **28**, each available in a single step from commercially available (*S*)-1,2,3,4-tetrahydroisoquinoline-3-carboxylic acid, were utilized (Scheme 10.7). Coupling with **26** using DCC and HOBt in dichloromethane afforded the penultimate compounds **29** and **30** as maleate salts. Cleavage of the *t*-butyl ester of **29** and treatment with HCl provided quinapril. Alternatively, hydrogenation of **30** under standard conditions cleanly removed the benzyl ester, and quinapril (**3**) was isolated after formation of the hydrochloride salt.

A recent patent application describes, in part, a multikilogram scale synthesis of quinapril (Jennings, 2004). The carboxyanhydride of **26** is prepared by treatment with phosgene (Scheme 10.8) (Youssefyeh et al., 1987). This is next coupled with tetrahydroisoquinoline subunit **27** in the presence of catalytic acid. Without isolation of the resultant quinapril *t*-butyl ester, the reaction solution is treated with acetic acid and anhydrous hydrogen chloride to deprotect the ester. Amorphous quinapril hydrochloride is obtained via treatment with acetonitrile.

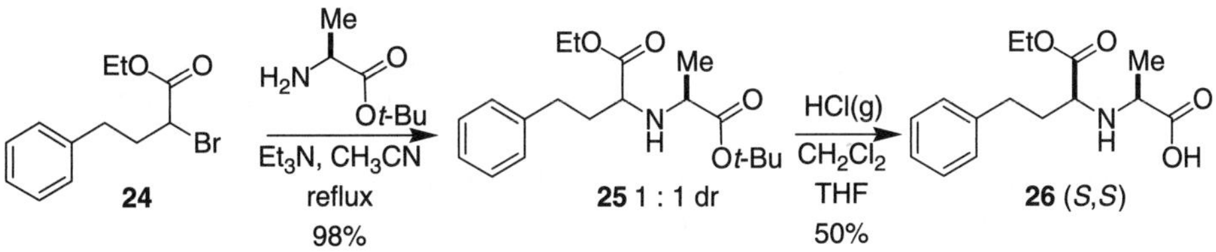

24 → (H_2N-CH(Me)-CO-O*t*-Bu; Et_3N, CH_3CN, reflux, 98%) → 25 1 : 1 dr → (HCl(g), CH_2Cl_2, THF, 50%) → 26 (*S*,*S*)

Scheme 10.6.

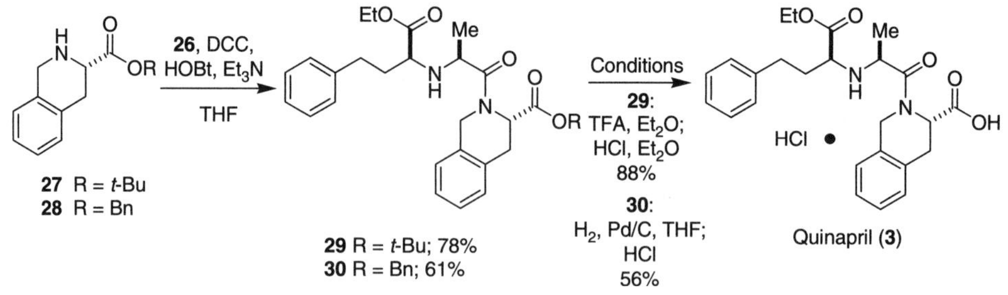

27 R = *t*-Bu
28 R = Bn

26, DCC, HOBt, Et_3N, THF

29 R = *t*-Bu; 78%
30 R = Bn; 61%

Conditions
29: TFA, Et_2O; HCl, Et_2O 88%
30: H_2, Pd/C, THF; HCl 56%

Quinapril (**3**)

Scheme 10.7.

26 Phosgene, THF/PhMe → **31**

1) **27**, AcOH, PhMe
2) AcOH, HCl, acetone
90%

Quinapril (**3**)

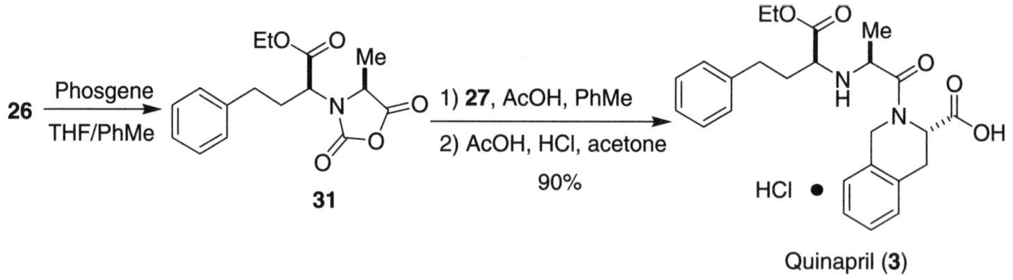

Scheme 10.8.

10.5 SYNTHESIS OF BENAZEPRIL

Benazepril (**4**, CGS-14824A) was discovered in the early 1980s by Ciba-Geigy (now Novartis) and was launched in 1991 as Lotensin®. Key to the discovery of benazepril was the recognition that cyclic captopril analog **32** possessed reasonable ACE inhibitory activity (Condon et al., 1982; Klutchko et al., 1981; Watthey et al., 1984). Optimization of **32** involved annulation of a benzene ring and expansion of the six-membered ring, resulting in the potent thiol-based inhibitor **33**. Replacement of the thiol with homophenylalanine as the zinc-binding moiety provided benazeprilat **34**. Not surprisingly, **34** exhibited marginal in vivo efficacy when dosed orally in animal models due to poor absorption. Consistent with the development of other bis-acid ACE inhibitors, the monoethyl ester proved to be a suitable prodrug for oral dosing (Watthey et al., 1985).

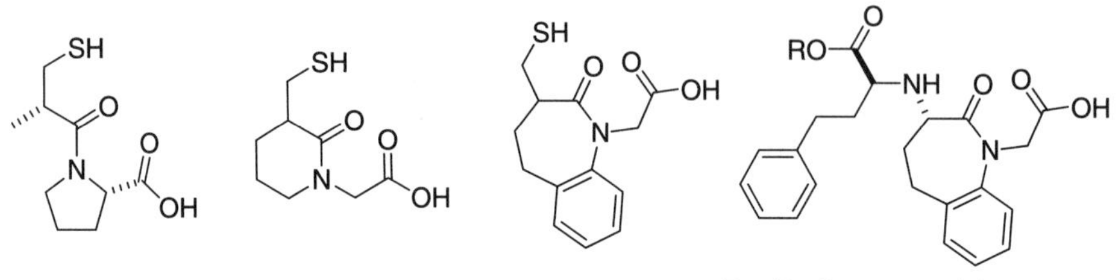

Captopril (**9**) 15 nM   **32** 1,500 nM   **33** 9 nM

R = Et: Benazepril (**4**)
R = H: Benazeprilat (**34**) 1.7 nM

The synthesis of the benzazepinone portion of benazepril began with monobromination of 1-tetralone (**35**), followed by oxime formation to give **36** (Scheme 10.9). A Beckmann rearrangement mediated by polyphosphoric acid provided the ring-expanded lactam **37**. Displacement of the α-bromine with potassium phthalimide installed the necessary

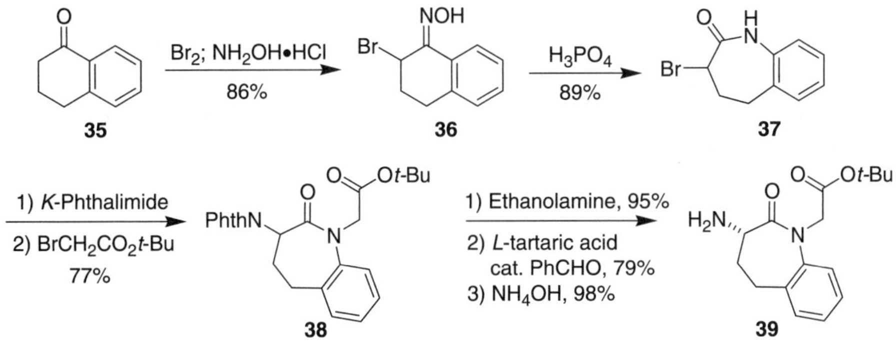

Scheme 10.9.

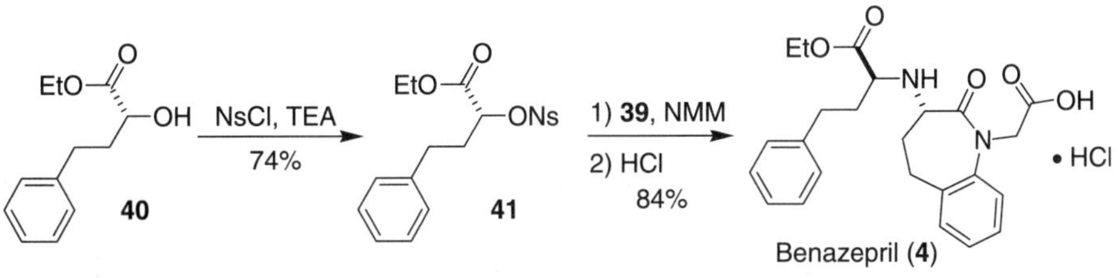

Scheme 10.10.

amine, and selective *N*-alkylation of the amide provided the *N*-carboxymethyl benzazepinone **38**. Ethanolamine was then employed to reveal the primary amine and subsequent tartrate resolution provided enantioenriched **39**. A crystallization-induced asymmetric transformation (CIAT, see Section 2.3.6 in Chapter 2 for more details) was applied to achieve >50% yields in this resolution (For a review see Anderson, 2005). The presence of a catalytic amount of benzaldehyde during crystallization effects racemization of the undesired, soluble enantiomer via the corresponding benzaldehyde imine. A 79% yield of the desired tartrate salt was obtained, and treatment with ammonium hydroxide liberated the free amine **39**.

The completion of the synthesis of benazepril involves the stereospecific installation of the homophenylalanine moiety (Scheme 10.10). (*R*)-2-Hydroxy-4-phenylbutyric acid (**40**, obtained in 82% *ee* from catalytic asymmetric hydrogenation of ethyl benzylpyruvate) was initially reported as the homophenylalanine precursor. A more recent publication from Ciba-Geigy details the reduction of ethyl benzylpyruvate to give **40** in >99% *ee* using *D*-lactic acid dehydrogenase derived from *Staphylococcus epidermis* (Ghisalba et al., 1999). The hydroxyl group is activated as its 4-nitrobenzenesulfonate to give **41**, which undergoes stereospecific displacement with amine **39**. Acid hydrolysis of the *tert*-butyl ester provides benazepril (**4**).

10.6 SYNTHESIS OF RAMIPRIL

Ramipril (**5**, HOE-498), discovered at Hoechst (now Sanofi-Aventis), was launched in the United States as Altace® in 1991 and is currently marketed by King Pharmaceuticals. Upon administration, rapid hepatic cleavage of the ethyl ester results in ramiprilat (**42**), the active drug substance. A detailed medicinal chemistry account of the discovery and development of ramipril has not been disclosed. Presumably the researchers were exploring extended, hydrophobic proline moieties to optimize interactions with the S2′ pocket. Ramipril contains the same (*S*)-2-((*S*)-1-ethoxy-1-oxo- 4-phenylbutan-2-ylamino) propanoic acid portion as enalapril and quinapril. This subunit raises the same challenges regarding an efficient diastereoselective synthesis. In addition, ramipril also contains the complex azabicyclo[3.3.0] octane-3-carboxylic acid subunit, introducing three additional chiral centers.

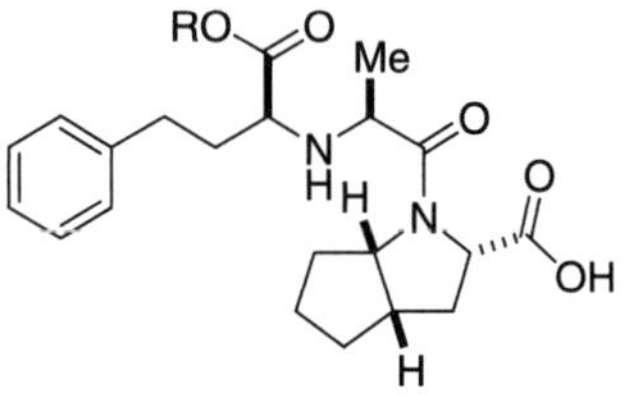

R = Et: Ramipril (**5**)
R = H: Ramiprilat (**42**)

The initial medicinal chemistry route to the azabicyclo[3.3.0]octane-3-carboxylic acid produced the azabicyclo system in a diastereoselective but racemic manner, and required a classical resolution to achieve enantioenriched material (Teetz et al., 1984a, b; 1988). Reaction of (*R*)-methyl 2-acetamido-3-chloropropanoate (**43**) and 1-cyclopentenylpyrrolidine (**44**) in DMF followed by an aqueous acidic work-up provided racemic keto ester **45** in 84% yield (Scheme 10.11). Cyclization of **45** in refluxing aqueous hydrochloric acid provided the bicyclic imine, which was immediately reduced under acidic hydrogenation conditions. The desired *cis-endo* product **46** was obtained upon recrystallization. The acid was protected as the benzyl ester using thionyl chloride and benzyl alcohol, providing subunit **47** as the racemate. Resolution of **47** was accomplished by crystallization with benzyloxy-carbonyl-*L*-phenylalanine or *L*-dibenzoyl-tartaric acid.

Hoechst has reported an enantioselective approach toward the key azabicyclo[3.3.0] octane-3-carboxylic acid **46** that preserves the stereochemistry of the *L*-serine-derived starting material (Urbach and Henning, 1991). *L*-Serine methyl ester (**48**) was alkylated

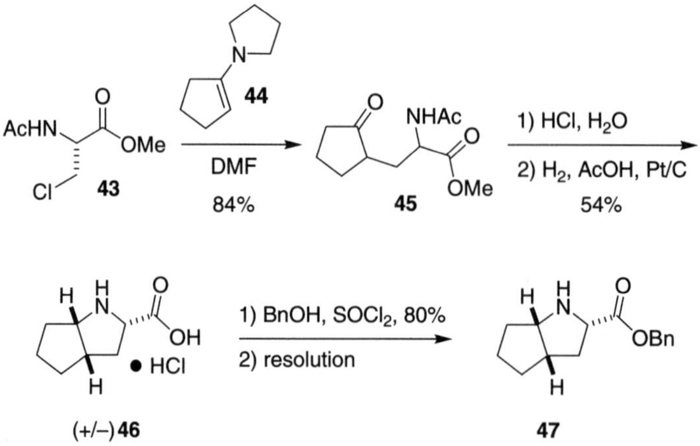

Scheme 10.11.

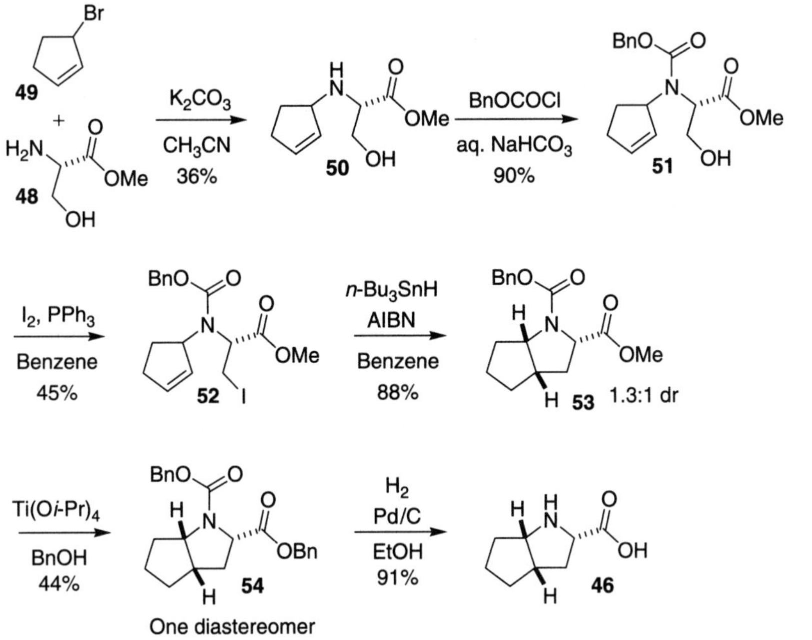

Scheme 10.12.

with 3-bromocyclopentene (**49**) to obtain a mixture of diastereomers (ratio not specified) (Scheme 10.12). This mixture was protected as the benzyl carbamate, transformed to iodide **52**, and then subjected to radical cyclization. The resulting *cis*-fused bicycle **53** was isolated as a 1.3 : 1 mixture of diastereomers resulting from the earlier nonselective alkylation step. Following transesterification to the benzyl ester, the diastereomers were separated to give stereochemically pure **54**. Hydrogenolysis afforded the desired ramipril subunit **46** in six synthetic steps overall.

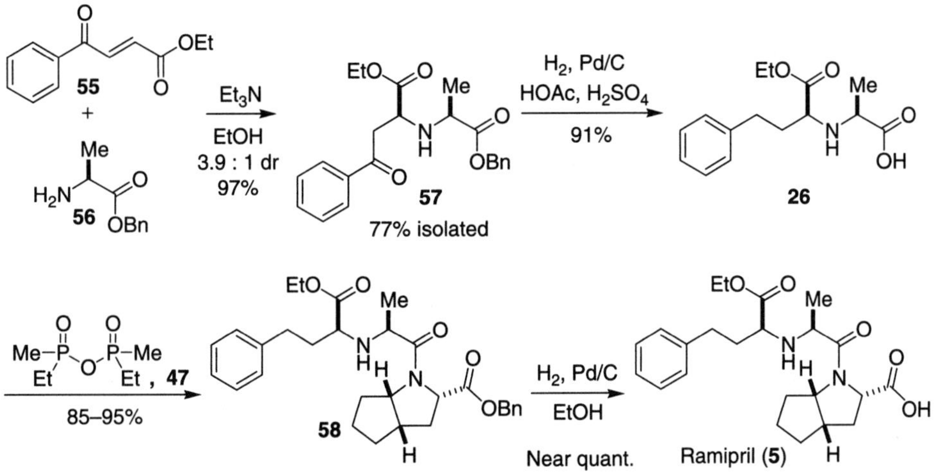

Scheme 10.13.

The Hoechst synthesis of the *N*-carboxyalkyl alanine intermediate **26** involves a diastereoselective conjugate addition (Teetz et al., 1984a; Urbach and Henning, 1984). Reaction of (*S*)-alanyl benzyl ester (**56**) with *trans*-ethyl-4-oxo-4-phenyl-2-butenoate (**55**) proceeded with 3.9 : 1 dr and provided the desired diastereomer **57** in 77% yield after recrystallization (Scheme 10.13). Reduction of the substituted acetophenone and cleavage of the benzyl ester were accomplished by catalytic hydrogenation to give intermediate **26**. Completion of the synthesis involves coupling of **47** and **26** followed by deprotection. Although both racemic and enantiopure **47** have been prepared in this regard, the synthesis of ramipril from the latter is described. Coupling of **47** and **26** with ethylmethylphosphinic acid anhydride provided ramipril benzyl ester **58** in 85–95% yield. Hydrogenolysis of the benzyl ester provided ramipril in near quantitative yield (Teetz et al., 1984a).

10.7 SYNTHESIS OF FOSINOPRIL SODIUM (MONOPRIL®)

Fosinopril sodium (**6**, SQ-28555) was discovered in the mid-1980s and launched by E.R. Squibb & Sons (now Bristol-Myers Squibb) in 1991 as Monopril®. Key to the discovery of fosinopril sodium was the hypothesis that 4-substituted proline analogs would be tolerated at P2′, and it was believed that these analogs would exhibit greater in vitro potency and a longer duration of action than the ACE inhibitors available at that time. It was subsequently found that substituted proline residues markedly improved the activity of a series of inhibitors containing a phosphinic acid as the zinc ligand. Cyclohexyl was determined to be ideal, and the *trans*-diastereomer **60** was judged to be slightly superior to the *cis*-diastereomer **61** (Table 10.2). Consistent with most bis-acid ACE inhibitors, oral absorption of the free phosphinic acid fosinoprilat **60** was judged to be insufficient for further development, but the (acyloxy)alkyl prodrug fosinopril (**6**) remedied this problem (Krapcho et al., 1988).

Two conceptually different approaches to the *trans*-4-cyclohexylproline fragment have been disclosed by the Squibb process group. The first route relies on a stereoselective alkylation for installation of the cyclohexyl substituent (Thottathil et al., 1986). *L*-Pyroglutamic acid (**62**) was reduced using literature methods to give the hydroxymethyl

TABLE 10.2. In vitro and In vivo Activities for Phosphinic Acids 59–61

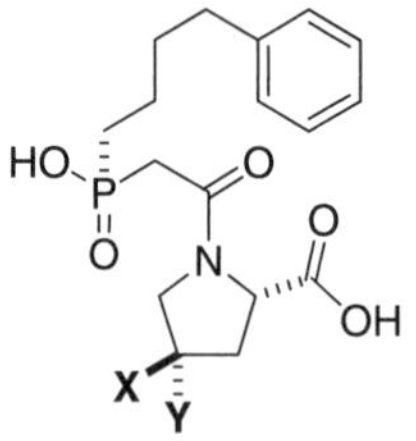

Compound	X	Y	$IC_{50}(nM)^a$	% Inhibition of Ang I Response[b]
59	H	H	180	56 ± 4
60 (fosinoprilat)	C_6H_{11}	H	7	87 ± 1
61	H	C_6H_{11}	11	84 ± 8

[a]Rabbit lung ACE; [b]Normotensive rats after 170 min (iv, 0.5 μmol/kg).

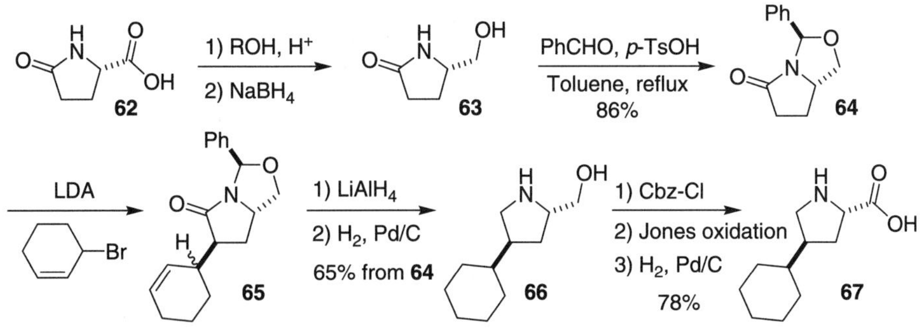

Scheme 10.14.

pyrrolidinone **63** (Scheme 10.14). Treatment of **63** with benzaldehyde and *p*-toluenesulfonic acid gave *N,O*-acetal **64** as a single diastereomer. Formation of the lithium enolate of **64** with LDA and alkylation with 3-bromocyclohexene installed the necessary cyclohexyl fragment with complete control of diastereoselectivity at the newly formed pyrrolidine stereocenter. Reduction of the lactam and *N,O*-acetal with $LiAlH_4$ was followed by hydrogenolysis to liberate the benzylated pyrrolidine, providing **66**. A three-step sequence involving carbamate formation, Jones oxidation, and deprotection gave *trans*-4-cyclohexylproline **67**.

An alternative approach to the *trans*-4-cyclohexylproline fragment **67** relied on a stereospecific arylation to install the precursor to the cyclohexyl group (Scheme 10.15). This route is likely used for manufacturing and begins with commercially available *L*-hydroxyproline. *N*-Benzoylation and esterification gives **69** (Kronenthal et al., 1990a). Mesylation of **69** with inversion of configuration was accomplished by a Mitsunobu approach employing methanesulfonic acid, triphenylphosphine, diisopropyl azodicarboxylate, and triethylamine. On a multikilogram scale this reaction provided **70** in 80–85% yield after saponification of the methyl ester and recrystallization (Anderson et al., 1996). A highly stereospecific Friedel–Crafts alkylation mediated by aluminum trichloride installed a 4-phenyl substituent with complete inversion (for an approach to *trans*-4-phenylproline involving tosylate displacement with retention of configuration, see Thottathil and Moniot, 1986). The unusually high level of inversion in this process is attributed to the electron-withdrawing nature of the adjacent nitrogen, which is believed to inhibit carbocation formation, favoring alkylation via an S_N2 pathway. On a 100-kg scale, a 75% yield of **71** is reported (Kronenthal et al., 1990b). The details regarding the conversion of **71** to *trans*-4-cyclohexylproline **67** have not been revealed. Cleavage of the

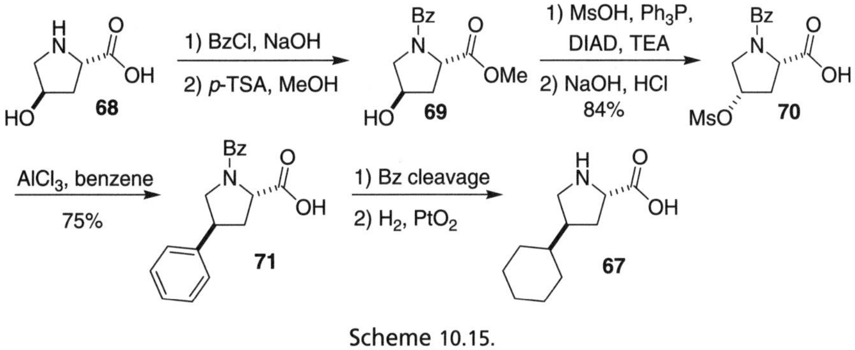

Scheme 10.15.

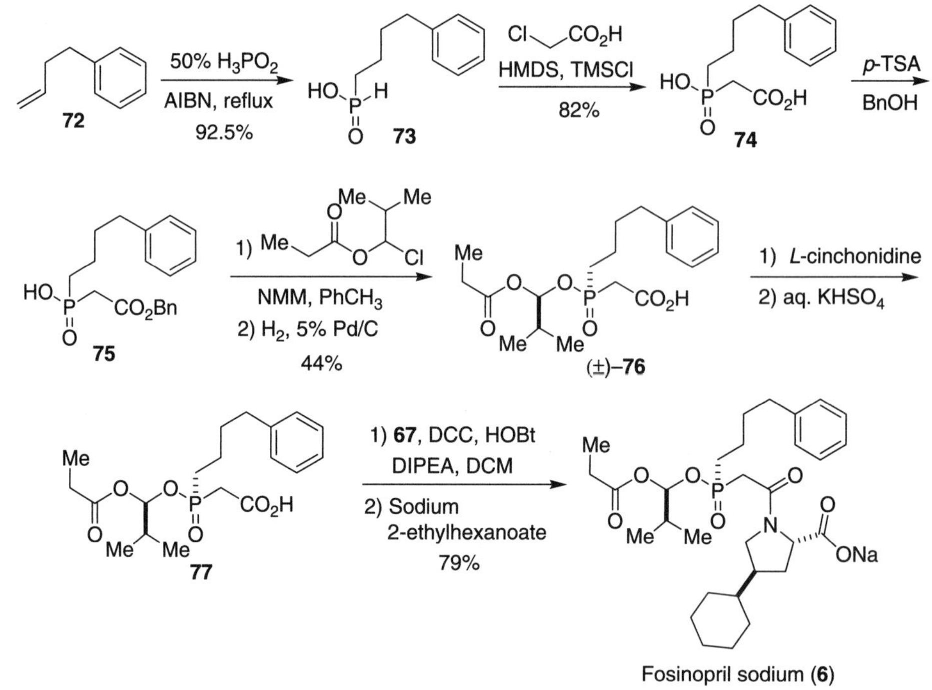

Scheme 10.16.

benzoyl group would provide *trans*-4-phenylproline, which is known to provide **67** via catalytic hydrogenation over platinum oxide (Krapcho et al., 1988).

The phosphinic acid fragment of fosinopril is constructed from terminal olefin **72**. A radical-initiated addition of hypophosphorous acid provided phosphinic acid **73** in approximately 98 : 2 regioselectivity (Scheme 10.16). The removal of the undesired regioisomer at this stage was problematic, but it was effectively reduced to $<0.1\%$ by recrystallization of later intermediates (Anderson et al., 1997a). A trimethylsilyl chloride-modified Arbuzov reaction (Thottathil et al., 1984) between **73** and chloroacetic acid provided the dialkyl phosphinic acid **74**. This reaction was the subject of extensive optimization for manufacturing, and is reported to provide **74** in 82% yield with 99.9% purity on a several hundred-kilogram scale (Anderson et al., 1997b). The acid was protected as its benzyl ester to give **75**, and alkylation with racemic 1-chloro-2-methylpropyl propionate gave a mixture of diastereomers from which the desired racemic product could be obtained by fractional crystallization. Hydrogenolysis of the benzyl ester provided racemic **76** in 44% yield over two steps. A classical resolution provided enantioenriched **77**, which was coupled to **67** under standard conditions. Conversion of the acid to its sodium salt provided fosinopril sodium (**6**) in 79% yield over the final two steps (Sedergran, 1991).

REFERENCES

Acharya, K. R., Sturrock, E. D., Riordan, J. F., and Ehlers, M. R. W. (2003). *Nat. Rev. Drug Discovery*, 2: 891–902.

Anderson, N. G. (2005). *Org. Proc. Res. Dev.*, 9: 800–813.

Anderson, N. G., Ciaramella, B. M., Feldman, A. F., Lust, D. A., Moniot, J. L., Moran, L., Polomski, R. E., and Wang, S. S. Y. (1997b). *Org. Process Res. Dev.*, 1: 211–216.

Anderson, N. G., Coradetti, M. L., Cronin, J. A., Davies, M. L., Gardineer, M. B., Kotnis, A. S., Lust, D. A., and Palaniswamy, V. A. (1997a). *Org. Process Res. Dev.*, 1: 315–319.

Anderson, N. G., Lust, D. A., Colapret, K. A., Simpson, J. H., Malley, M. F., and Gougoutas, J. Z. (1996). *J. Org. Chem.*, 61: 7955–7958.

Blacklock, T. J., Shuman, R. F., Butcher, J. W., Shearin, W. E., Jr., Budavari, J., and Grenda, V. J. (1988). *J. Org. Chem.*, 53: 836–844.

Boyer, S. K., Pfund, R. A., Portmann, R. E., Sedelmeier, G. H., and Wetter, H. F. (1988). *Helvetica Chimica Acta*, 71: 337–343.

Brown, N. J. and Vaughan, D. E. (1998). *Circulation*, 97: 1411–1420.

Condon, M. E., Petrillo, E. W., Ryono, D. E., Reid, J. A., Neubeck, R., Puar, M., Heikes, J. E., Sabo, E. F., Losee, K. A., Cushman, D. W., and Ondetti, M. A. (1982). *J. Med. Chem.*, 25: 250–258.

Cushman, D. W. and Ondetti, M. A. (1999). *Nature Med.*, 5: 1110–1112.

Cushman, D. W., Cheung, H. S., Sabo, E. F., and Ondetti, M. A. (1977). *Biochemistry*, 16: 5484–5491.

Frazier, C. G., Shah, S. H., Armstrong, P. W., Bhapkar, M. V., McGuire, D. K., Sadowski, Z., Kristinsson, A., Aylward, P. E., Klein, W. W., Weaver, W. D., and Newby, L. K. (2005). *Am. Heart. J.*, 150: 1260–1267.

Friedman, D. I. and Amidon, G. L. (1989). *J. Am. Pharm. Sci.*, 78: 995–998.

Gavras, H., Brunner, H. R., Laragh, J. H., Sealey, J. E., Gavras, I. and Vukovich, R. A. (1974). *New Engl. J. Med.*, 291: 817–821.

Ghisalba, O., Gygax, D., Lattmann, R., Schär, H.-P., Schmidt, E., and Seelmeier, G. (1992). U.S. Patent 5,098,841 (to Ciba-Geigy).

Harris, E. E., Patchett, A. A., Tristram, E. W., and Wyvratt, M. J. (1980). U.S. Patent 4,374,829 (to Merck & Co.).

Harris, E. E., Patchett, A. A., Tristram, E. W., and Wyvratt, M. J. U.S. Patent 4,374,829 (1983 to Merck & Co.).

Hoefle, M. L. and Klutchko, S. (1982). U.S. Patent 4,344,949 (to Warner-Lambert).

Huffman, M. A. and Reider, P. J. (1999). *Tetrahedron Lett.*, 40: 831–834.

Huffman, M. A., Reider, P. J., Leblond, C., and Sun, Y. (2000) U.S. Patent 6,025,500 (2000 to Merck & Co.).

Jennings, S. M. (2004). U.S. Patent Application 2004/0192613A1 (to Warner-Lambert).

Kaltenbronn, J. S., DeJohn, D., and Krolls, U. (1983). *Org. Prep. Proced. Int.*, 15: 35–40.

Klutchko, S., Blankley, C. J., Fleming, R. W., Hinkley, J. M., Werner, A. E., Nording, I., Holmes, A., Hoefle, M. L., Cohen, D. M., Essenburg, A. D., and Kaplan, H. R. (1986). *J. Med. Chem.*, 29: 1953–1961.

Klutchko, S., Hoefle, M. L., Smith, R. D., Essenburg, A. D., Parker, R. B., Nemeth, V. L., Ryan, M. J., Dugan, D. H., and Kaplan, H. R. (1981). *J. Med. Chem.*, 24: 104–109.

Krapcho, J., Turk, C., Cushman, D. W., Powell, J. R., DeForrest, J. M., Spitzmiller, E. R., Karanewsky, D. S., Duggan, M., Rovnvak, G., Schwartz, J., Natarajan, S., Godfrey, J. D., Ryono, D. E., Neubeck, R., Atwal, K. S., and Petrillo, E. W., Jr. (1988). *J. Med. Chem.*, 31: 1148–1160.

Kronenthal, D., Kuester, P. L., and Mueller, R. H. (1990a). U.S. Patent 4,912,231 (to E. R. Squibb & Sons, Inc.).

Kronenthal, D. R., Mueller, R. H., Kuester, P. L., Kissick, T. P., and Johnson, E. J. (1990b). *Tetrahedron Lett.*, 31: 1241–1244.

Natesh, R., Schwager, S. L. U., Sturrock, E. D., and Acharya, K. R. (2003). *Nature*, 421: 551–554.

Ondetti, M. A., Williams, N. J., Sabo, E. F., Pluscec, J., Weaver, E. R., and Kocy, O. (1971). *Biochemistry*, 10: 4033–4039.

Patchett, A. A. (1993). "Enalapril and Lisinopril". *Chronicles of Drug Discovery*, 3: 125–162.

Patchett, A. A. (2002). *J. Med. Chem.*, 45: 5609–5616.

Patchett, A. A., Harris, E., Tristram, M. J., Wyvratt, M. J., Wu, M. T., Taub, D., Peterson, E. R., Ikeler, T. J., ten Broeke, J., Payne, L. G., Ondeyka, D. L., Thorsett, E. D., Greenlee, W. J., Lohr, N. S., Hoffsommer, R. D., Joshua, H., Ruyle, W. V., Rothrock, J. W., Aster, S. D., Maycock, A. L., Robinson, F. M., and Hirschmann, R. (1980). *Nature*, 288: 280–283.

Ram, C. V. S. (2002). *Am. J. Cardiovasc. Drugs*, 2: 77–89.

Ranadive, S. A., Chen, A. X., and Serajuddin, A. T. M. (1992). *Pharm. Res.*, 9: 1480–1486.

Sedergran, T. C. (1991). U.S. Patent 5,008,399 (to E. R. Squibb & Sons, Inc.).

Smith, C. G. and Vane, J. R. (2003). *FASEB*, 17: 788–789.

Teetz, V., Geiger, R., and Gaul, H. (1984b). *Tetrahedron Lett.*, 25: 4479–4482.

Teetz, V., Geiger, R., Henning, R., and Urbach, R. (1984a). *Arzneim.-Forsch*, 34: 1399–1401.

Teetz, V., Geiger, R., Urbach, H., Becker, R., and Schölkens, B. (1988). U.S. Patent 4,727,160 (to Hoechst).

Thottathil, J. K. and Moniot, J. L. (1986). *Tetrahedron Lett.*, 27: 151–154.

Thottathil, J. K., Moniot, J. L., Mueller, R. H., Wong, M. K. Y., and Kissick, T. P. (1986). *J. Org. Chem.*, 51: 3140–3143.

Thottathil, J. K., Przybyla, C. A., and Moniot, J. L. (1984). *Tetrahedron Lett.*, 25: 4737–4740.

Urbach, H. and Henning, R. (1984). *Tetrahedron Lett.*, 25: 1143–1146.

Urbach, H. and Henning, R. (1991). U.S. Patent 5,011,940 (to Hoechst).

Vertas, V. and Haynie, R. (1992). *Am. J. Cardiol.*, 69: 8C–16C.

Watthey, J. W. H., Gavin, T., and Desai, M. (1984). *J. Med. Chem.*, 27: 816–818.

Watthey, J. W. H., Stanton, J. L., Desai, M., Babiarz, J. E., and Finn, B. M. (1985). *J. Med. Chem.*, 28: 1511–1516.

Wu, M. T., Douglas, A. W., Ondeyka, D. L., Payne, L. G., Ikeler, T. J., Joshua, H., and Patchett, A. A. (1985). *J. Pharm. Sci.*, 74: 352–354.

Wyvratt, M. J., Tristram, E. W., Ikeler, T. J., Lohr, N. S., Joshua, H., Springer, J. P., Arison, B. H., and Patchett, A. A. (1984). *J. Org. Chem.*, 49: 2816–2819.

Youssefyeh, R. D., Skiles, J. W., Suh, J. T., and Jones, H. (1987). U.S. Patent 4,686,295 (to USV Pharmaceutical Corporation).

Zaman, M. A., Oparil, S., and Calhoun, D. A. (2002). *Nat. Rev. Drug Discovery*, 1: 621–636.

11

DIHYDROPYRIDINE CALCIUM CHANNEL BLOCKERS FOR HYPERTENSION

Daniel P. Christen

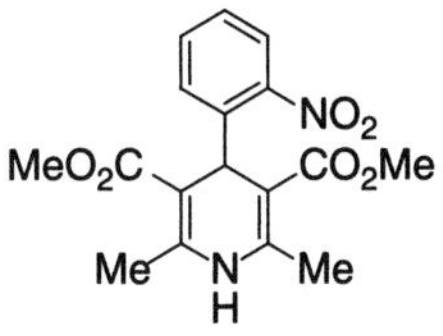

1

USAN: Nifedipine
Trade name: Adalat®
Bayer
Launched: 1975
M.W. 346.34

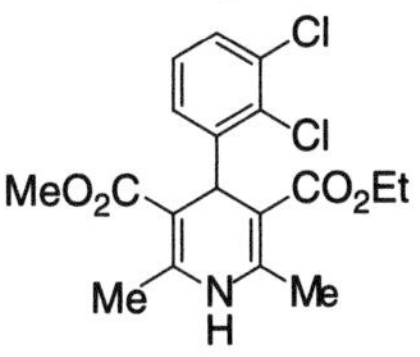

2

USAN: Felodipine
Trade name: Plendil®
Astra-Zeneca
Launched: 1988
M.W. 384.26

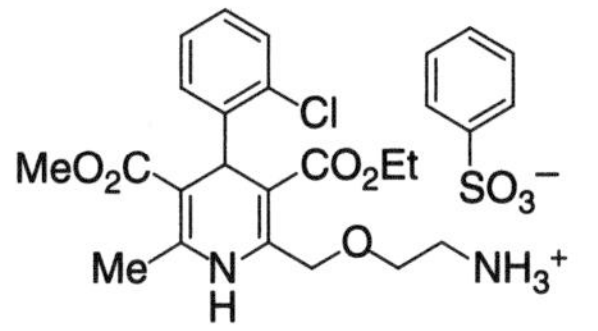

3

USAN: Amlodipine besylate
Trade name: Norvasc®
Pfizer
Launched: 1990
M.W. 567.06 (parent 408.88)

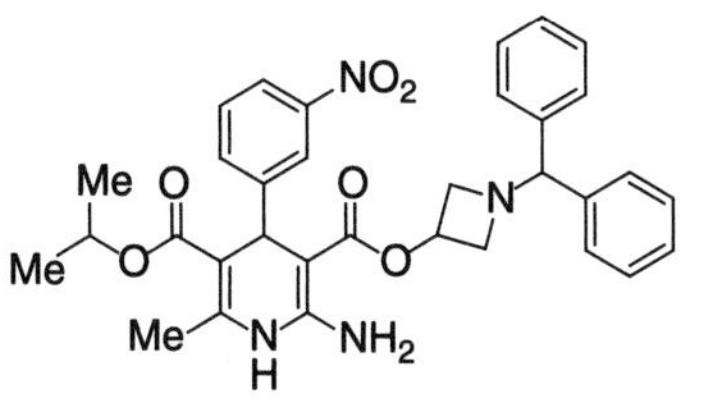

4

USAN: Azelnidipine
Trade name: Calblock®
Sankyo, Ube.
Launched: 2003
M.W. 582.65

11.1 INTRODUCTION

Hypertension, elevated blood pressure indicated by systolic or diastolic pressures exceeding 140 mmHg or 90 mmHg, respectively, is becoming an increasingly important medical problem as the general population becomes older and more overweight. A recent summary of research in the area of hypertension indicates that 50 million Americans and as many as one billion people globally suffer from hypertension (NHLBI, www.nhlbi.nih.gov, 2007). More importantly, the treatment of hypertension is becoming more complicated as the population ages, because many patients will take drugs to treat conditions such as diabetes, rheumatoid arthritis, atherosclerosis, and so on. As such, there is a continuing evolution of treatment regimens for hypertension (Chobanian et al., 2001). In general terms, once a patient is diagnosed with hypertension, the only treatment methods are diet, exercise, and drug intervention. Although the former options can lead to some reduction in blood pressure, most patients are treated with a mono or dual drug treatment regimen. Typically, the drugs for either therapy are drawn from the diuretics, β-blockers, angiotensin-converting enzyme (ACE) inhibitors, angiotensin II antagonists, and/or calcium channel blockers. In some cases, two of the drugs are sold in a single pill. Other treatment drugs include aldosterone-receptor blockers, combined α- and β-blockers, α_1-blockers, central α_2-agonists, and direct vasodilators. The actual drug treatment plan depends on factors such as age, cardiac health, other health issues, and how severe the hypertension is. In addition, renin-angiotensin inhibitors are expected to become members of the treatment arsenal within the next two years.

The calcium channel blockers (CCBs) are an important class of hypertension drugs incorporating three distinctive compound subtypes. Nifedipine (**1**), first discovered in the 1960s at Bayer, remains the prototypical representative of the dihydropyridine variant of the calcium channel blockers. Verapamil (**5**) and diltiazem (**6**) are representatives of the other two major structural types of calcium channel blockers used for the treatment of hypertension. This review will focus on the biological properties and chemical synthesis of the dihydropyridines **1–4** shown above. During the last 25 years, other dihydropyridine CCBs have entered the market here and abroad (Fig. 11.1). These compounds include nitrendipine (Baypress®, **7**), nilvadipine (Nivadil®, **8**), aranidpine (Bec®, **9**), nisoldipine (Syscor®, **10**), nimodipine (Nimotop®, **11**), cinaldipine (Siscrad®, **12**), lacidipine (Lacirex®, **13**), nicardipine hydrochloride (Cardene SR®, **14**), barnidipine hydrochloride (Hypoca®, **15**), benidipine hydrochloride (Coniel®, **16**), lercanidipine hydrochloride (Zanedip®, **17**), franidipine hydrochloride (Calslot®, **18**), and efonidipine hydrochloride ethanol (Landel®, **19**).

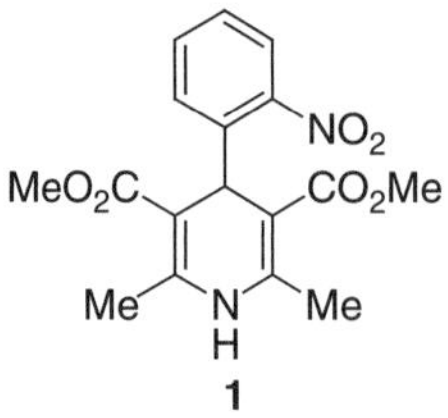

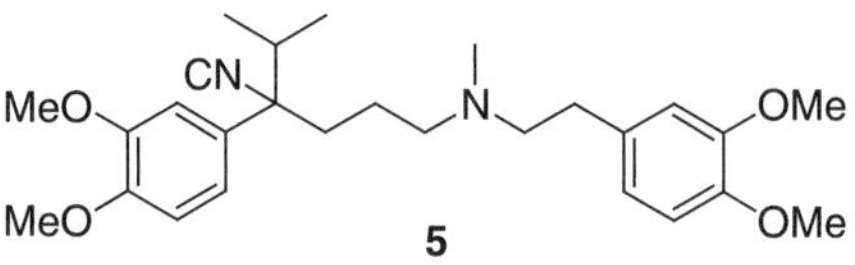

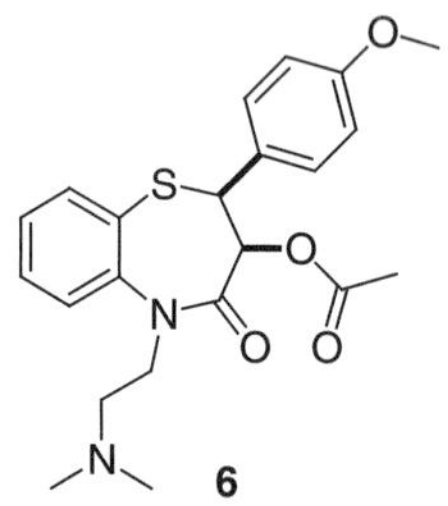

Generally, the dihydropyridine CCBs have evolved into three distinct subclasses based on their pharmacokinetics and pharmacodynamics. Early dihydropyridines such as nifedipine and nicardipine are characterized by a rapid onset of action and short duration of action due to limited half-life lives, thus requiring twice-daily dosing (Bayer, 2004). In addition, these short-acting compounds may have potential detrimental effects

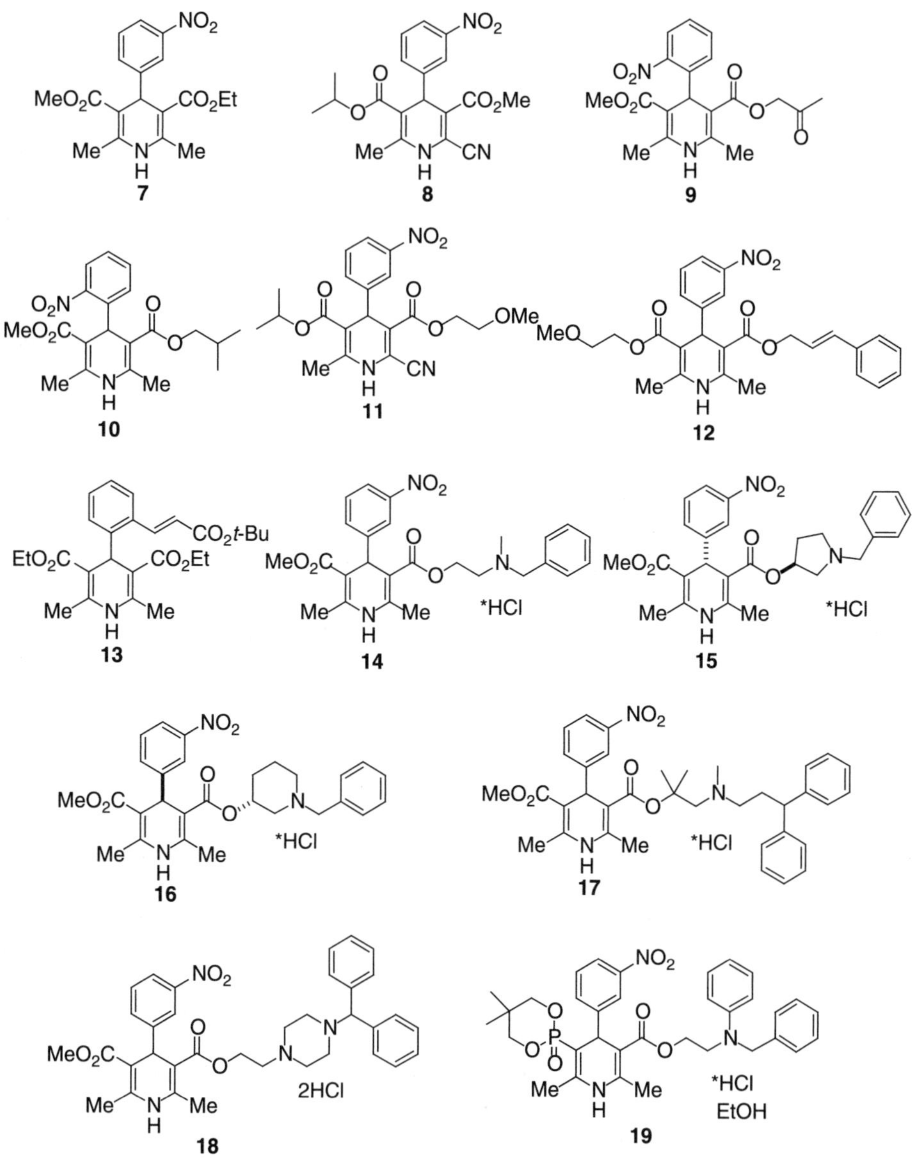

Figure 11.1. Structures of dihydropyridine calcium channel blockers (CCBs) (**7–19**).

on the patient (Kizer and Kimmel, 2000). These compounds are generally no longer in use. The second generation of dihydropyridine CCBs includes the extended or slow release (SR or ER) variants of nifedipine, felodipine, and nicardipine, as well as benidipine, isradipine, manidipine, nilvadipine, nimodipine, nisoldipine, and nitrendipine (Freedman and Waters, 1987). The extended release variants of the first-generation drugs are differently formulated to lead to a more gradual onset and extend the release time and half-life of the drugs. The newer analogs in the second group show improved pharmacokinetic properties such as half-life and potentially greater vascular selectivity. The third-generation compounds improve over the pharmacokinetics of the second generation even more.

The dihydropyridine CCBs act as antagonists of the voltage-dependent L-type calcium channel (Rampe and Kane, 1994; Triggle, 2003; Zamponi, 1997). These channels, a subclass of the calcium channel family, are structurally very complex, with multiple membrane spanning regions (Perez-Reyes and Schneider, 1994). Extensive molecular biology work has established that the $\alpha 1$ subunit of the channel hosts the ion-conducting unit; the role of the $\alpha 2$, β, γ, and δ associated regions are less well established. Tritium-labeled 1,4-dihydropyridine derivatives have shown specific membrane binding for vascular and cardiac smooth muscle from rabbits, rats, and dogs. The selectivity of the dihydropyridines for vascular or cardiac tissue varies from compound to compound; this can lead to differences in peripheral vasodilation and the concomitant lowering of vascular resistance and blood pressure. Greater selectivity for the vascular calcium channels appears to be preferred. Further refinement of these studies show that there are several dihydropyridine binding sites on the $\alpha 1$ subunit of the L-type calcium channel. The diltiazem and verapamil antagonists appear to have different binding sites. The drugs inhibit the influx of calcium ions through the channel, but much additional work is needed in this area to gain a better understanding of how these drugs interact with the ion channels.

A few other general observations can be made about the marketed dihydropyridine drugs. First, most of them are reported to be highly protein bound in plasma (see the prescription information for each drug). In addition, the drugs generally suffer from extensive hepatic metabolism, primarily due to significant activation of CYP3A4 during first-pass clearance. Interestingly, grapefruit juice counteracts this effect (Bailey et al., 1998; Paine et al., 2006). Experiments in which grapefruit juice was used to mask the taste of ethanol in a study to determine if there were any interactions between the alcohol and felodipine indicated that the mixture led to longer duration of action of the drug's effect. Additional work showed that the grapefruit juice or components thereof inhibit CYP3A4 isozymes and lead to a two- or three-fold increase in AUC and C_{max} of felodipine, thus leading to more available drug and a longer lasting lowering of blood pressure. This "grapefruit effect" is not limited to the dihydropyridine CCBs and is expected to come into play for other drugs exhibiting high activation of CYP3A4. Importantly, patients taking dihydropyridine CCBs need to be aware if other drugs they are taking also interact with CYP3A4, as such interactions change the effective concentration of some or all of their medications. Third, many members of this drug family show an age-dependent PK profile, with older patients achieving higher C_{max} and average plasma concentrations than younger patients (see the prescription information for each drug). This observation is probably attributable to changes in liver enzyme composition for older patients. Lastly, many of the early dihydropyridines are also used as antianginal medications; the mechanism of action for this is not understood.

11.2 SYNTHESIS OF NIFEDIPINE (ADALAT®)

As mentioned before, nifedipine was the first marketed dihydropyridine CCB. Initially, the drug indication was exclusively for angina, but the more recently developed extended release (ER) formulations are used off-label primarily for hypertension (Bayer, 2004). The extended-release tablets use a cellulose coat that extends their release time. The half-life of the ER formulation is reported as 7 h, whereas the immediate-release formulation has a half-life of 2 h (Bayer).

The preparation of nifedipine and other simple symmetric analogs is very straightforward (Scheme 11.1). 4-Aryldihydropyridine-3,5-dicarboxylates are isolable intermediates

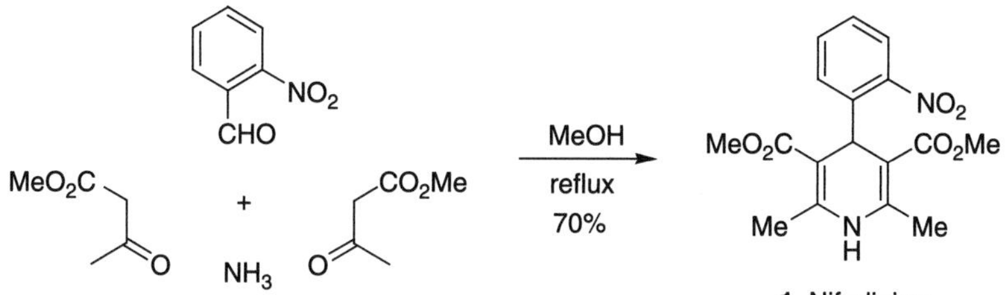

Scheme 11.1. Synthesis of nifedipine (1).

in the well-known Hantzsch pyridine synthesis (Bossert et al., 1981; Eisner and Kuthan, 1972; Stout and Meyers, 1982). This reaction, in its most simple form, is a one-pot reaction of an aryl aldehyde, acetoacetic acid alkyl ester, and ammonia heated in refluxing methanol for a few hours. If side reactions or by-products become troublesome, a stepwise process works as well. The original preparation of nifedipine uses 2-nitrobenzaldehyde, acetoacetic acid methyl ester, methanol, and ammonia (Bossert and Vater, 1969). The mixture is heated for several hours, cooled back down, and the product is filtered off as a yellow crystalline material.

11.3 SYNTHESIS OF FELODIPINE (PLENDIL®)

Plendil®, launched in 1988, is another dihydropyridine CCB antagonist formulated as an extended-release tablet (Astrazeneca, 2003). Felodipine, unlike nifedipine, has two different ester substituents; this breaks the symmetry of the compound and produces two possible enantiomers. Although one of the enantiomers is 1000-fold more active, the commercially available compound is sold exclusively as a racemic mixture because the manufacturing process for a pure enantiomer would be significantly more cumbersome and expensive. Felodipine has been shown to reversibly compete with labeled nitrendipine for the dihydropyridine binding sites in direct competition experiments. In addition, felodipine preferably blocks voltage-gated L-type calcium channels in vascular tissue, thus showing a preference for the cell type thought to play a more important role in blood pressure reduction. Like other dihydropyridines, felodipine shows pronounced plasma protein binding, first-pass hepatic clearance, and high activation of CYP3A4.

The synthesis of racemic felodipine (Scheme 11.2) utilizes a stepwise Hantzsch dihydropyridine reaction (Berntsson et al., 1981). 2,3-Dichlorobenzylidineacetyl acetic acid

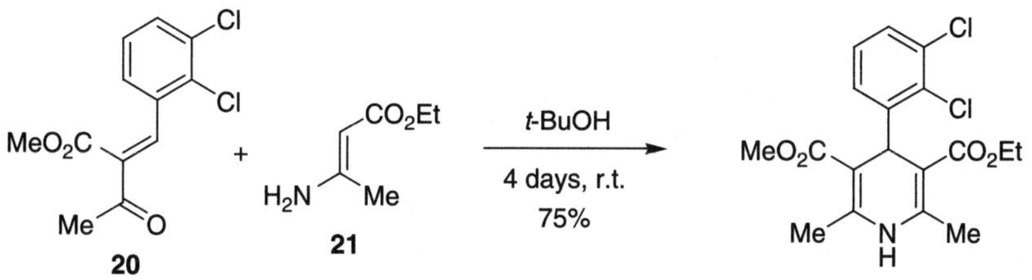

Scheme 11.2. Synthesis of felodipine (2).

methyl ester (**20**) was prepared from 2,3-dichlorobenzaldehyde and acetyl acetic acid methyl ester via a Knoevenagel condensation. 3-Aminocrotonic acid ethyl ester (**21**) was prepared from methyl acetoacetate and ammonium acetate. Felodipine was prepared by reacting fragments **20** and **21** in *t*-butanol for four days at room temperature. After evaporation, the product was recrystallized from isopropyl ether.

11.4 SYNTHESIS OF AMLODIPINE BESYLATE (NORVASC®)

Norvasc®, Pfizer's entry into the crowded CCB market, was launched in 1990. Like most of the more recently launched compounds, this drug also has two different ester substituents and thus is sold as a mixture of isomers. Initial work on the analysis of the two possible isomers suggested that the *R* isomer **22** was more potent than the corresponding *S* isomer **23**; this suggested that amlodipine differed in this respect from other dihydropyridine CCBs in which the *S* isomer is usually the more active (Goldmann and Stoltefuss, 1991). A more recent preparation of the (−)-(1*S*)-camphanic derivatives of amlodipine and the subsequent X-ray of its derivatives clearly showed that it is the *S* derivative that is more active (Arrowsmith et al., 1986; Goldmann et al., 1992).

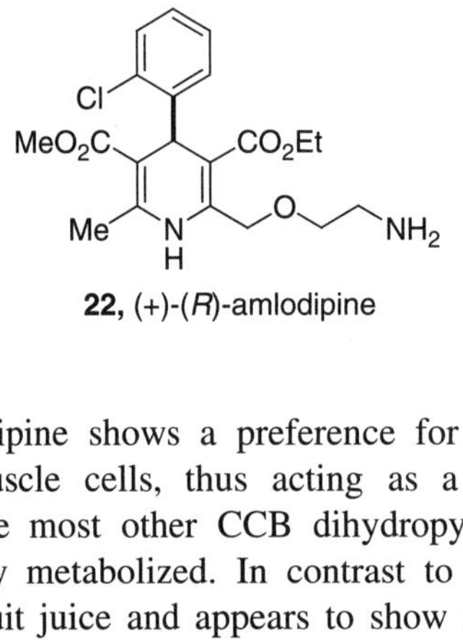

22, (+)-(*R*)-amlodipine

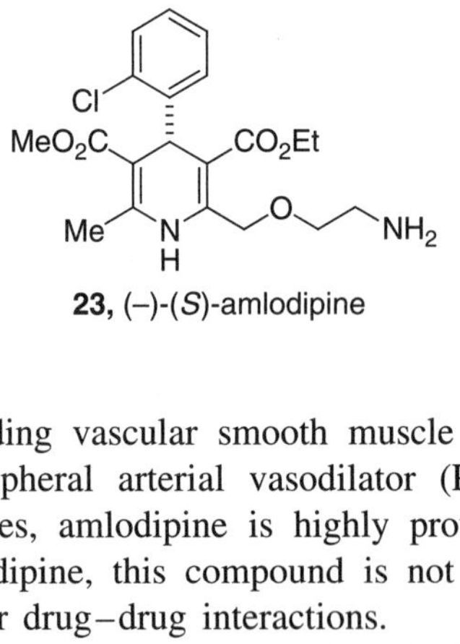

23, (−)-(*S*)-amlodipine

Amlodipine shows a preference for binding vascular smooth muscle cells over cardiac muscle cells, thus acting as a peripheral arterial vasodilator (Pfizer, Inc. 2005). Like most other CCB dihydropyridines, amlodipine is highly protein bound and heavily metabolized. In contrast to felodipine, this compound is not influenced by grapefruit juice and appears to show fewer drug–drug interactions.

The synthesis of amlodipine is outlined in Scheme 11.3 (Campbell et al., 1986). Once again, a precursor to the final drug is prepared from the Hantzsch condensation of two fragments. In this case, methyl-2-(2-chlorobenzylidene)acetoacetate (**24**) was prepared via Knoevenagel condensation of methyl acetoacetate and 2-chlorobenzaldehyde. Ethyl-4-(2-azidoethoxy)acetoacetate (**25**) was prepared from the sodium salt of 2-azidoethanol and ethyl-4-chloroacetoacetate. Intermediate **25** was then reacted with ammonium acetate to give **26**, followed by addition of **24** to furnish the dihydropyridine **27**. After recrystallization, the azide functionality of **27** was reduced to the corresponding amine by using palladium on calcium carbonate or zinc and acetic acid. A number of salts of amlodipine have been prepared, but the final commercial compound was obtained by treating the intermediate amine with benzenesulfonic acid to prepare the corresponding besylate salt (Davison and Wells, 1989).

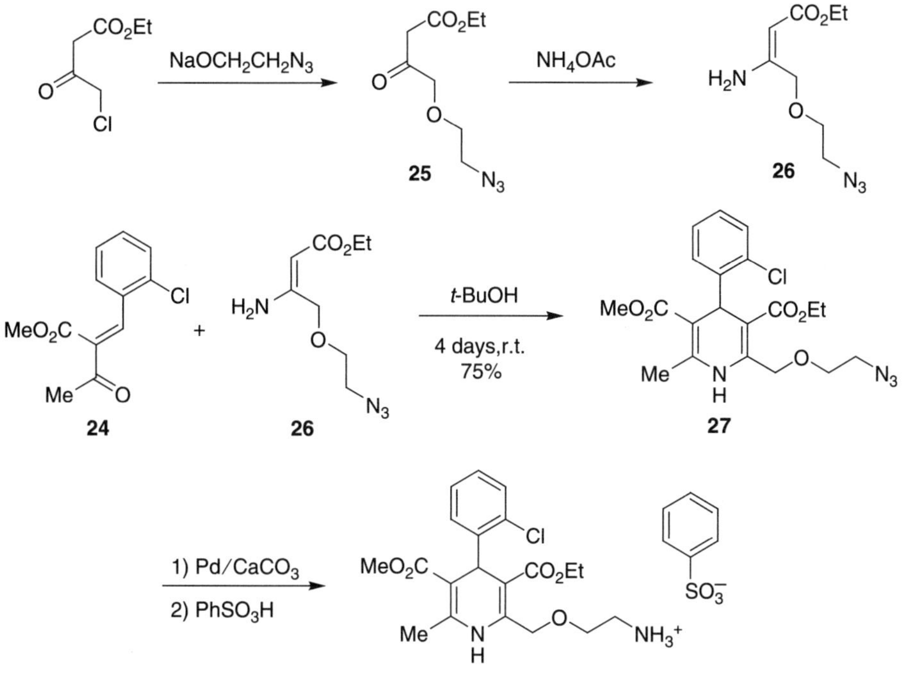

Scheme 11.3. Synthesis of amlodipine besylate (3).

11.5 SYNTHESIS OF AZELNIDIPINE (CALBLOCK®)

Azelnidipine, first described in the 1980s, took more than 15 years to get onto the market in Japan; the compound is currently not sold in the United States. This calcium channel antagonist, like the other congeners in its class, selectively blocks voltage-dependent Ca^{2+} influx through membrane-bound L-type calcium channels; this receptor binding is reversible (Wellington and Scott, 2003). Azelnidipine is an orally bioavailable drug that is taken once daily as an 8–16-mg dose, preferably after a meal. Clinical trials in mild-to-moderate hypertensive patients showed that azelnidipine effectively controls blood pressure, with a mean reduction in blood pressure of 27.8/16.6 mmHg (1 year duration). Additional double-blind, randomized studies in comparison with amlodipine or nitrendipine have indicated that azelnidipine is equipotent to those calcium channel blockers. Azelnidipine appears to be well tolerated without the tachychardiac effects associated with some of the earlier calcium channel blockers.

The reported synthesis of azelnidipine is straightforward (Scheme 11.4) (Koike et al., 1988). Benzhydrylamine and epichlorohydrin are mixed together without solvent to give 1-benzhydryl-3-hydroxyazetidine (**28**) in 57% yield after a long reaction time. Cyanoacetic acid and an equimolar amount of **28** are mixed together in THF, and 1.1 equiv. of 1,3-dicyclohexylcarbodiimide (DCC) was added. After 11 h of reaction at 55°C, the reaction was worked up and purified by flash chromatography to give the desired

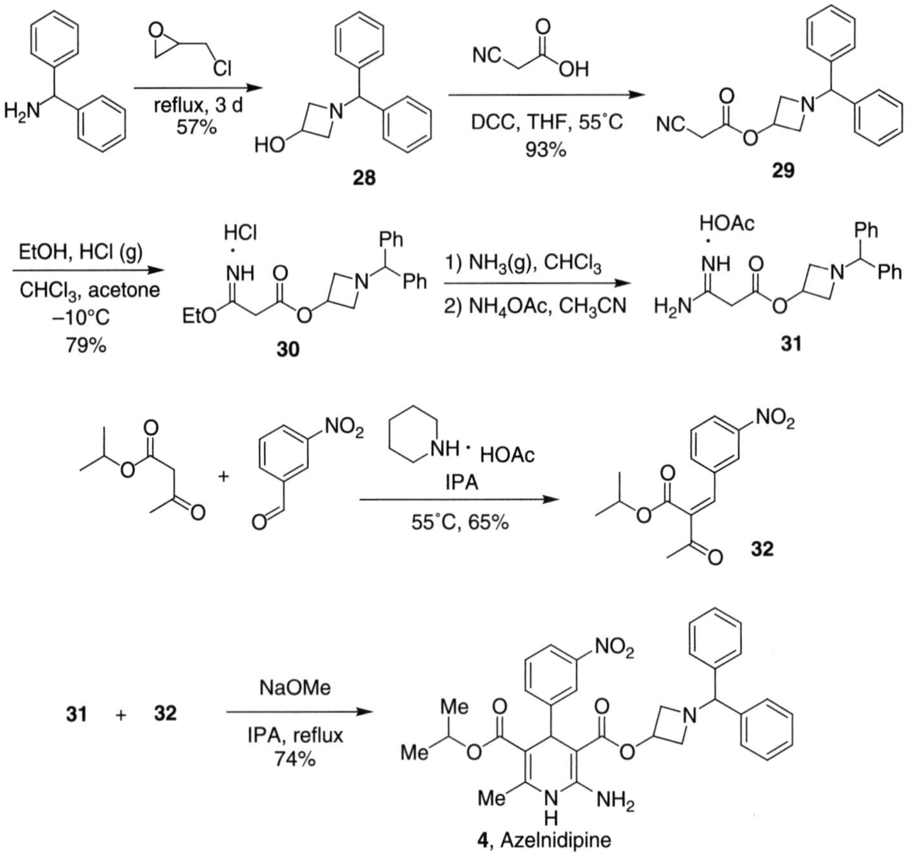

Scheme 11.4. Synthesis of azelnidipine (**4**).

1-benzhydryl-3-azetidinyl cyanoacetate **29**. The cyanoacetate derivative was then converted into the HCl salt of amidate **30**; this HCl salt was neutralized with ammonia and the resulting ammonium chloride salt by-product was removed by filtration. The resulting amidine was converted into the corresponding acetate salt **31** by heating with ammonium acetate for 1 h. After precipitation of the acetate salt **31** with diethyl ether, the stage was set for the final formation of azelnidipine via a Hantzsch dihydropyridine synthesis. Coupling of fragments **31** and **32** gave azelnidipine in good yield.

REFERENCES

Arrowsmith, J. E., Campbell, S. F., Cross, P. E., Stubbs, J. K., Burges, R. A., Gardiner, D. G., and Blackburn, K. J. (1986). *J. Med. Chem.*, 29: 1696–1702.

AstraZeneca. Plendil®. (2003). Prescription Information. AstraZeneca.

Bailey, D. G., Malcolm, J., Arnold, O., and Spence, D. J. (1998). *Br. J. Clin. Pharmacol.*, 46: 101–110.

Bayer. Adalat® (2004). Prescription information. Bayer, Inc.

Berntsson, P. B., Carlsson, S. A. I., Gaarder, J. O., and Ljung, B. R. (1981). "2,6-Dimethyl-4-2, 3-disubstituted phenyl-1,4-dihydro-pyridine-3,5-dicarboxylic acid-3,5-asymmetric diesters having hypotensive properties as well as a method for treating hypertensive conditions and pharmaceutical preparations containing same," US 4,264,611 (to Aktiebolaget Hassle).

Bossert, F., Meyer, H., and Wehinger, E. (1981). *Angew. Chem. Int. Ed. Engl.*, 20: 762–769.

Bossert, F. and Vater, W. (1969). "4-Aryl-1,4-dihydropyridines," US 3,485,847 (to Bayer, Inc).

Campbell, S. F., Cross, P. E., and Stubbs, J. K. (1986). "2-(Secondary aminoalkoxymethyl)dihydropyridine derivatives as anti-ischaemic and antihypertensive agents," US 4,572,909 (to Pfizer, Inc.).

Chobanian, A. V., Bakris, G. L., Black, H. R., Cushman, W. C., Green, L. A., Izzo, J. L., Jones, D. W., Materson, B. J., Oparil, S., Wright, J. T., Jr., Roccella, E. J., and the National High Blood Pressure Education Program Coordinating Committee. (2001). *JAMA* 289: 2560–2572.

Davison, E. and Wells, J. I. (1989). "Pharmaceutically acceptable salts," US 4,879,303 (to Pfizer, Inc.).

Eisner, U. and Kuthan, J. (1972). *Chem. Rev.*, 72: 1–42.

Freedman, D. D. and Waters, D. D. (1987). *Drugs*, 34: 578–598.

Goldmann, S. and Stoltefuss, J. (1991). *Angew. Chem. Int. Ed. Engl.*, 30: 1559–1578.

Goldmann, S., Stoltefuss, J., and Born, L. (1992). *J. Med. Chem.*, 35: 3341–3344.

Kizer, J. R. and Kimmel, S. E. (2000). *Pharmacoepidemiology and Drug Safety*, 9: 25–36.

Koike, H., Nishino, H., and Yoshimoto, M. (1988). "Dihydropyridine derivatives, their preparation and their use," US 4,772,596 (to Sankyo Company Limited and Ube Industries Limited).

NHLBI, "The seventh report of the joint national committee on the prevention, detection, evaluation and treatment of High Blood Pressure" (JNC VII). Available at http://www.nhlbi.nih.gov/guidelines/hypertension/ last accessed March 4, 2007.

Paine, M. F., Widmer, W. W., Hart, H. L., Pusek, S. N., Beavers, K. L., Criss, A. B., Brown, S. S., Thomas, B. F., and Watkins, P. B. (2006). *Am. J. Clin. Nutr.*, 83: 1097–1105.

Perez-Reyes, E. and Schneider, T. (1994). *Drug Development Research*, 33: 295–318.

Pfizer, Inc. Norvasc®. (2005). Prescription information. Pfizer, Inc.

Rampe, D. and Kane, J. M. (1994). *Drug Development Research*, 33: 344–363.

Stout, D. M. and Meyers, A. I. (1982). *Chem. Rev.*, 82: 223–243.

Triggle, D. J. (2003). *Drug Development Research*, 58: 5–17.

Wellington, K. and Scott, L. J. (2003). *Drugs*, 63: 2613–2621.

Zamponi, G. W. (1997). *Drug Development Research*, 42: 131–143.

12

SECOND-GENERATION HMG-CoA REDUCTASE INHIBITORS

Jeffrey A. Pfefferkorn

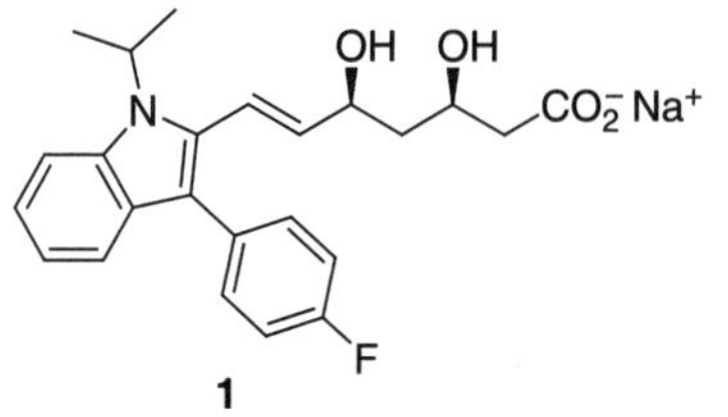

1

USAN: Fluvastatin sodium
Trade name: Lescol®
Novartis (Sandoz)
Launched: 1994
M.W. 433.45 (parent: 411.5)

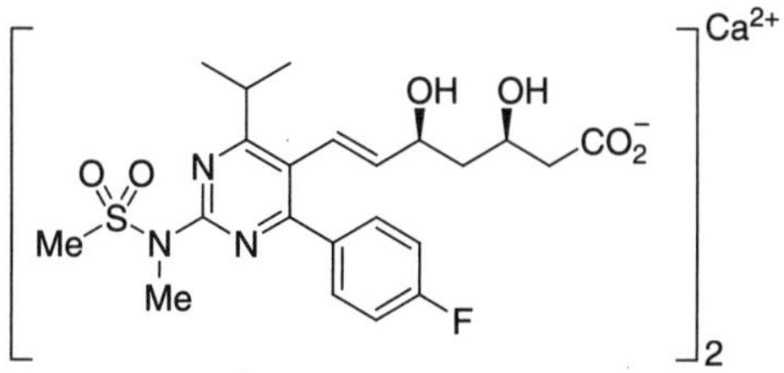

2

USAN: Rosuvastatin calcium
Trade name: Crestor®
AstraZeneca
Launched: 2003
M.W. 1001.13 (parent: 481.5)

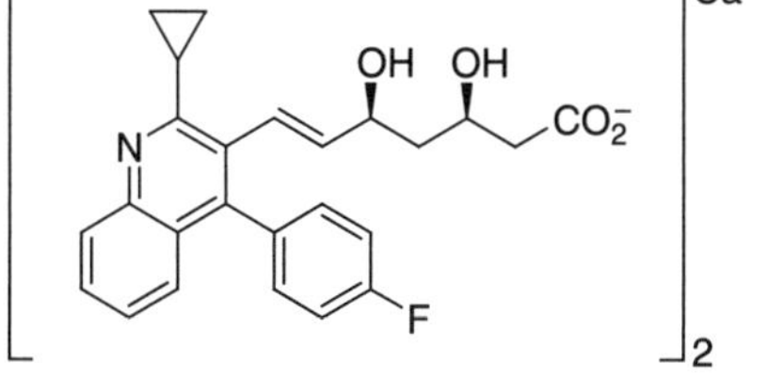

3

USAN: Pitavastatin calcium
Trade name: Livalo®
Kowa/Nissan Chemical/Sankyo
Launched: 2003
M.W. 880.98 (parent: 421.5)

The Art of Drug Synthesis. Edited by Douglas S. Johnson and Jie Jack Li

12.1 INTRODUCTION

Coronary heart disease (CHD) is the leading cause of death worldwide. It is estimated that in the United States alone, over 12 million people suffer from this disease, at an annual cost of 50 to 100 billion dollars (Grundy, 1997). Epidemiological evidence indicates that there is a strong correlation between hypercholesterolemia (elevated serum cholesterol levels) and the risk of CHD; consequently, substantial efforts have been undertaken to mitigate this key risk factor (Castelli, 1984). One of the most effective methods of reducing serum cholesterol is through the use of HMG-CoA reductase inhibitors. This class of drugs, commonly referred to as statins, inhibit the conversion of 3-hydroxy-3-methylglutaryl CoA (HMG-CoA) to mevalonate (Fig. 12.1), which is the rate-limiting step of cholesterol biosynthesis (McKenney, 2003).

The first generation of statins to be approved for the treatment of hypercholesterolemia was based upon the natural product compactin, an HMG-CoA reductase inhibitor originally isolated from cultures of *Penicillium*. The most widely prescribed of these first-generation statins are simvastatin (**4**, Fig. 12.2), marketed as Zocor®, and pravastatin (**5**), marketed as Pravacol®. Based in large part on the clinical effectiveness of these early

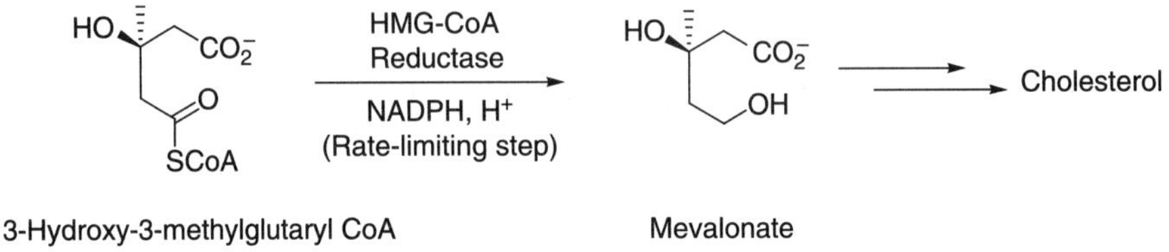

HO, CO_2^-, O, SCoA — HMG-CoA Reductase, NADPH, H^+ (Rate-limiting step) → HO, CO_2^-, OH → Cholesterol

3-Hydroxy-3-methylglutaryl CoA — Mevalonate

Figure 12.1. HMG-CoA catalyzed conversion of 3-hydroxy-3-methylglutaryl CoA into mevalonate as the rate-limiting step of cholesterol biosynthesis.

4, Simvastatin (Zocor®)

5, Pravastatin (Pravacol®)

6, Atorvastatin (Lipitor®)

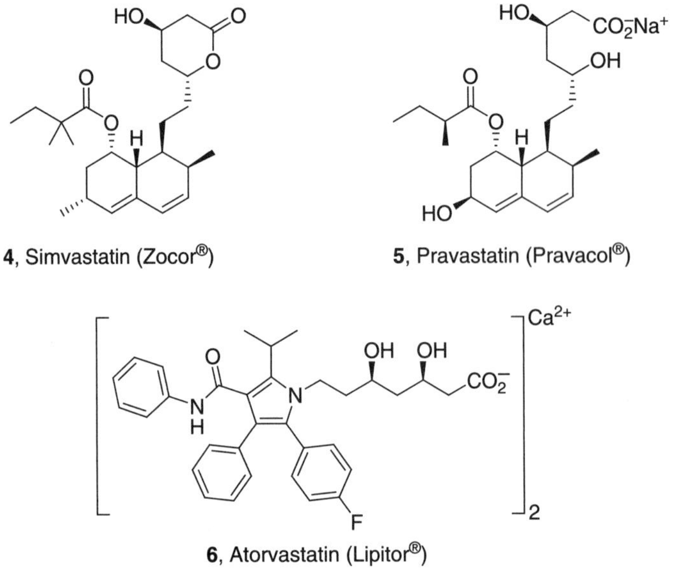

Figure 12.2. Leading HMG-CoA reductase inhibitors.

statins, pharmaceutical companies worldwide began searching for the second-generation HMG-CoA inhibitors with improved efficacy and tolerability. One such investigation led to the identification of atorvastatin (**6**, Lipitor®) as a completely synthetic inhibitor with remarkable efficacy. Atorvastatin is currently the most widely prescribed treatment for hypercholesterolemia, and a detailed account of its discovery and synthesis can be found in *Contemporary Drug Synthesis* (Li et al., 2004: Chapter 9).

In this chapter we will consider the synthesis of several other second-generation HMG-CoA reductase inhibitors that have been approved for the treatment of patients at risk for CHD due to elevated cholesterol levels. We will first consider fluvastatin (**1**), marketed as Lescol®, which was launched in 1994. Subsequently, we will examine the synthetic stories of rosuvastatin (**2**), marketed as Crestor®, and pitavastatin (**3**), marketed as Livalo®.

12.2 SYNTHESIS OF FLUVASTATIN (LESCOL®)

Fluvastatin, launched in 1994, was the first synthetic HMG-CoA reductase inhibitor to be approved for the treatment of hypercholesterolemia. This statin was developed by Novartis (Sandoz), and it is marketed as a racemate under the trade name Lescol® (Asberg and Holdaas, 2004). In laboratory studies, fluvastatin was shown to potently inhibit HMG-CoA reductase ($IC_{50} = 8$ nM) and effectively block cholesterol biosynthesis in hepatocyte cells ($IC_{50} = 52$ nM) (Parker et al., 1990). Clinically, fluvastatin is prescribed less frequently than simvastatin, atorvastatin, or rosuvastatin, but it nevertheless remains an important medication for the treatment of hypercholesterolemia; moreover, as the first completely synthetic statin to be approved, it offers a compelling synthetic story (Repic et al., 2001).

An instructive early discovery synthesis of the fluvastatin template is outlined in Scheme 12.1 for the *N*-Me analog of fluvastatin (that is, compound **19**) (Kathawala, 1994; Repic et al., 2001). 4-Fluorobenzyl chloride (**7**) was converted to the corresponding Grignard by treatment with magnesium, and the resulting organometallic reagent was reacted with diethyl oxalate to generate α-ketoester **8** (Schuster and Coppola, 1994). The heterocyclic core of fluvastatin was next constructed through a Fischer indole synthesis that involved treatment of α-ketoester **8** with phenyl hydrazine followed by HCl to afford indole **13**. Mechanistically, this transformation can be understood as having proceeded through formation and subsequent protonation of imine **9**, which tautomerized to ene-hydrazine **10** (Robinson, 1963, 1968). A [3,3]-sigmatropic rearrangement and subsequent proton shift of intermediate **10** provided imine **11**, which underwent intramolecular cyclization with expulsion of NH_3 to generate the indole **13**. Alkylation of the indole **13** was accomplished by treatment with sodium hydride and methyl iodide to afford *N*-methyl indole **14**. The ester of **14** was converted to aldehyde **15** via a reduction/oxidation sequence. This newly installed aldehyde was then homologated to enal **16** by nucleophilic attack of LiCH=CHOEt (prepared via transmetallation of tri-*n*-butylstannylvinylethoxide with *n*-BuLi) followed by acid-mediated hydrolysis of the resulting vinyl ether. In the next step, the dianion of methylacetoacetate was generated through double deprotonation and reacted with enal **16** to generate β-hydroxy ketone **17** as a racemic mixture (*3R*/*3S*). Racemic **17** was reduced with borane-*t*-butylamine complex to generate diol **18** as a mixture of four stereoisomers: (*3R*, *5R*), (*3S*, *5S*), (*3R*, *5S*), and (*3S*, *5R*). The two diastereomeric pairs of enantiomers (*3R*, *5R*)/(*3S*, *5S*) and (*3R*, *5S*)/(*3S*, *5R*) were chromatographically separated from one another and the enantiomeric *syn*-diols **18**: (*3R*, *5R*)/(*3S*, *5S*) were finally saponified to potassium carboxylate salt **19**.

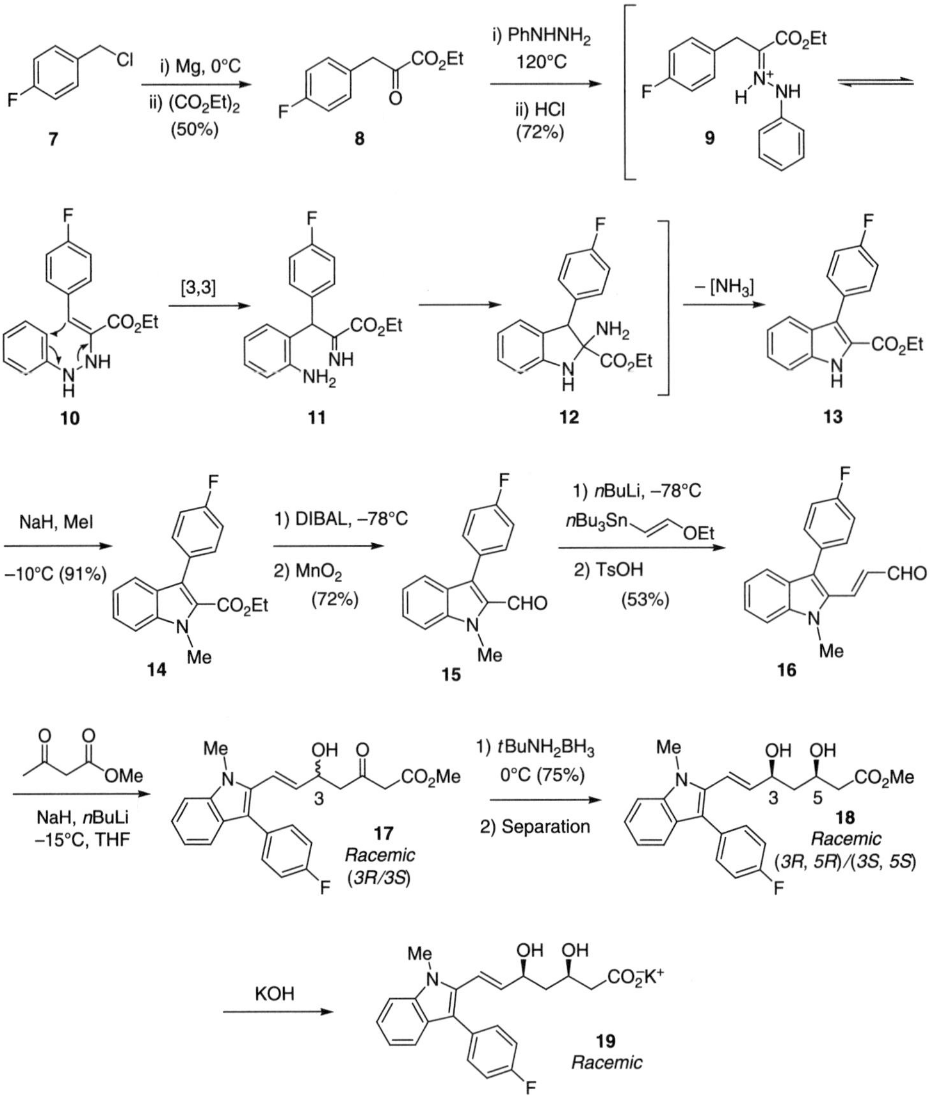

Scheme 12.1. Early racemic synthesis of the fluvastatin template.

Although this early synthesis of the fluvastatin template (Scheme 12.1) was important for structure–activity studies, a second synthesis (Scheme 12.2) was later developed to facilitate large-scale production of fluvastatin (Repic et al., 2001). As is typical in process development work, this second approach sought to shorten the reaction sequence, eliminate toxic reagents (such as tri-*n*-butylstannylvinylethoxide), and improve the yield and/or selectivity of reactions in order to reduce the amount of chromatography required to isolate the products.

As outlined in Scheme 12.2, the process route to fluvastatin (**1**) commenced with Friedel–Crafts acylation of fluorobenzene (**20**) with chloroacetyl chloride (**21**) in the presence of $AlCl_3$ to prepare α-chloroketone **22** (Repic et al., 2001). Reaction of **22** with *N*-*i*-Pr aniline at elevated temperature generated tertiary amine **23**, which was engaged

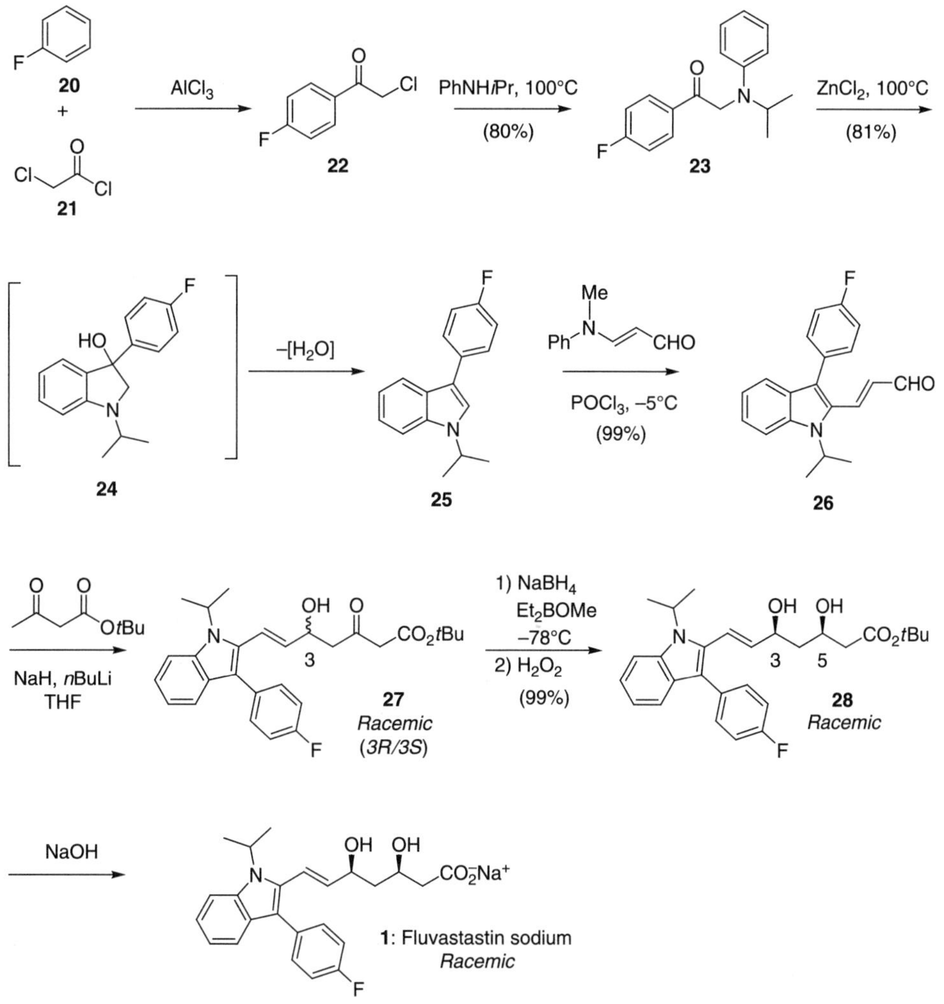

Scheme 12.2. Second generation racemic synthesis of fluvastatin (**1**).

in a $ZnCl_2$-mediated Bischler indole synthesis to afford indole **25** (Walkup and Linder, 1985). Next, installation of the 3,5-dihydroxyheptanoic acid side-chain began with an innovative variation of the traditional Vilsmeier–Haack formylation reaction. In this event, 3-(*N*-methyl-*N*-phenylamino)acrolein was first treated with $POCl_3$ to generate an iminium salt, which was then reacted with indole **25** to give enal **26** in excellent yield (Chen et al., 1993). This modification of the earlier route is notable, because it facilitated installation of the requisite enal moiety in a single step as opposed to the four-step sequence utilized in the original synthesis (Scheme 12.1). Moreover, the current approach eliminated the need to use the toxic tri-*n*-butylstannylvinylethoxide reagent. The remainder of the side-chain was installed by generation of the dianion of *t*-butylacetoacetate followed by reaction with **26** to afford β-hydroxy ketone **27** as a racemic mixture (*3R/3S*). The researchers noted that use of *t*-butylacetoacetate as opposed to the original methylacetoacetate in this reaction was important, because the resulting *t*-butyl ester on the side-chain was more robust and less prone to side-chain lactonization, a downstream

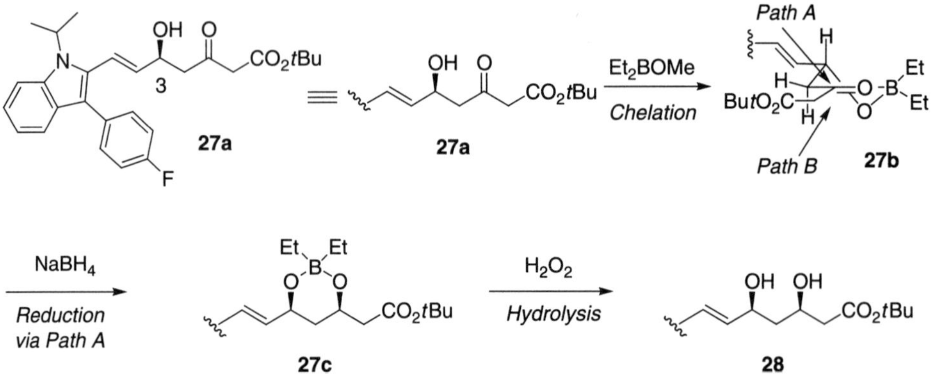

Scheme 12.3. Stereochemical rationalization of *syn*-reduction of β-hydroxy ketoester **27a**.

side reaction that resulted in erosion of stereochemical integrity, thus creating purification difficulties. Returning to Scheme 12.2, racemic β-hydroxy ketone **27** was subjected to a chelation controlled *syn*-reduction with $NaBH_4$ and Et_2BOMe to give diol **28** as a racemic mixture of *syn*-1,3-diols [(*3R*, *5R*)/(*3S*, *5S*)] (Chen et al., 1987). Finally, the *t*-butyl ester of **28** was saponified with aqueous NaOH to generate racemic fluvastatin (**1**) as a sodium salt.

The penultimate step of this second generation synthesis, namely the sterocontrolled reduction of β-hydroxy ketone **27**, is notable in that it exclusively afforded only the two desired *syn*-1,3-diols as compared to the original synthesis (Scheme 12.1), which generated four stereoisomers in the reduction operation. The observed stereoselectivity of this reduction was rationalized by the researchers as outlined in Scheme 12.3 (Chen et al., 1987). Treatment of β-hydroxy ketone **27a** with Et_2BOMe initially afforded the cyclic boronate chelate **27b** shown in a chair-like conformation. Examination of **27b** reveals that the two potential trajectories (that is, *Path A* or *Path B*) for the incoming reducing agent ($NaBH_4$) were sterically differentiated, resulting in nucleophilic attack via the less hindered *Path A* to provide, after hydrolysis of cyclic boronate **27c**, the desired *syn*-reduction product **28**. As we will see later in the chapter, this method of chelation-controlled reduction of β-hydroxy ketones will prove to be important in the side-chain synthesis of several other HMG-CoA reductase inhibitors.

12.3 SYNTHESIS OF ROSUVASTATIN (CRESTOR®)

Rosuvastatin (Crestor®) is the most recent HMG-CoA reductase inhibitor to be widely approved for the treatment of hypercholesterolemic patients. It has established an impressive record of clinical efficacy and is often prescribed to patients that do not respond well to earlier HMG-CoA reductase inhibitors (Schuster, 2003). This fully synthetic statin, developed by Astra-Zeneca and Shionogi Research Laboratories, has been shown to potently inhibit cholesterol biosynthesis in rat hepatocytyes ($IC_{50} = 1.12$ nM). By comparison, pravastatin (**5**) was >100-fold less active in the same assay ($IC_{50} = 198$ nM) (Hirai et al., 1993; Watanabe et al., 1997).

Structurally, rosuvastatin contains a central pyrimidine core substituted with a methanesulfonamide, a 4-fluorophenyl ring, an isopropyl group, and the 3,5-dihydroxyheptanoic acid side-chain that is common to all drugs of this class. In 1997, Watanabee and co-workers reported the synthesis of rosuvastatin as outlined in Scheme 12.4 (Hirai et al., 1993; Watanabe et al., 1997). This route commenced with the condensation of 4-fluorobenzaldehyde (**29**) and ethyl isobutrylacetate (**30**) to provide β-ketoester **31** as an inconsequential mixture of stereoisomers. Condensation of β-ketoester **31** with *S*-methylisothiourea hemisulfate in the presence of HMPA at elevated temperature followed by dehydrogenation with DDQ delivered the pyrimidine core of the rosuvastatin template. To facilitate installation of the 2-amino substituent, the thioether of **32** was oxidized with *m*CPBA to the corresponding sulfone **33**. Displacement of this sulfone with methyl amine resulted in the formation of 2-methylamino-pyrimidine **34**. The newly installed amine was then converted to methanesulfonamide **35** by treatment with sodium hydride and methanesulfonyl chloride.

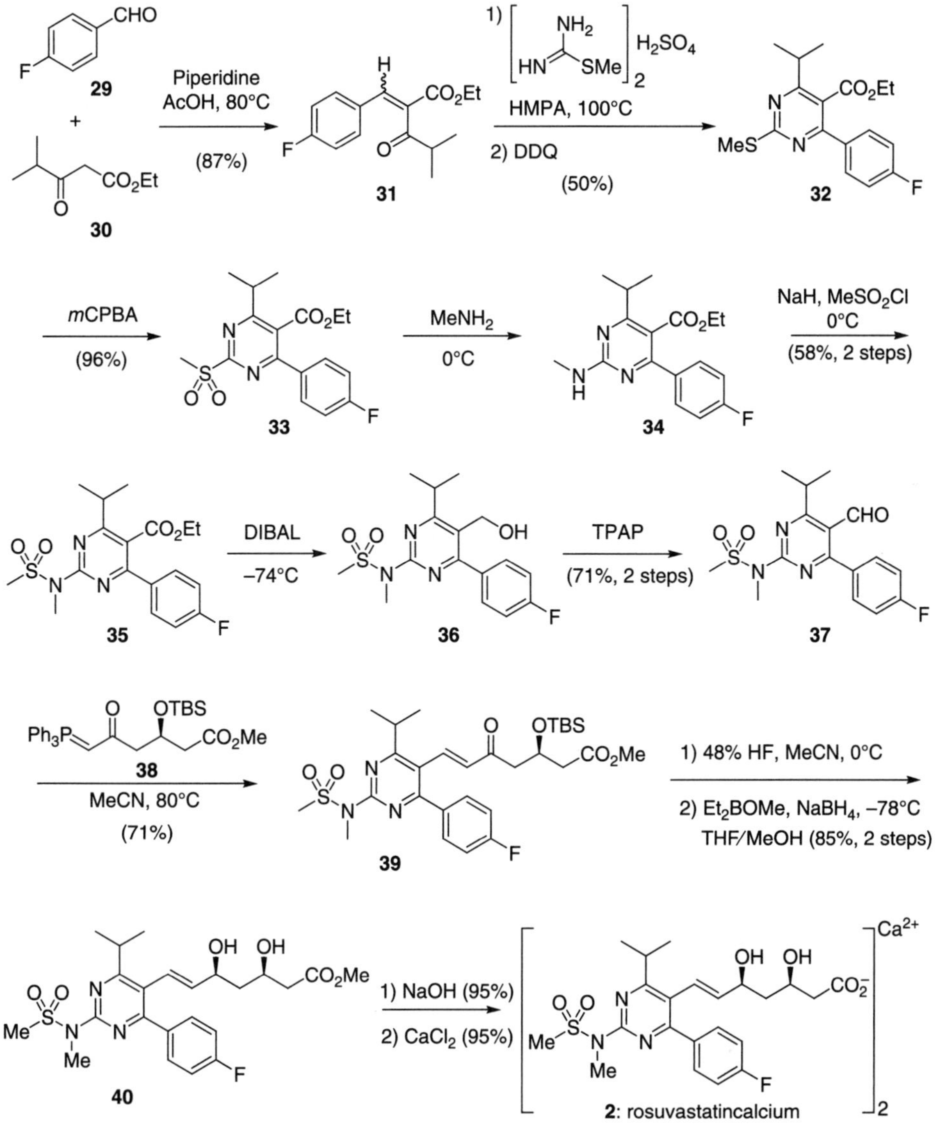

Scheme 12.4. Stereoselective synthesis of rosuvastatin (**2**).

With the fully functionalized heterocyclic core completed, synthetic attention next focused on introduction of the 3,5-dihydroxyheptanoic acid side-chain. This required initial conversion of the ethyl ester of **35** to the corresponding aldehyde through a two-step reduction/oxidation sequence. In that event, a low-temperature DIBAL reduction of **35** provided primary alcohol **36**, which was then oxidized to aldehyde **37** with TPAP. Subsequent installation of the carbon backbone of the side-chain was accomplished using a Wittig olefination reaction with stabilized phosphonium ylide **38** resulting in exclusive formation of the desired *E*-olefin **39**. The synthesis of phosphonium ylide **38** will be examined in Scheme 12.5 (Konoike and Araki, 1994).

Continuing with the rosuvastatin synthesis, completion of the 3,5-dihydroxyheptanoic acid side-chain required removal of the TBS protecting group from the secondary alcohol. This was accomplished by treatment of **39** with hydrofluoric acid to afford a β-hydroxy ketone that was then subjected to chelation-controlled *syn*-reduction with Et_2BOMe and $NaBH_4$ to deliver diol **40** as a single stereoisomer (see Scheme 12.3, of fluvastatin synthesis, for additional explanation of this reaction). Finally, with diol **40** in hand, completion of the synthesis was straightforward. The side-chain methyl ester was first hydrolyzed to the corresponding carboxylic acid sodium salt, which was then treated with $CaCl_2$ to complete the synthesis of rosuvastatin calcium (**2**).

A notable element of the rosvastatin synthesis was installation of the chiral 3,5-dihydroxyheptanoic acid side-chain through a Wittig olefination reaction with phosphonium ylide **38**. A stereoselective synthesis of this key ylide reagent was reported by Konoike and co-workers (Konoike and Araki, 1994) in 1994. The Konoike synthesis started from prochiral anhydride **44** (Scheme 12.5), which was first prepared in three steps from diethyl 3-hydroxygluturate (**41**) according to the method of Heathcock (Theisen and Heathcock, 1988). Thus, as outlined in Scheme 12.5, diethyl 3-hydroxygluturate (**41**) was TBS-protected to give **42**, which was then exhaustively saponified to dicarboxylate **43**. Subsequent reaction of **43** with acetic anhydride at elevated

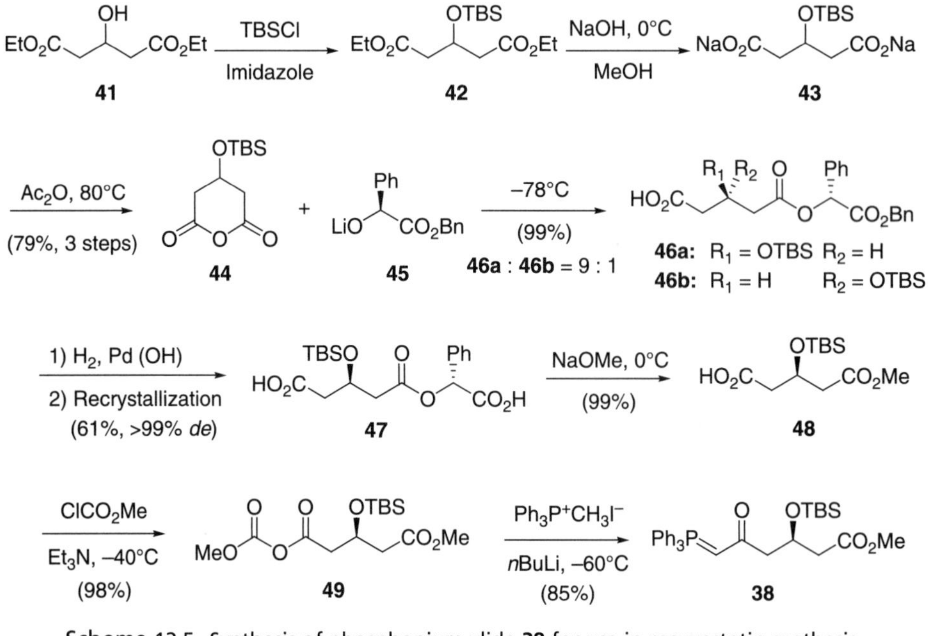

Scheme 12.5. Synthesis of phosphonium ylide **38** for use in rosuvastatin synthesis.

temperature resulted in formation of cyclic anhydride **44**, which is prochiral due to the presence of a plane of symmetry. As described by Konoike, prochiral anhydride **44** was stereoselectively desymmetrized by reaction with the lithium salt of benzyl-(*R*)-(−)-mandelate (**45**) as a "chiral inducer" to generate a 9 : 1 mixture of **46a** and **46b**. Although the desired stereoisomer **46a** was the major product of the reaction, the researchers were unable to satisfactorily separate it from the undesired **46b** at this stage, so the mixture (**46a** + **46b**) was debenzylated via hydrogenolysis to afford the corresponding carboxylic acids. Recrystallization then provided the desired carboxylic acid **47** as a single stereoisomer (*de* > 99%) in 61% yield. Having served its purpose, the chiral inducer moiety was cleaved by transesterification to generate methyl ester **48**. Finally, the free carboxylic acid of **48** was activated as the mixed anhydride (that is, **49**) and reacted with $Ph_3P{=}CH_2$ (prepared from $Ph_3P^+CH_3I^-$ and *n*-BuLi) to generate phosphonium ylide **38**, which was used in the synthesis of rosuvastatin as described in Scheme 12.4.

12.4 SYNTHESIS OF PITAVASTATIN (LIVALO®)

Pitavastatin (**3**) was launched in 2003 and is currently marketed in Japan under the trade name Livalo®. Like rosuvastatin and fluvastatin, pitavastatin is a completely synthetic HMG-CoA reductase inhibitor that was developed by Kowa, Nissan Chemical, and Sankyo (Sorbera et al., 1998). Multiple syntheses of pitavastatin have been reported and an exhaustive review of these efforts is beyond the scope of this text (Hiyama et al., 1995a, b; Minami and Hiyama, 1992; Miyachi et al., 1993; Takahashi et al., 1993, 1995; Takano et al., 1993). Instead, we will focus our discussion on two related and innovative synthetic approaches that differ strategically from the routes we have previously examined for rosuvastatin and fluvastatin. These routes to pitavastatin employed palladium-mediated coupling reactions to install the 3,5-dihydroxyheptanoic acid side-chain. This key retrosynthetic disconnection is highlighted in Scheme 12.6, in which a suitable functionalized side-chain (**52** or **53**) is attached to the heterocyclic core of pitavastatin (**51**) through palladium-mediated coupling.

Both of the potential cross-coupling routes highlighted in Scheme 12.6 require heteroaryl iodide **51** as a coupling partner, and its preparation is outlined in Scheme 12.7 (Takahashi et al., 1993). Hence, anthranilic acid (**54**) was first tosylated and then treated with PCl_5 to generate acid chloride **55**. Friedel–Crafts acylation of **55** with fluorobenzene (**20**) in the presence of $AlCl_3$ provided 2-aminobenzophenone **56**. Separately, cyclopropyl methyl ketone (**57**) was treated with diethyl carbonate and a suitable base to generate β-ketoester **58**, which was then condensed with 2-aminobenzophenone **56** under acid-catalyzed conditions, resulting in the formation of 2-cyclopropyl-4-(4-fluoro-phenyl)-quinoline-3-carboxylic acid ethyl ester (**59**). Finally,

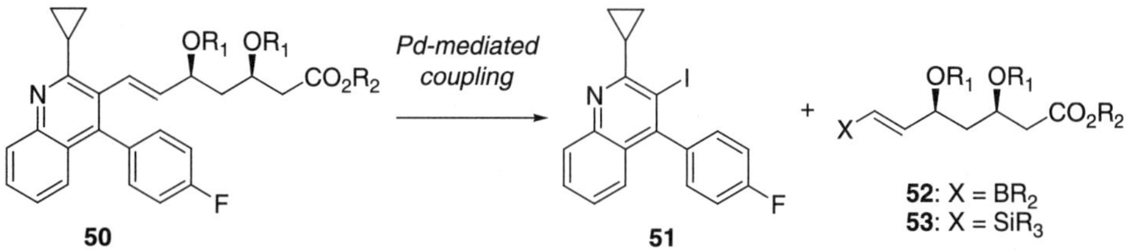

Scheme 12.6. Key retrosynthetic disconnection in the pitavastatin synthesis.

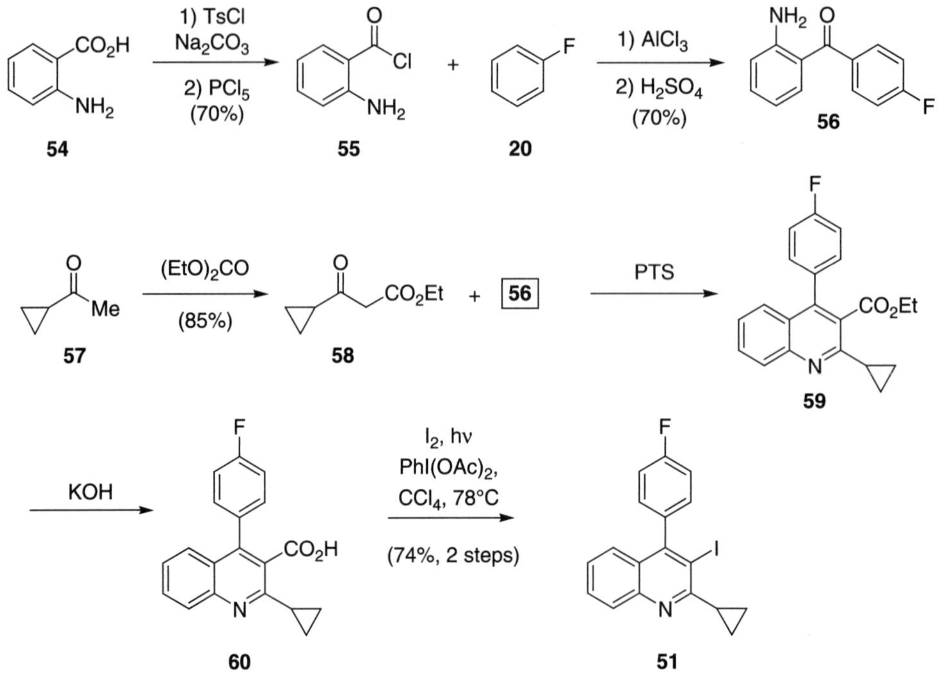

Scheme 12.7. Synthesis of heteroaryl iodide **51** for the synthesis of pitavastatin.

59 was converted to heteroaryl iodide **51** by saponification and then treatment with I_2, $PhI(OAc)_2$, and UV light.

As illustrated in Schemes 12.8 and 12.9, heteroaryl iodide **51** was engaged in palladium-mediated couplings with both a vinyl silane functionalized side-chain (Scheme 12.8) and vinyl borane functionalized side-chain (Scheme 12.9). We will consider these routes separately, beginning with the vinyl silane approach illustrated in Scheme 12.8. Takahashi and co-workers (Takahashi et al., 1993, 1995) reported the preparation of the requisite vinyl silane coupling partner **73** from chiral epoxide **68**, which itself could be synthesized according to the method of Lopp and co-workers (Lopp et al., 1991). Hence, (*R*,*R*)-diethyl tartrate (**61**) was first converted to acetonide **62**, and this intermediate was exhaustively reduced with $LiAlH_4$ to diol **63** (Feit, 1964). The researchers noted that attempted mono-chlorination of diol **63** was not successful, so it was instead selectively mono-benzylated and then treated with CCl_4/Ph_3P to provide chloride **64** (Lopp et al., 1991). Treatment of **64** with $LiNH_2$ in ammonia effected a base-induced elimination reaction that generated propargylic alcohol **65** with no loss of stereochemical integrity. Protection of both the alcohol and terminal acetylene of **65** was accomplished with *n*-BuLi and TBSCl to produce intermediate **66**. Treatment of **66** with BBr_3 at low temperature resulted in conversion of the TBS-protected secondary alcohol to its corresponding bromide with inversion of configuration along with concomitant cleavage of the benzyl protecting group to afford bromohydrin **67**. Base-mediated closure of bromohydrin **67** provided epoxide **68**, which was subsequently opened by addition of cyanide to generate β-hydroxy nitrile **69** with no erosion of stereochemical integrity (Takahashi et al., 1993, 1995). Using a Blaise reaction, the organozinc derivative of *t*-butyl bromoacetate was generated and then added to β-hydroxy nitrile **69** to provide δ-hydroxy-β-ketoester **70**. A stereoselective *syn*-reduction of **70** was accomplished with

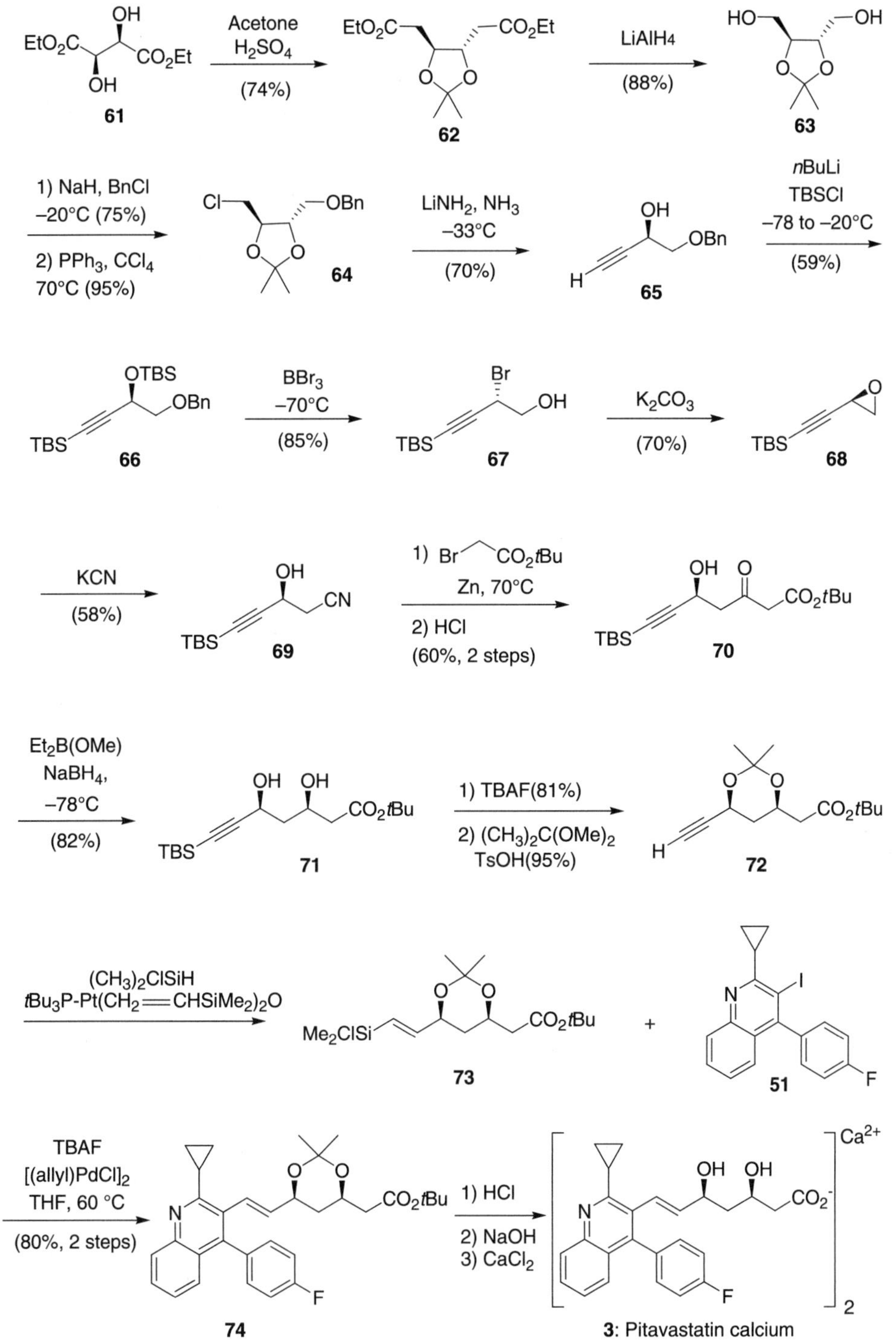

Scheme 12.8. Synthesis of pitavastatin (3) through a hydrosilylation cross-coupling reaction.

$NaBH_4$ and Et_2BOMe to generate *syn*-1,3-diol **71**. Subsequently, the acetylinic TBS-protecting group of **71** was cleaved with TBAF and the 1,3-diol was protected as the corresponding acetonide to give **72**, which was ready to be utilized in the hydrosilylation cross-coupling reaction. In that event, **72** was initially hydrosilylated with chlorodimethylsilane in the presence of catalytic tBu$_3$P-Pt(CH$_2$=CHSiMe$_2$)$_2$O to afford vinyl silane **73**, which was then reacted with heteroaryl iodide **51** in the presence of TBAF and catalytic [(allyl)PdCl]$_2$ to provide **74** in good overall yield. Finally, **74** was converted to pitavastatin (**3**) through the standard deprotection and hydrolysis sequence.

As an alternative to the hydrosilylation cross-coupling strategy described in Scheme 12.8, a separate synthesis of pitavastain was reported in which the key side-chain connection was accomplished through a hydroboration cross-coupling reaction (Scheme 12.9) (Miyachi et al., 1993). As illustrated in Scheme 12.9, this approach again began with

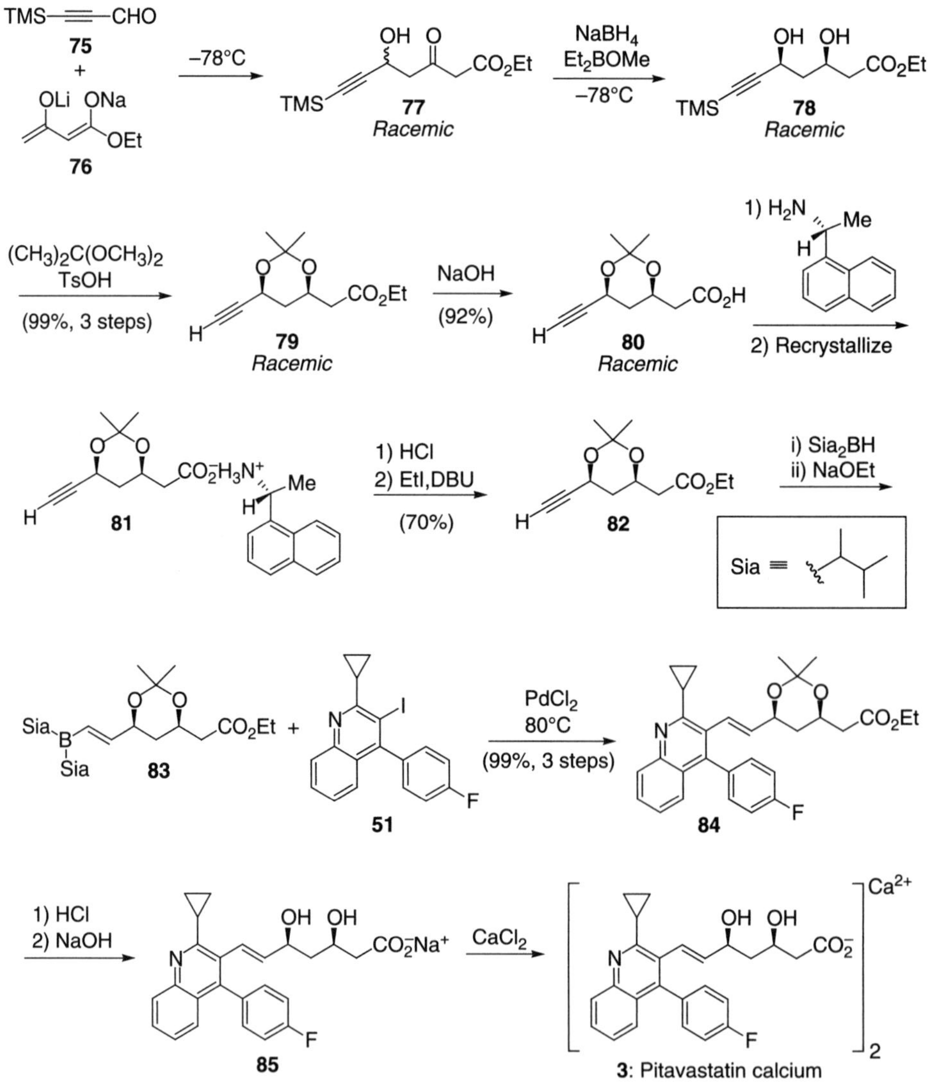

Scheme 12.9. Synthesis of pitavastatin (**3**) through a hydroboration cross-coupling reaction.

construction of an appropriately functionalized side-chain component. In this case, the side-chain was prepared in racemic form and resolved at an advanced stage. As shown, the dianion of 3-oxo-butyric acid ethyl ester (**76**, Scheme 12.9) was first added to trimethylsilanyl-propynal (**75**) to construct racemic propargylic alcohol **77**. A stereoselective *syn*-reduction of **77** was accomplished with $NaBH_4$ and Et_2BOMe to produce racemic *syn*-1,3-diol **78**, which was subsequently protected as the corresponding acetonide **79** by treatment with 2,2-dimethoxypropane and catalytic acid. Treatment of **79** with NaOH resulted in saponification of the ethyl ester to carboxylic acid **80**, which was then treated with (*R*)-naphthylethylamine to produce two diastereomeric quaternary amine salts that were separable by recystallization. This classical resolution provided the desired stereoisomer **81** in 31% yield. After separation, treatment of quaternary amine salt **81** with aqueous HCl resulted in the liberation of the carboxylic acid, which was then esterified to provide ester **82** for use in the palladium-mediated coupling reaction. In that event, treatment of **82** with disiamylborane was followed by base-generated vinyl borane **83**, which was then reacted with heteroaryl iodide **51** in the presence of catalytic $PdCl_2$ to provide **84** in excellent overall yield. Finally, **84** was converted to pitavastatin (**3**) through a typical reaction sequence.

REFERENCES

Asberg, A. and Holdaas, H. (2004). *Expert. Rev. Cardiovasc. Ther.*, 2: 641–652.

Castelli, W. P. (1984). *Am. J. Med.*, 76: 4–12.

Chen, K.-M., Hardtmann, G. E., Prasad, K., Repic, O., and Shapiro, M. J. (1987). *Tetrahedron Lett.*, 28: 155–158.

Chen, K.-M., Kapa, P. K., Lee, G. T., Repic, O., Hess, P., and Crevoisier, M. (1993). EP 0363934.

Feit, P. W. (1964). *J. Med. Chem.*, 7: 14–17.

Grundy, S. M. (1997). *J. Intern. Med.*, 241: 295–306.

Hirai, K., Ishiba, T., Koike, H., and Watanabe, M. (1993). U.S. Patent 5260440.

Hiyama, T., Minami, T., and Takahashi, K. (1995a). *Bull. Chem. Soc. Jpn.*, 68: 364–372.

Hiyama, T., Minami, T., Yanagawa, Y., and Ohara, Y. (1995b). PCT WO 9511898.

Kathawala, F. G. (1994). U.S. Patent 5354772.

Konoike, T. and Araki, Y. (1994). *J. Org. Chem.*, 59: 7849–7854.

Li, J.-J., Johnson, D. S., Sliskovic, D. R., and Roth, B. D. (2004). *Contemporary Drug Synthesis*, 9: 113–124, Wiley & Sons, Inc. Hoboken (NJ).

Lopp, M., Kanger, T., Muraus, A., Pehk, T., and Lille, U. (1991). *Tetrahedron: Asymmetry*, 2: 943–944.

McKenney, J. M. (2003). *Clin. Cardiol. (Suppl. III).*, 26: 32–38.

Minami, T. and Hiyama, T. (1992). *Tetrahedron Lett.*, 33: 7525–7526.

Miyachi, N., Yanagawa, Y., Iwasaki, H., Ohara, Y., and Kiyama, T. (1993). *Tetrahedron Lett.*, 34: 8267–8270.

Parker, R. A., Clark, R.W., Sit, S.-Y., Lanier, T. L., Grosso, R. A., and Wright, J. J. K. (1990). *J. Lipid Res.*, 31: 1271–1282.

Repic, O., Prasad, K., and Lee, G. T. (2001). *Org. Process. Res. Dev.*, 5: 519–527.

Robinson, B. (1963). *Chem. Rev.*, 63: 373–401.

Robinson, B. (1968). *Chem. Rev.*, 69: 227–250.

Schuster, H. (2003). *Cardiology*, 99: 216–139.

Schuster, H. F. and Coppola, G. M. (1994). *J. Heterocyclic Chem.*, 31: 1381–1384.

Sorbera. L. A., Leeson, P. A., and Castaner, J. (1998). *Drugs Future*, 23: 847–859.

Takahashi, K., Minami, T., Ohara, Y., and Hiyama, T. (1993). *Tetrahedron Lett.*, 34: 8263–8266.

Takahashi, K., Minami, T., Ohara, Y., and Hiyama, T. (1995). *Bull. Chem. Soc. Jpn.*, 68: 2649–2656.

Takano, S., Kamikubo, T., Sugihara, T., Suzuki, M., and Ogasawara, K. (1993). *Tetrahedron: Asymmetry*, 4: 201–204.

Theisen, P. D. and Heathcock, C. H. (1988). *J. Org. Chem.*, 53: 2374–2378.

Walkup, R.E. and Linder, J. (1985). *Tetrahedron Lett.*, 26: 2155–2158.

Watanabe, M., Koike, H., Ishiba, T., Okada, T., Seo, S., and Hirai, K. (1997). *Bioorg. Med. Chem.*, 5: 437–444.

13

CHOLESTEROL ABSORPTION INHIBITORS: EZETIMIBE (ZETIA®)

Stuart B. Rosenblum

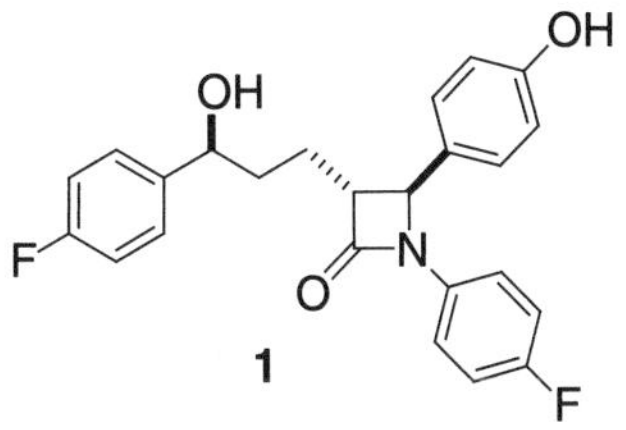

OH
OH
F
O
N
1
F

USAN: Ezetimibe
Trade names: Zetia®, Ezetrol®
Schering-Plough Corporation
Launched: 2002

M.W. (free form) 407.43
CAS: [163222-33-1]

Crystalline solid; Mp: 164–166°C; $[\alpha]_D^{22}$ = 33.9° (c = 3 MeOH)

13.1 INTRODUCTION

Cholesterol is critical for proper cellular and organ function in mammals, but elevated plasma cholesterol levels (hypercholesterolemia) is a major risk factor for the development of cardiovascular disease. The increase in hypercholesterolemia over the past 50 years is partly due to significant dietary changes in most developed countries and diminished daily physical activity. Based on pooled epidemiological studies (Grundy et al., 2004), for every 1% reduction in the serum low-density lipoprotein fraction (LDL-C, so called "bad" cholesterol), at least a 1% reduction in cardiovascular risk is realized. For patients with two or more additional risk factors (e.g., smoking, hypertension, or diabetes mellitus) the benefit of LDL-C reduction is significantly greater. In 2005, it was

The Art of Drug Synthesis. Edited by Douglas S. Johnson and Jie Jack Li

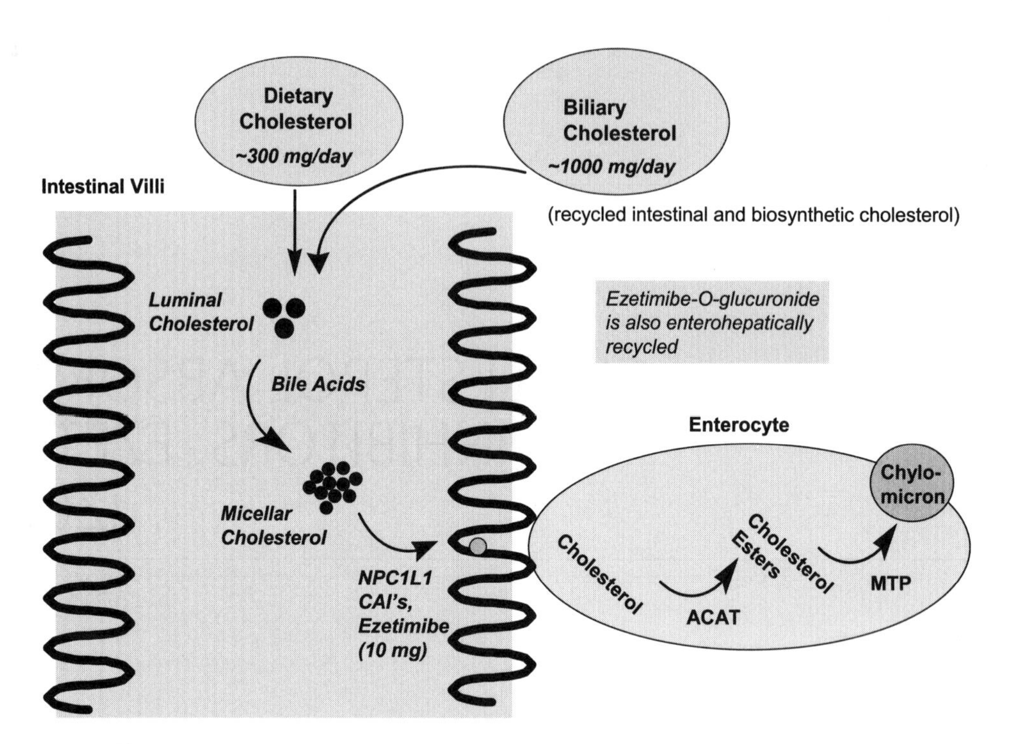

Figure 13.1. Ezetimibe effects absorption of both dietary and biliary cholesterol.

estimated that more than 56 million people would benefit from some form of cholesterol reduction, that is lifestyle change, medication, or both (Persell et al., 2006).

Drug discovery targets for LDL-C reduction include the biosynthesis, esterification, intracellular transport and regulation, plasma transport, and intestinal absorption of cholesterol. The "statin" class of therapeutics, typified by simvastatin and atorvastatin, intervene in de novo cholesterol synthesis by inhibition of the ER membrane-bound enzyme 3-hydroxy-3-methylglutaryl coenzyme A (HMG-CoA) reductase, which reduces HMG to mevalonate. Additional modalities that target either synthesis or cholesterol processing include fibric acid derivatives ("fibrates"), cholesterol ester transfer protein (CETP) inhibitors, perioxisome proliferators activator receptor (PPAR) agonists, nicotinamide receptor agonists (NAR) and acyl CoA: cholesterol acyltransferase (ACAT) inhibitors (Krause et al., 1995). In 2002, the first of a new class of LDL-C lowering agents, the intestinal cholesterol absorption inhibitor (CAI) ezetimibe (**1**), was introduced (Fig. 13.1). Coincident with the development of successive generations of mechanistically novel and improved compounds targeting cholesterol lowering, has been the growing acceptance to treat asymptomatic patients determined to be at risk of cardiovascular disease. The successful discovery and commercialization of safe and effective compounds that lower LDL-C, coupled with identification of patients at risk, have benefited millions of individuals.

13.2 DISCOVERY PATH TO EZETIMIBE

Ezetimibe (**1**) is the first in a new class of cholesterol absorption inhibitors that reduce plasma LDL-C levels by direct inhibition of the uptake of free cholesterol from the small intestine. In addition to its mechanistic novelty, ezetimibe (**1**) has a long duration of action, low systemic exposure, and an excellent safety profile. The discovery path to ezetimibe encompassed the common themes of scientific achievement–inspiration, perspiration, and serendipity.

During the early phases of an ACAT inhibitor discovery program at Schering-Plough, conformationally restricted analogs based on the 2-azetidinone backbone were targeted by Burnett and co-workers (Burnett et al., 1994). Early in the biological evaluation, it became apparent that even though the in vitro ACAT inhibitory activity of these analogs was modest (e.g., IC_{50} values of 2–50 μM), they exhibited significant activity in a cholesterol-fed hamster model (CFH). The discovery of the prototypical 2-azetidinone CAI, SCH-48461 (ACAT IC_{50} ~26 μM, ED_{50} of CE reduction in hamsters ~2.2 mpk) and the details of the first-generation SAR have been described in detail (Clader et al., 1996).

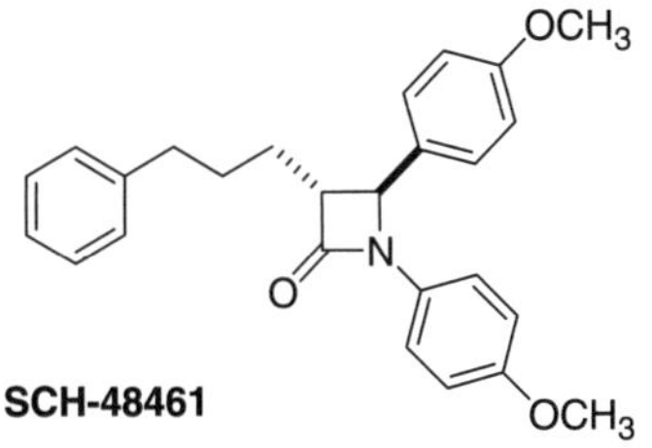

OCH_3 N O OCH_3

SCH-48461

SCH-48461 reduced LDL-C by −13% at a daily dose of 25 mg [~0.36 mg/kg (mpk)] in patients adhering to the American Heart Association's low cholesterol diet. Surprisingly, even at doses ten-fold greater, the response was not significantly larger (−16% LDL-C @ 5.7 mpk). In addition to the clinical results, data indicating rapid metabolism and biliary excretion of the metabolites of the first-generation compound were emerging. Speculation on the existence of active metabolites was further supported by demonstration of significant cholesterol absorption reduction when "SCH-48461 metabolite" bile was dosed into a recipient rat (Van Heek et al., 1997). Continued SAR exploration based on the SCH-48461 structure and using the CFH model as the only guide had been accomplished, but with little gain in potency. Since SCH-48461 reduced LDL-C to a greater degree than diet alone, clinical proof-of-concept was achieved. This spurred the continued search for analogs with improved activity in the CFH model, simpler metabolic profiles, and with improved CYP450 selectivity.

Our initial efforts focused on methodology development to prepare the putative metabolites and thereby expand the SAR. The synthetic details are described fully below. The speculated modes of primary metabolism of SCH-48461 were dealkylation of the C4- and N1-methoxyphenyl groups, hydroxylation of the pendent C3 side-chain phenyl group, benzylic hydroxylation (potentially two stereoisomers), and 2-azetidinone ring opening (Fig. 13.2). A directed effort toward the most probable metabolites led to the discovery of hydroxylated analogs (3′-OH diastereomeric mix and racemate) with a six-fold improvement in in vivo CAI activity. By late 1993, the ten most probable metabolites of SCH-48461 had been synthesized, evaluated in the CFH model, and the impact of the complexity of the metabolism on CAI activity and pharmacokinetics was outlined. More significantly, hydroxylated and metabolically blocked targets were proposed and speculated to have increased potency and simpler metabolic and pharmacokinetic profiles. The knowledge that metabolic liberation of a 4-hydroxy phenyl at N1 (by demethylation) or at the pendent phenyl (by hydroxylation) did not improve activity promoted strategic installation of fluorine atoms at these metabolic soft spots. Activity enhancements with a 3′-benzylic hydroxyl or a 4-hydroxyphenyl moiety at

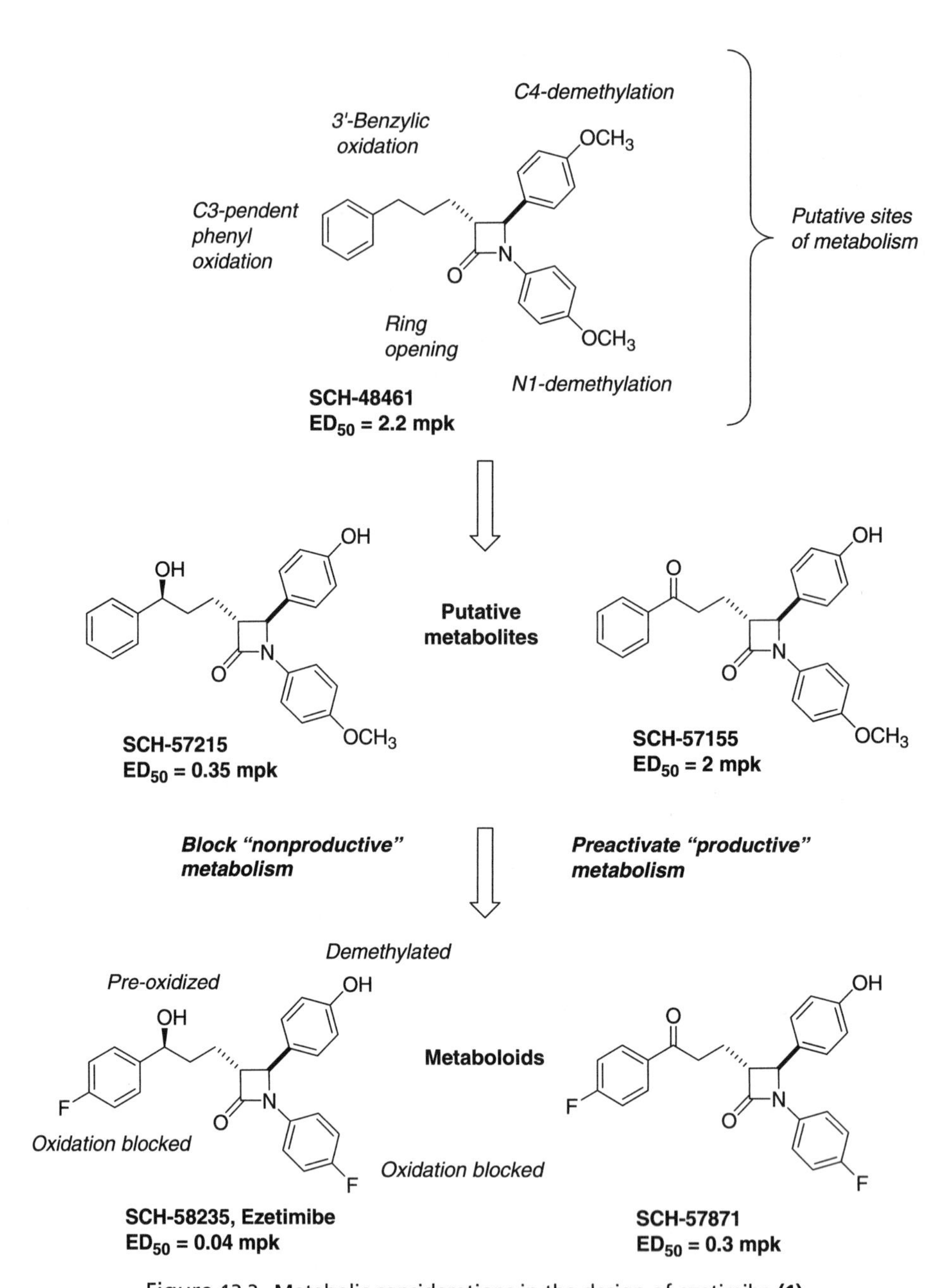

Figure 13.2. Metabolic considerations in the design of ezetimibe (**1**).

C4 promoted the preinstallation of these metabolically "productive" functionalities. Coupling the above observations, it became clear to us to simultaneously optimize all regions of the structure and target what is now known as ezetimibe (**1**) (Rosenblum et al., 1998). As the SAR was developed with an in vivo assay, improvement in activity reflected a complex balance between intrinsic receptor affinity and ADME considerations. We now appreciate that the site of action is the enterocyte on the intestinal villi, and thus systemic absorption is not required for activity. The above outlined combination of synthetic organic chemistry, medicinal chemistry, pharmacology, and drug metabolism led to the design of ezetimibe.

Ezetimibe (**1**) has been shown to be a highly potent inhibitor of cholesterol absorption in several animal models, with ED_{50} values of 0.04 mpk, 0.007 mpk, and 0.0005 mpk in the cholesterol-fed seven-day hamster, seven-day dog and three-week monkey models, respectively. The increased activity in the dog and monkey is speculated to be due to more efficient intestinal *O*-glucuronidation and enterohepatic recirculation in higher mammals. In a clinical trial with patients who were pre-equilibrated with a low-cholesterol diet, treatment with 10 mg ($\sim$0.13 mpk) ezetimibe reduced LDL-C levels an additional -18.5% (Bays and Moore, 2001). Ezetimibe and its *O*-glucuronide exhibit low systemic absorption in man, with 48 h AUC values of <800 ng $\cdot$ h/mL (Patrick et al., 2002). No significant effects on liver enzyme levels have been observed in animals or during clinical trials (Bays and Moore, 2001). As endogenous cholesterol also undergoes recirculation via the bile (Fig. 13.1), ezetimibe manifests an effect on plasma LDL-C greater than would be expected from the inhibition of dietary cholesterol alone. Additionally, ezetimibe (**1**) exhibits desirable effects on triglyceride, apolipoprotein B, and C-reactive protein (CRP) levels. Combination therapy of ezetimibe with agents that act by complementary mechanisms has also been demonstrated (Ballantyne and Blazing, 2004).

Not until 2004, almost 14 years after the inception of the CAI program and after marketing approval, was the molecular target of ezetimibe elucidated (Garcia-Calvo et al., 2005). Use of radio-labeled ezetimibe localized the molecular target at or near the intestinal brush-border membrane. After a number of false leads from membrane isolation studies, genomic sleuthing implicated Nieman-Pick C1-like 1 (NPC1L1) protein as the likely target. This was strongly supported by the unresponsiveness of NPC1L1 knock-out mice to ezetimibe (**1**) and to cholesterol elevation when fed a high-cholesterol diet. Ezetimibe has been shown to bind to NPC1L1 in vitro, but the observed in vivo activity is still surprising based on its apparent moderate NPC1L1 affinity.

In summary, based on the observation of divergent in vitro and in vivo SAR, a new class of cholesterol-lowering compounds was discovered. Subsequent biological characterization of the first generation lead compound SCH-48461, limited the site of action to be at or near the intestinal villi, with the most probable mechanism being the inhibition of luminal absorption. An SAR optimization based solely on an in vivo assay and without biochemical characterization of the molecular target led to the design of ezetimibe (**1**).

13.3 SYNTHESIS OF EZETIMIBE (ZETIA®)

Ezetimibe (**1**) contains three *para*-substituted phenyl rings, a chiral benzylic hydroxyl and two additional stereogenic centers decorating a rigid 2-azetidinone scaffold. The three chiral centers give rise to eight stereoisomers that have been individually characterized and shown to exhibit significantly different CAI profiles (Rosenblum et al., 1998). The stereocontrol of the two azetidinone stereocenters (abs. 3*R*, 4*S*) and the installation of the benzylic hydroxyl with absolute 3′*S* stereochemistry, were the most significant synthetic challenges.

The development of flexible, high-yielding synthetic methodology played a critical role for the progression of the CAI program. As the seven-day cholesterol-fed hamster model was the only means available for determination of cholesterol absorption inhibition, the early SAR development required in excess of 150 mg of each analog. Use of the (a) ester enolate-imine and (b) ketene-imine 2-azetidinone ring construction methodologies

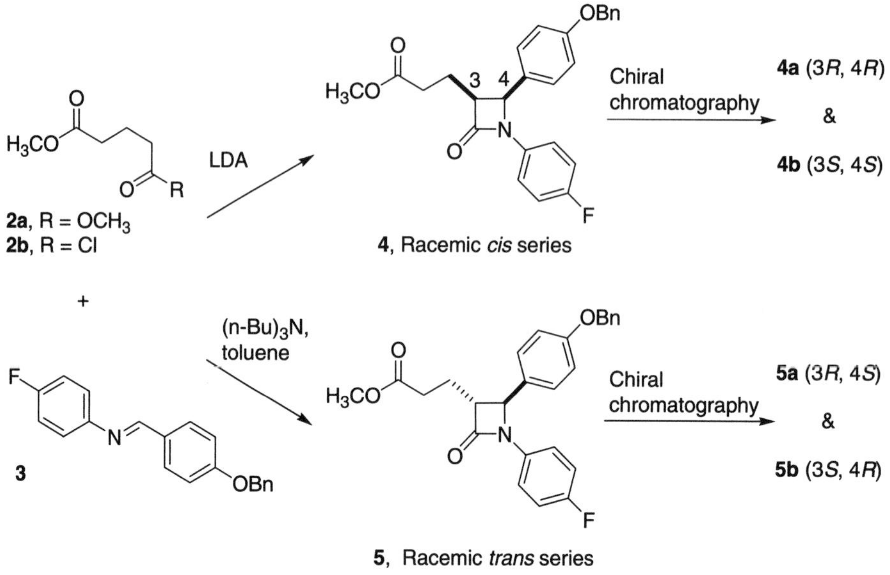

2a, R = OCH_3
2b, R = Cl
+
3
LDA
$(n\text{-}Bu)_3N$, toluene
4, Racemic *cis* series
5, Racemic *trans* series
Chiral chromatography
4a (3*R*, 4*R*) & 4b (3*S*, 4*S*)
5a (3*R*, 4*S*) & 5b (3*S*, 4*R*)

Scheme 13.1.

(Burnett et al., 1994) allowed efficient preparation of gram quantities of advanced intermediates (Scheme 13.1). Functionalized side-chain intermediates such as **4** and **5** were envisioned and proved synthetically useful for the late introduction of the 3′-hydroxyl and pendent aryl moieties.

Our initial route to C3-side-chain hydroxylated analogs focused on exploiting esters **4** and **5**. Mono-lithiation of dimethyl glutarate (**2a**) with LDA at −78°C in THF followed by addition of imine **3**, afforded a 1 to 4 racemic mixture of *cis*- and *trans*-substituted 2-azetidinones in moderate yield. In contrast, when acid chloride **2b** was added slowly to a heated (45–80°C) solution of imine **3** and tri-*n*-butylamine in heptane (bp = 80°C), racemic *trans* 2-azetidinone **5** was produced in good yield (>15 : 1 *trans* : *cis*). Interestingly, generation of the putative ketene intermediate in low concentration at high temperature in the presence of a large excess of imine **3** gave rise to rapid and stereoselective cycloaddition with few side products (Browne et al., 1995). Room-temperature generation of ketene (or acylaminium) species, followed by imine addition resulted in poorer stereocontrol and increased amounts of self-condensation by-products. Both preferred methods were successfully scaled to afford multigram quantities of racemic material; however, a resource-intensive separation via chiral chromatography (CHIRACEL OD column; hexane : 2-propanol eluent) was required to advance the active enantiomeric series.

The first route envisioned (Scheme 13.2) to the 3′-benzylic hydroxyl analogs was via aldehyde **6**, obtained in good yield by DIBAL reduction of the ester **5** in toluene at −20°C. Reaction of the commercially available 4-fluorophenylmagnesium bromide (1 M in THF) with aldehyde **6** afforded an equal mixture of alcohols **7a** and **7b** in 20% yield. Side reactions included 2-azetidinone ring opening and products derived from C3 proton abstraction. The lack of stereocontrol at the 3′-benzylic center, the low yield, and the excellent CAI activity of **7a** and **7b**, prompted an aggressive search for alternative methodologies. Gratifyingly, reaction of acid chloride **9** with 4-fluorophenylzinc bromide under the aegis of catalytic tetrakis(triphenylphosphonium)-palladium (Negishi cross-coupling conditions) (Negishi et al., 1983) afforded the arylketone **10** in 80% yield. Acid chloride **9**

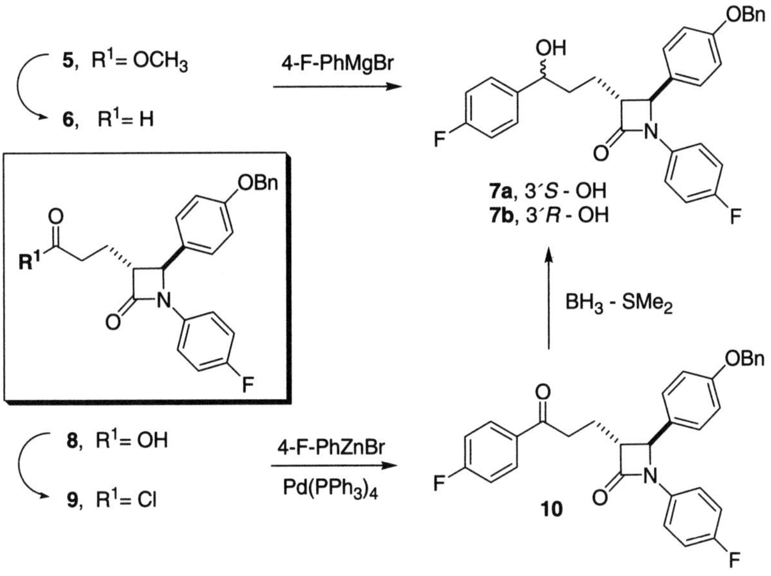

Scheme 13.2.

was prepared from ester **5** via lithium hydroxide hydrolysis and subsequent treatment with one equivalent of oxalyl chloride. Reduction of **10** with borane-dimethyl sulfide afforded an equal mixture of separable 3′*S* (**7a**) and 3′*R* (**7b**) alcohols in high yield. Initial attempts to promote 1,3 chiral induction by low-temperature "precomplexation" were not promising. Treatment of arylketone **10** with a catalytic (*R*)-diphenyloxazolidine and borane (Corey–Bakshi–Shibata, (CBS) Reduction) afforded 3′*S* alcohol (**7a**) in excellent yield, and use of the (*S*)-CBS catalyst provides the 3′*R* alcohol (**7b**). Hydrogenolysis of **7a** using 10% Pd/C in ethanol under a 60 psi hydrogen atmosphere, afforded ezetimibe (**1**) as a crystalline white solid (mp: 164–166°C).

Early in the synthetic program we observed ring opening of the 2-azetidinone ring or C3 deprotonation under conditions that liberated a free alkoxide or enolate anion. Judicious use of the CBS methodology, which masks the alkoxide as the boronate ester until work-up, and the palladium catalyzed arylation reaction, which uses the relatively nonbasic arylzinc, affords the desired adducts in high yield. The application of the Negishi coupling reaction followed by the chemoselective chiral-borane mediated reduction methodology proved reproducible and scalable, and allowed variation of the pendent aryl substituents with excellent stereochemical control at the 3′ center. As the later steps of the eight-step route also proved scalable, the synthetic bottleneck was the chiral chromatographic purification of key ester intermediate **5a**.

During the scale-up of a first-generation CAI (SCH-48461), process methodology was developed and refined for the enantioselective construction of 2-azetidinones by the Schering-Plough Development Group (Fu et al., 2003; Thiruvengadam et al., 1996, 1999). Access to their extensive experience allowed for rapid optimization of an enantioselective synthesis of key intermediate **5a**. Evans–Aldol condensation (Evans et al., 2002) of the boron or titanium enolate of chiral oxazolidinone **11** with benzaldehyde (**12**) at −78°C, afforded crystalline *syn*-aldol products **13** (*RRR* with minor amounts of *RSS* isomer) with moderate diastereoselectivity (Scheme 13.3) (Thiruvengadam et al., 1999).

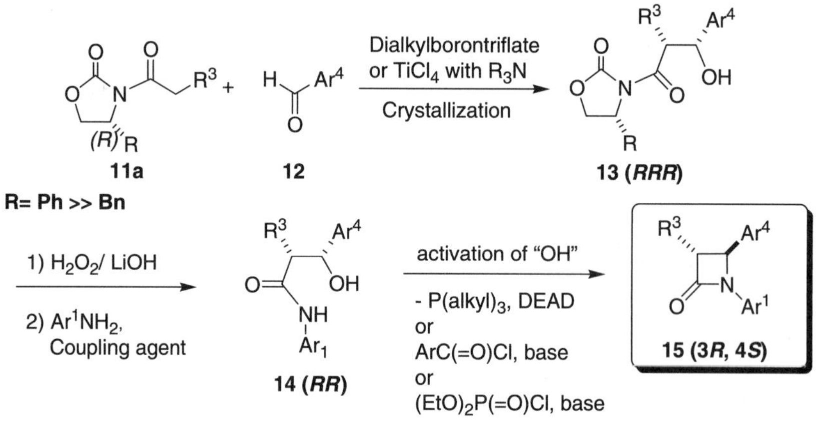

Scheme 13.3. Aldol route.

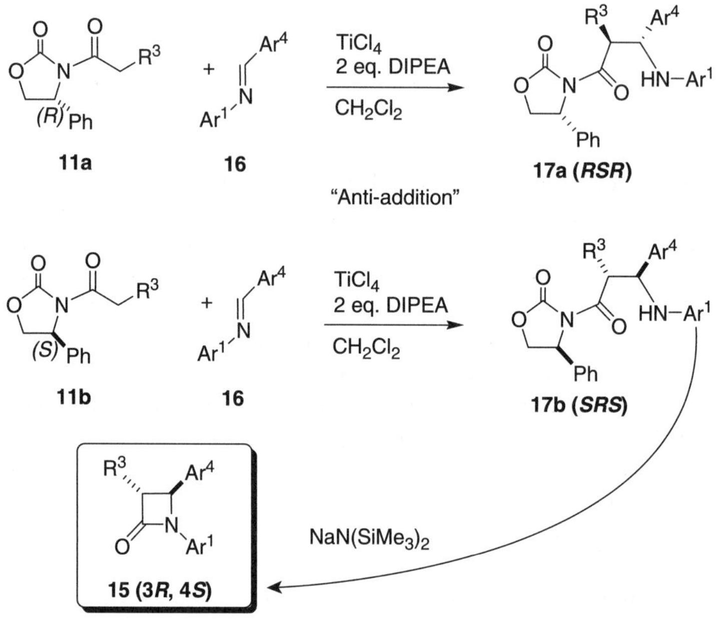

Scheme 13.4. Imine-addition route.

Two equivalents of the tertiary amine base are required, and a significant improvement in the diastereoselectivity was observed with TMEDA over DIPEA. Purification and further enrichment of the desired *RRR* isomer to >98% *ee* was achieved by crystallization. Oxidative removal of the chiral auxiliary followed by carbodiimide mediated amide formation provides β-keto carboxamide **14** in good yield. Activation of the benzylic hydroxyl via PPh_3/DEAD, acylation, or phosphorylation, effects 2-azetidinone ring-closure with inversion of stereochemistry at the C4 position. Unfortunately, final purification could not be effected by crystallization and the side products and or residual reagents could only be removed by careful chromatography on silica.

Further process optimization by Thiruvengadam and co-workers (Thiruvengadam et al., 1999), led to a novel, stereoselective, scalable two-step process devoid of chromatography for chiral 2-azetidinone construction (Scheme 13.4). As above, the titanium-enolate of chiral oxazolidinone **11a** was preformed, but now when reacted with well behaved imines of type **16**, affords the *unexpected* anti-addition product. This surprising result was further supported by careful comparison to minor antiproducts obtained in the earlier aldol-addition methodology and determination that the major product was indeed **17a** (undesired *RSR* series). Adjustment of the oxazolidinone absolute stereochemistry to the fortuitously less expensive 2*S*-series afforded the desired diastereomer **17b** in 95% *de* and in 50–70% yield. Recrystallization improved the stereochemical purity to >99% *de*.

The chiral enolate-imine addition methodology was examined in detail (Thiruvengadam et al., 1999). Enolate formation proceeds to completion within an hour at temperatures from −30 to 0°C with either 1 equiv. $TiCl_4$ or $TiCl_3O$-iPr (preformed or prepared in the presence of substrate by addition of $TiCl_4$ and followed by a third of an equivalent $Ti(O\text{-}iPr)_4$ and two equivalents of a tertiary amine base). Unlike the aldol process with the same titanium enolate, the nature of the tertiary amine base had no effect on the diastereoselectivity. The diastereoselectivity is maximized by careful control of the internal temperature to below −20°C during the imine addition (2 equiv.) as well as during the acetic acid quench. The purity of the crude 2-amino carboxamide derivatives (**17a** or

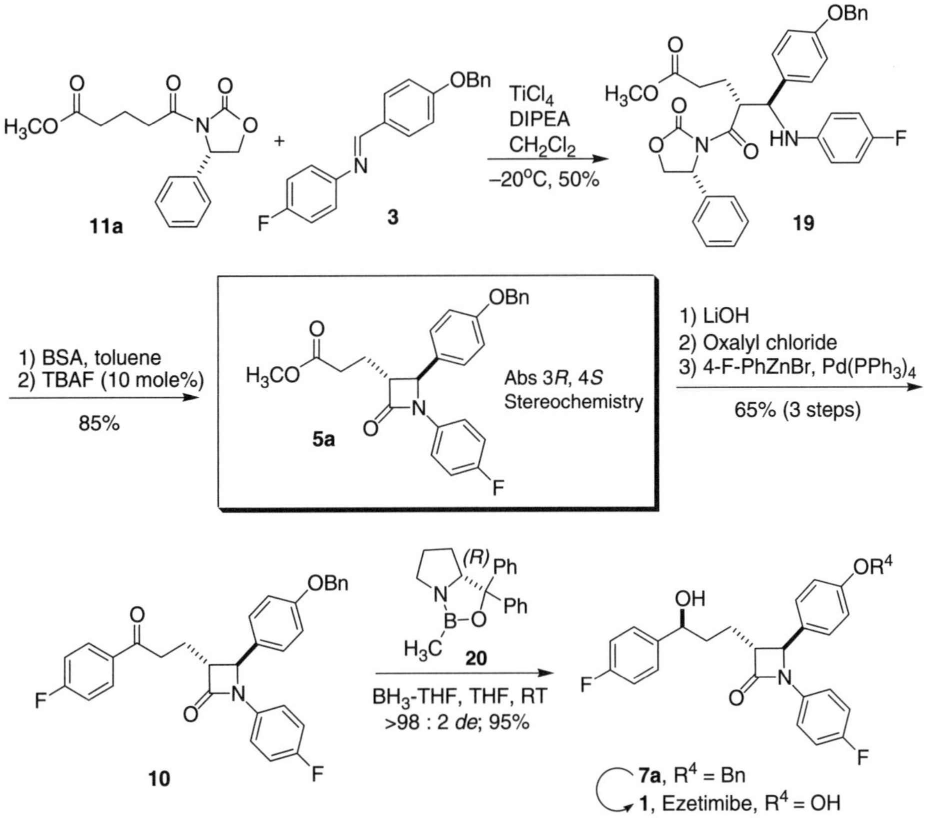

11a + 3 → 19 ($TiCl_4$, DIPEA, CH_2Cl_2, −20°C, 50%)

19 → 5a (1) BSA, toluene; 2) TBAF (10 mole%); 85%) — Abs 3*R*, 4*S* Stereochemistry

5a → 10 (1) LiOH; 2) Oxalyl chloride; 3) 4-F-PhZnBr, $Pd(PPh_3)_4$; 65% (3 steps))

10 → 7a (20, BH_3-THF, THF, RT; >98 : 2 *de*; 95%)

7a, R^4 = Bn
1, Ezetimibe, R^4 = OH

Scheme 13.5.

17b) was enriched to greater than 99% by crystallization with isolated yields typically in the 50–70% range.

Based on the early racemic synthesis of **4** (*cis* series), it had already been demonstrated that 2-azetidinone ring closure could be achieved via nucleophilic attack of a lithium amine anion on a β-ester. Cyclization could be accomplished with other strong bases, but sodium bistrimethylsilylamide was found to effect efficient cyclization without significant racemization at C3. During the search for experimentally convenient bases, it was noted that Noyori (Nakamura et al., 1983) reported that tetrabutylammonium fluoride (TBAF) as well as LiF, KF, and CsF could serve as the base in Aldol reactions. Treatment of **17a** or **17b** with TBAF trihydrate in THF did not affect cyclization. After much experimentation it was found that addition of *N,O*-bistrimethylacetamide (BSA) to **19** followed by TBAF addition, effected 2-azetidinone ring closure. Further optimization found that use of catalytic TBAF (<1%) in methylene chloride afforded near quantitative cyclization.

In order to meet an aggressive time-line, application of the optimized methodology led to the seven-step process shown in Scheme 13.5. The final benzylic deprotection conditions (Parr-shaker, 60 psi H_2 gas) were less than optimal from a safety perspective and alternatives were sought. Hydrogenolysis of **7a** was cleanly accomplished in under 2 h by treatment with ammonium formate (5 equiv.) and 10% Pd/C catalyst (0.1 equiv. by weight) in methanol at mild reflux. Use of higher palladium loads and prolonged reaction times produced unwanted side-products derived from 2-azetidinone cleavage. All the synthetic steps after the enolate-imine addition step (slightly variable, 50–70% yield)

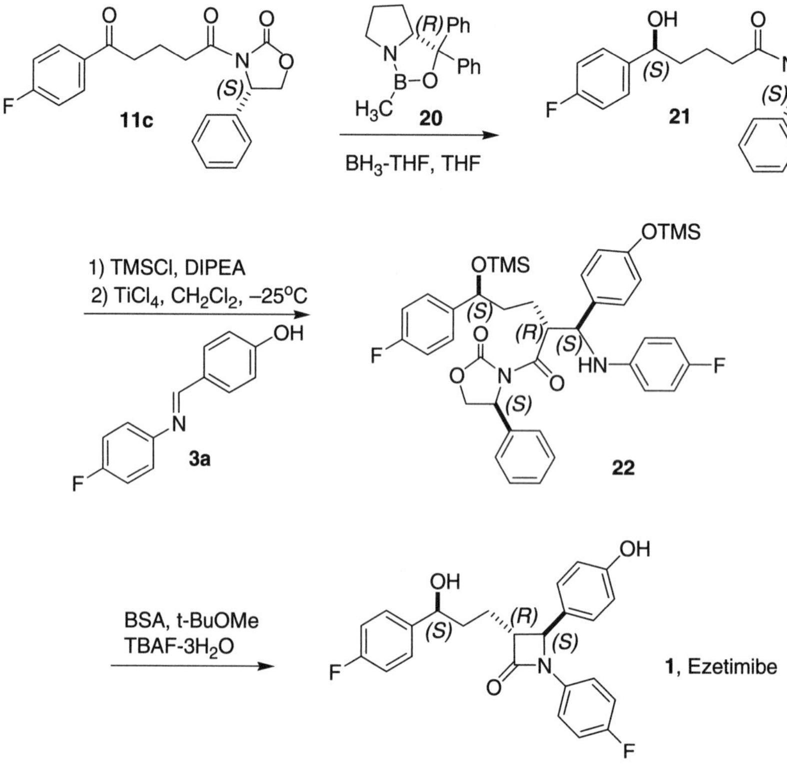

11c

20

BH_3-THF, THF

21

1) TMSCl, DIPEA
2) $TiCl_4$, CH_2Cl_2, –25°C

3a

22

BSA, t-BuOMe
TBAF-$3H_2O$

1, Ezetimibe

Scheme 13.6.

proceeded in better than 85% yields. The first 5 kg GMP batch of ezetimibe (**1**) was produced by this route (Rosenblum, 1995, 1997).

During the successful progression of ezetimibe (**1**) through clinical trials, the Schering-Plough Process group was further challenged to improve the research synthesis into a commercial process. This was accomplished by fully elaborating the latent C3 side chain before the enolate-imine condensation. Thus, CBS reduction (5% catalyst load) of ketone **11c**, afforded chiral alcohol **21** in >95% yield (Scheme 13.6). Judicious choice of the trimethylsilyl protecting group allowed clean in-situ protection of both the benzylic and phenolic hydroxyl groups with TMS-Cl. Alcohol **21** was treated with two equiv. of TMS-Cl, followed by titanium enolate formation ($TiCl_4$), and then addition of phenolic imine **3a**. Excess TMS-Cl present in the reaction reacts with the C4-phenol to afford crystalline **22**. Cyclization mediated by BSA and TBAF proceeded smoothly, but two equiv. of TBAF were required to effect complete deprotection of the benzylic and phenolic hydroxyl groups. Minor modifications of this process have been used to produce ezetimibe on a commercial scale.

Alternative syntheses of ezetimibe (**1**) have been demonstrated. The three approaches outlined below proceed through penultimate intermediate **10a** and differ with respect to the number of carbons and functionality on the side-chain before attachment to the 2-azetidinone fragment. These alternative strategies illustrate the remarkable stability of *N*-aryl 2-azetidinones to a wide variety of conditions and allowed access to a wide variety of ezetimibe analogs.

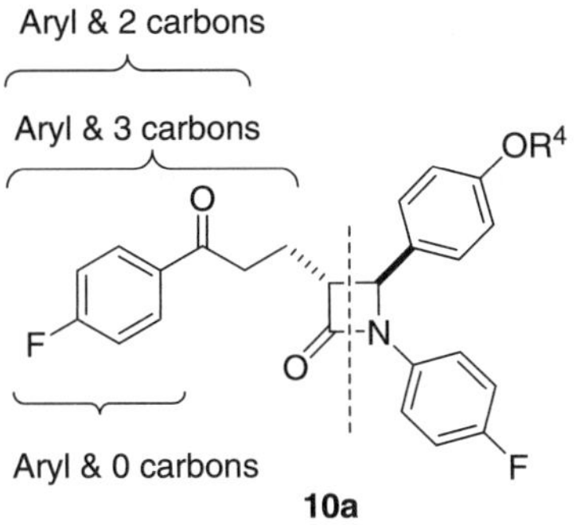

Aryl & 2 carbons

Aryl & 3 carbons

OR^4

O

F

O

N

F

Aryl & 0 carbons

10a

In the course of examining the CAI effect of conformational restriction of the C3-side-chain, intermediate **24** was prepared. Shankar and co-workers (Shankar et al., 1996) demonstrated that **10**, a key intermediate in the research synthesis could be accessed by Wacker oxidation of olefin **24** (Scheme 13.7). Additionally, an alternative chiral variant of the well-precedented addition of zinc enolates to imines was demonstrated. Treatment of the bromoacetate **25**, derived from 8-phenylmenthol with zinc and sonication followed by imine addition afforded **26** in 55% yield with greater than 99% *de*. Ethyl magnesium promoted ring-closure followed by C3 alkylation with **28**, intercepts the previously demonstrated route through formation of olefin **24** (Shankar et al., 1996).

Access to a variety of substituted 2-azetidinones analogs via palladium-mediated arylation of versatile intermediate **30** has been reported by Rosenblum (Rosenblum et al., 2000). Specifically, the reaction of **30** with substituted aryl halides to form compounds of type **24a** was investigated (Scheme 13.8). The use of $PdOAc_2/PPh_3$ was equivalent to $PdCl_2(PPh_3)_2$ but superior to $Pd(PPh_3)_4$, and addition of both a tetraalkylammonium salt and an alkali metal acetate (e.g., KOAc) were required. Although the yields for

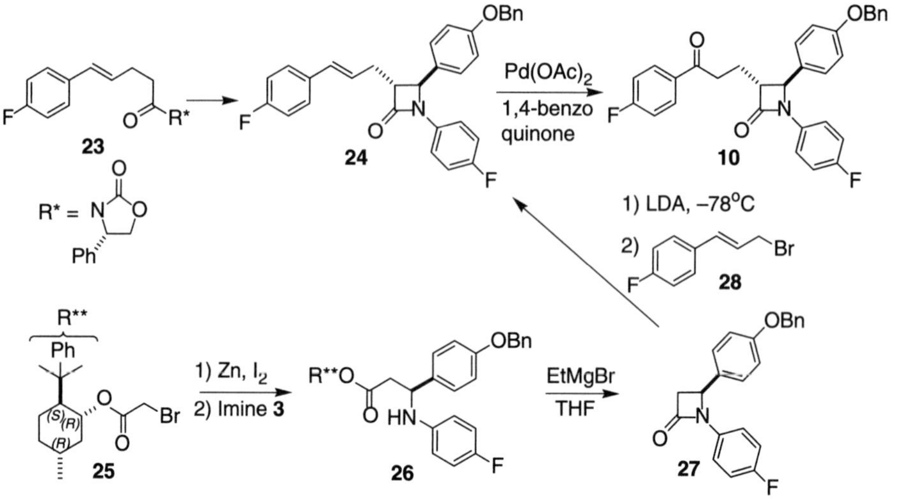

Scheme 13.7.

arylation of **30** were modest (20–70%), this methodology was convenient and versatile for SAR exploration.

In an analogous late-stage arylation approach, terminal alkyne **31** was envisioned as a versatile intermediate. Slow addition of 4-pentynoyl chloride to imine **3** and $(n\text{-Bu})_3N$ at reflux (efficient condenser, 100°C, 12 h, 1 : 1 toluene : heptane) afforded only trace amounts of **31**. Reaction of 4-pentynoyl chloride with triethylamine in methylene chloride under preformed ketene conditions (−78°C, 1 h), followed by addition of **3** and warming to −10°C over 4 h, afforded a complex mixture of products. Since high-yield preparation of **31** remained elusive, access to internal alkynyl analogs (type **33**) was accomplished by preassembly of the appropriate arylalkynyl acid substrate for the ketene-imine cycloaddition reaction (Scheme 13.9).

Wu and co-workers (Wu et al., 1999) have demonstrated a novel chiral lactone enolate-imine process to access 2-azetidinone diols such as **35** (Scheme 13.10). Treatment of **34** with LDA at −25°C in THF followed by addition of imine **3**, afforded only trace product. Addition of HMPA or the less toxic DMPU during the lithium enolate formation step improved the yield and the *trans* : *cis* diastereoselectivity (~90 : 10). Recrystallization improved the purity to >95 : 5 *trans* : *cis* 2-azetidinone. Addition of an equivalent of lithium bromide accelerates the rate of ring closure, presumably by destabilizing the intermediate lithium aggregates. Side-chain manipulation of **35** was accomplished by sodium

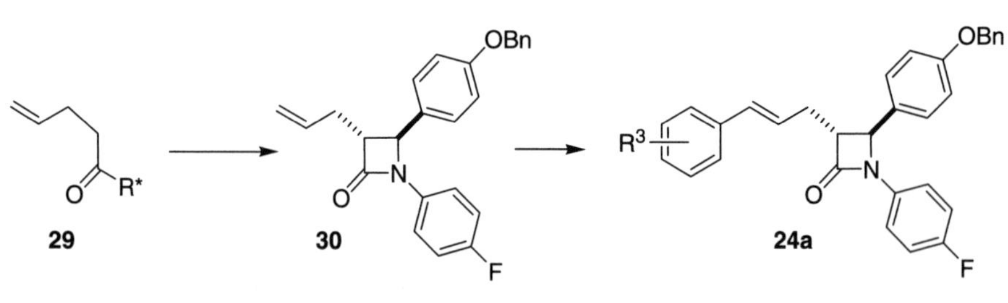

Scheme 13.8.

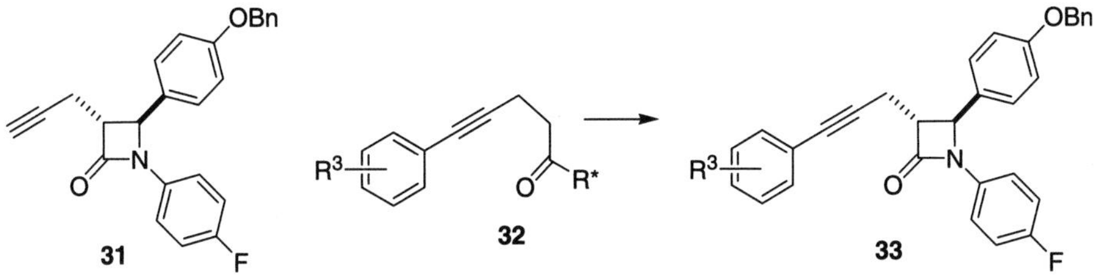

Scheme 13.9.

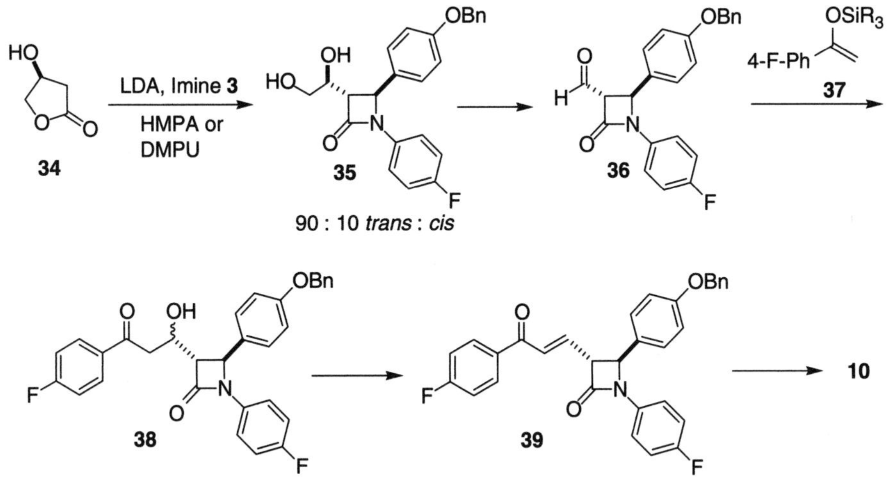

Scheme 13.10.

periodate oxidation to aldehyde **36**, followed by Mukaiyama aldol condensation. PTSA-mediated acid elimination of **38** followed by either homogeneous ($RhCl(PPh_3)_3$) or heterogeneous (Pd/C) hydrogenation afforded **10**, which can be manipulated into ezetimibe (**1**) as illustrated in Scheme 13.10.

The discovery history of cholesterol treatments illustrates how the understanding of biological systems has led to drugs, and how these drugs have served as tools to further elucidate the biological systems. A critical component of the success of this noble pursuit has been the creativity of chemists in the practice of the art of drug synthesis.

REFERENCES

Ballantyne, C. M. and Blazing, M. A. (2004). *American J. Cardiology*, 93: 1487–1494.

Bays, H. E. and Moore, P. B. (2001). *Clinical Therapeutics*, 23: 1209–1230.

Browne, M., Burnett, D. A., and Caplen, M. A. (1995). *Tetrahedron Lett.*, 36: 2555–2558.

Burnett, D. A., Caplan, M. A., Davis, H. R. Jr., Burrier, R. E., and Clader, J. W. (1994). *J. Med. Chem.*, 37: 1733–1736.

Clader, J. W., Burnett, D. A., Caplen, M., and Domalski, M. S. (1996). *J. Med. Chem.*, 39: 3684–3693.

Evans, D. A., Tedrow, J. S., Shaw, J. T., and Downey, C. W. (2002). *J. Am. Chem. Soc.*, 124: 392–393.

Fu, X., McAllister, T. L., Thiruvengadam, T. K., and Tann, C.-H. (2003). U.S. Patent 6,627,757 B2.

Garcia-Calvo, M., Lisnock, J., and Bull, H. G. (2005). *Proc. Natl. Acad. Sci. USA*, 102: 8132–8137.

Grundy, S. M., Cleeman, J. I., and Merz, C. N. B. (2004). *Circulation*, 110: 227–239.

Krause, B. R., Sliskovic, D. R., and Bocan, T. M. A. (1995). *Exp. Opinion. Invest. Drugs*, 4: 353–387.

Nakamura, E., Shimizu, M., Kuwajima, I., and Noyori, R. (1983). *J. Org. Chem.*, 48: 932–945.

Negishi, E., Bagheri, V., Chatterjee, S., Luo, F. T., Miller, J. A., and Stoll, A. T. (1983). *Tetrahedron Lett.*, 24: 5181–5184.

Patrick, J. E., Kosoglou, T., Stauber, K. L., and Alton, K. B. (2002). *Drug Metabolism and Disposition*, 30: 430–437.

Persell, S. D., Lloyd-Jones, D. M., and Baker, D. W. (2006). *J. Gen. Intern. Med.*, 21: 171–176.

Rosenblum, S. B. (1995). World (PCT) Patent WO9508532-A1.

Rosenblum, S. B. (1997). U.S. Patent 5,631,365.

Rosenblum, S. B., Huynh, T., Afonso, A., Davis, H. R., and Yumibe, N. (1998). *J. Med. Chem.*, 41: 973–980.

Rosenblum, S. B., Huynh, T., Afonso, A., and Davis, H. R. (2000). *Tetrahedron*, 56: 5735–5742.

Shankar, B. B., Kirkup, M. P., McCombie, S. W., Clader, J. W., and Ganguly, A. K. (1996). *Tetrahedron Lett.*, 37: 4095–4098.

Thiruvengadam, T. K., Sudhakar, A. R., and Wu, G. (1999). Process Chemistry in the Pharmaceutical Industry, Dekker, New York, NY, pp 221–242.

Thiruvengadam, T. K., Tann, C.-H., and McAllister, T. L. (1996). U.S. Patent 5,561,227.

Van Heek, M., France, C. F., Compton, D. S., Mcleod, R. L., Yumibe, N. P., Alton, K. B., Sybertz, E. J., and Davis, H. R., Jr. (1997). *J. Pharm. Exp. Ther.*, 283: 157–163.

Wu, G., Wong, Y., Chen, X., and Ding, Z. (1999). *J. Org. Chem.*, 10: 3714–3718.

III

CENTRAL NERVOUS SYSTEM DISEASES

14

DUAL SELECTIVE SEROTONIN AND NOREPINEPHRINE REUPTAKE INHIBITORS (SSNRIs) FOR DEPRESSION

Marta Piñeiro-Núñez

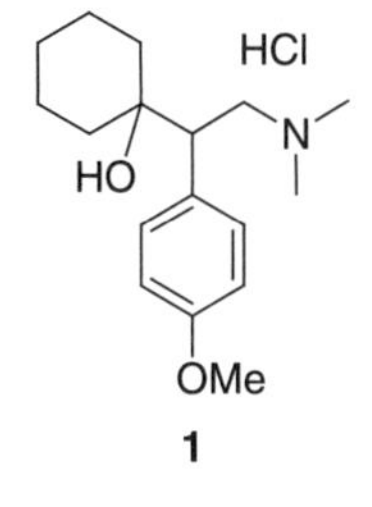

USAN: Venlafaxine hydrochloride
Trade name: Effexor®
Wyeth-Ayerst
Launched: 1994 (US)
M.W. (freebase): 277.41
Racemate

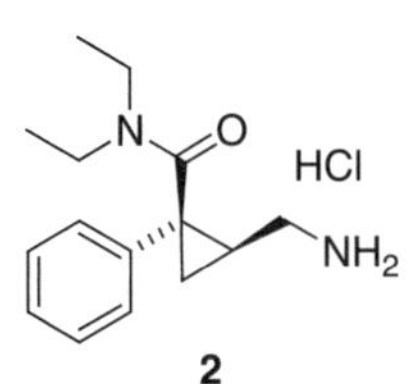

USAN: Milnacipran hydrochloride
Trade name: Ixel,® Dalcipran®
Pierre Fabre
Launched: 1995 (France)
M.W. (freebase): 246.36
Racemate

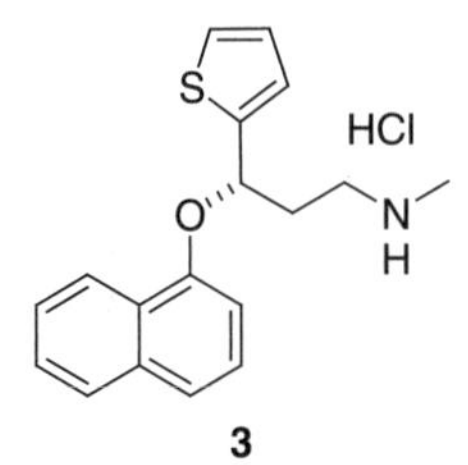

USAN: Duloxetine hydrochloride
Trade name: Cymbalta®
Eli Lilly and Company
Launched: 2004 (US)
M.W. (freebase): 297.42
Single Enantiomer

The Art of Drug Synthesis. Edited by Douglas S. Johnson and Jie Jack Li

14.1 INTRODUCTION

The burden of mental illness has traditionally been underestimated worldwide. Despite treatment advances, major depressive disorder (MDD) is still a significant cause of morbidity and mortality. In fact, depression was the fourth leading cause of disease burden in the world in 1990, and is projected to be the second leading cause of disability by 2020. In the United States alone, it causes billions of dollars annually in direct and indirect medical costs and losses in productivity. It is now known that patients with MDD may experience both psychological and medical complaints, including pain, which underscores the severity of impact of MDD on the health-care system.

It is generally accepted that the goals of antidepressant therapy are three-fold: first, producing symptomatic improvement (response); second, inducing symptom resolution and optimal functioning (remission); and third, preventing relapse or recurrence. Currently known therapies frequently fall short of providing full remission; in fact, best response rates hover at 60–70%, and remission generally stays below 50%. This fact, coupled with limitations deriving from delayed onset of action, tolerability problems, as well as treatment resistance, has continued to spur research for improved antidepressants that better address the needs of millions of patients worldwide (Briely, 1998; Ferrier, 1999; Farvolden et al., 2003; Gorman and Kent, 1999; Jain, 2004; Kirwin and Goren, 2005; Koch et al., 2003; Lantz et al., 2003; Stahl, 1998; Stahl et al., 2005; Thomson Micromedex Database; Walter, 2005; Westerberg, 1999; Wong and Bymaster, 2002).

Ultimately, the effects of virtually all existing antidepressants can be traced to the improvement of neurotransmission in the brain by one or more monoamine neurotransmitters, that is serotonin (5-HT, **4**), norepinephrine (NE, **5**), and dopamine (DA, **6**). By blocking monoamine transporters, which remove the neurotransmitter from the synapse and extracellular space by uptake processes, the drugs increase extracellular levels of the transmitter and cause a cascade of intracellular events leading to the desired CNS effect.

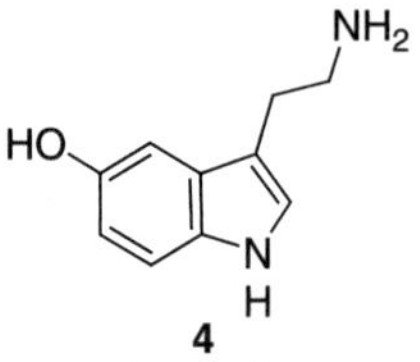

4 Serotonin **5** Norepinephrine **6** Dopamine

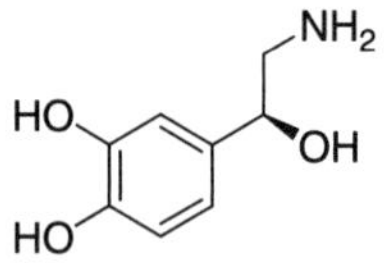

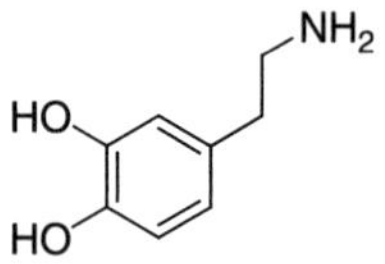

Despite its usefulness, this picture simply provides a partial explanation of the mechanism of MDD. Indeed, it is likely that other neurobiological systems are also involved in the pathogenesis of depression. In this regard, a number of novel agents, including corticotrophin-releasing factor antagonist, substance P antagonists and antiglucocorticoids, show considerable promise for refining treatment options. The current consensus is that depression is not a single entity, but a complex phenomenon encompassing many underlying causes. In addition, it is now known that the brain is not a static organ, but a plastic architecture affected by environmental influences, including learning, stress, and medication. For instance, there is evidence that prolonged stress can permanently damage neurons in the hippocampus, and that antidepressants actually stimulate the growth of hippocampal nerve cells. In summary, full appreciation of observed treatment response will require the development of more sophisticated theory-driven typologies of MDD.

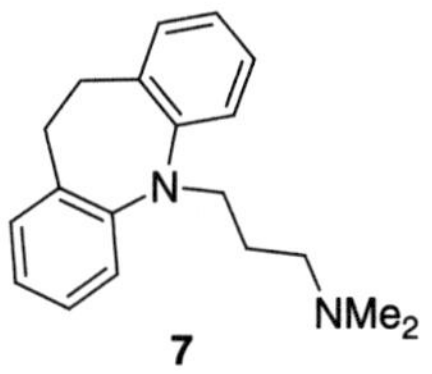

7 Imipramine

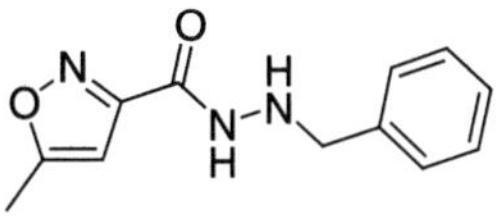

8 Isocarboxazid

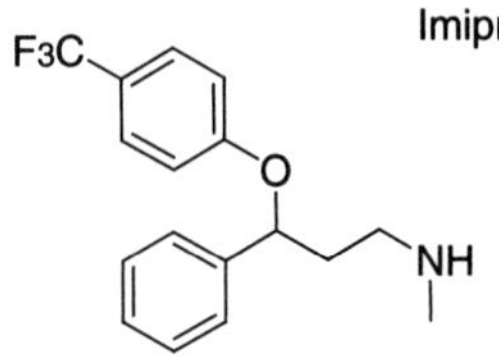

9 Fluoxetine

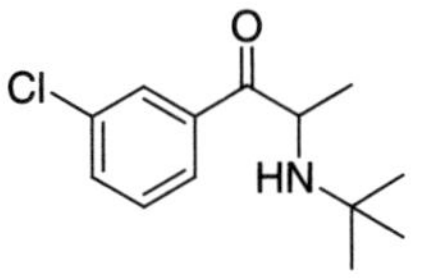

10 Bupropion

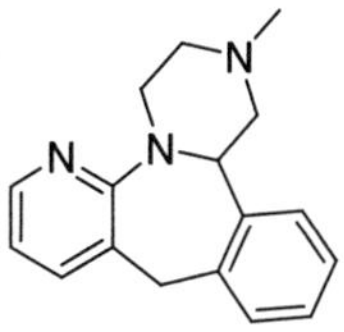

11 Mirtazapine

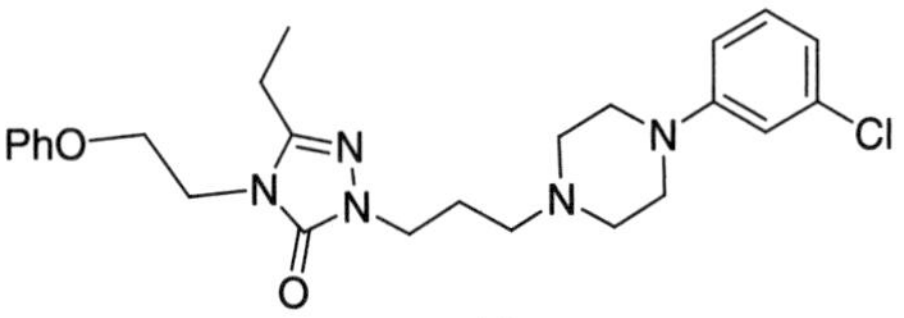

12 Nefazodone

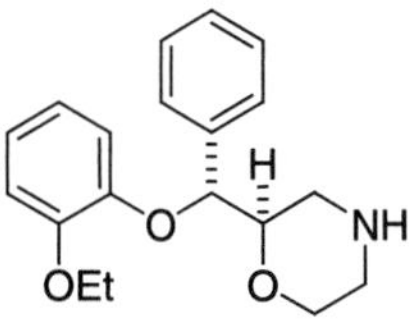

13 Reboxetine

Our understanding of the mechanism of antidepressant action has evolved over time. In the late 1950s, the first molecules introduced for the treatment of MDD were the so-called tricyclic antidepressants (TCAs), represented by imipramine (**7**). Subsequent experience with TCAs supported the role of both 5-HT and NE, although these drug molecules act on other neuronal systems as well. Despite their effectiveness, the use of TCAs was limited due to poor tolerability and safety concerns, in particular, severe toxicity when taken in overdose.

The next option became available in the 1960s with the introduction of monoamine oxidase inhibitors (MAOIs), exemplified by isocarboxazid (**8**). The MAOIs are multiple-action medications that exert their antidepressant effects by increasing levels of many monoamines, including 5-HT, NE, and DA. Unfortunately, MAOIs also suppress tyramine reuptake, which has serious unintended consequences. Indeed, in response to a diet-related surge of tyramine, some patients experienced sudden increases in blood pressure that caused fatal brain hemorrhaging. As a result, strict dietary restrictions had to be incorporated into standard MAOI therapy shortly after introduction, significantly reducing its usefulness. In addition, these drugs caused sedation and decreased alertness.

The field of antidepressant research was revolutionized in the late 1980s by the introduction of selective serotonin reuptake inhibitors (SSRIs), exemplified by fluoxetine (**9**). In addition to their antidepressant action, SSRIs have also proven effective for a broad range of psychiatric illnesses, and, more importantly, they demonstrated an improved tolerability profile as compared to TCAs and MAOIs due to their increased selectivity. On the other hand, SSRIs proved inferior to TCAs and MAOIs in their reduced antidepressant effects, slower onset of action, lower remission rates, and decreased ability to control the physical symptoms associated with depression.

TABLE 14.1. Diverse Mechanisms of Action Antidepressants from the 1990's

Drug	Launch Date	Mechanism
Bupropion (**10**)	1989	NE/DA dual reuptake inhibitor
Mirtazapine (**11**)	1994	Adrenergic/5-HT dual antagonist
Nefazodone (**12**)	1994	5-HT antagonist/reuptake inhibitor
Reboxetine (**13**)	1997	Selective NE reuptake inhibitor (SNRI)

In the wake of the SSRI revolution, the 1990s saw the emergence of a number of antidepressant drugs with diverse mechanisms of action. Again, the common thread behind the different approaches was an increase of synaptic monoamine levels (Table 14.1).

Clinical experience resulting from decades of application of the therapies represented by molecules **7–13** has provided evidence that both 5-HT and NE neuronal pathways contribute independent antidepressant mechanisms, but a number of preclinical studies have also demonstrated an interaction between 5-HT and NE neuronal systems that may play a role in the therapy of depression, beyond a simple additive effect. This interdependence, as well as the encouraging clinical experience with SSRIs and TCAs, led to the hypothesis that increasing 5-HT and NE availability by blocking both transporters could result in better pharmaceuticals for the treatment of MDD. This approach led to the emergence since the mid-1990s of dual selective serotonin norepinephrine reuptake inhibitors (SSNRIs), a class currently comprising venlafaxine (**1**), milnacipran (**2**), and duloxetine (**3**). There are important advantages to the use of SSNRIs over their SSRI predecessors in that they maintain favorable tolerability profiles, but they are also capable of reducing the symptoms of depression more effectively due to their dual action over separate neuronal pathways. Therefore, all three drugs block 5-HT and NE transporters (SERT and NET), but they do so with differing selectivity (Table 14.2). Thus, venlafaxine (**1**) is about 30 times more selective for 5-HT, milnacipran (**2**) is essentially equipotent, and duloxetine (**3**) is about 10 times more selective for 5-HT. When used at doses that block both reuptake systems, all three drugs produce higher rates of response and remission from MDD than the SSRIs. Their effect as anxiolytics is comparable to SSRIs, but, in addition, they are also effective in the treatment of chronic pain.

Beyond their action upon SERT and NET, venlafaxine (**1**), milnacipran (**2**) and duloxetine (**3**) are remarkably selective molecules. All three of them have displayed very low in vitro affinity ($Ki > 3000$ nM) for α_1- and α_2-adrenergic, histamine H_1, muscarinic, and DA D_2 receptors, consistent with favorable side-effect profiles. Venlafaxine (**1**) and duloxetine (**3**) also have low affinity for a number of serotonergic receptors, and do not inhibit monoamine oxidase A or B. An expanded in vitro receptor profile of >50 receptors and binding sites

TABLE 14.2. Transporter Binding Profiles of 1–3. In Vitro Inhibition of Radioligand Binding to Human Monoamine Transporters

	Ki (nM)		
	Venlafaxine (**1**)	Milnacipran (**2**)	Duloxetine (**3**)
5-HT transporter (SERT)	82	123	0.8
NE transporter (NET)	2,480	200	7.5
DA transporter (DAT)	7,647	>10,000	240
NET/SERT ratio	30	1.6	9.4

Sources: Koch et al., 2003; Stahl et al., 2005; Walter, 2005; Wong and Bymaster, 2002.

TABLE 14.3. Comparison of Clinical Parameters for 1–3 (Briley, 1998; Lantz et al., 2003; Thomson Micromedex Healthcare Series; Cymbalta US Package Insert and European Summary of Product Characteristics)

	Venlafaxine (**1**)	Milnacipran (**2**)	Duloxetine (**3**)
Route of administration	PO	PO	PO
Number of doses/day	2–3	2	1–2
Total dose/day (mg)	75–375	100	40–120
Onset (initial) (weeks)	2	1–3	3
TC_{max} (h)	1–2	0.5–4	6–10
%*F*	12.6	85	>70
%Protein binding	30	13	95
Volume of distribution (L)	420–490	371	1640
Elimination route	Renal	Renal/Hepatic	Renal
Half-life (h)	5	8	11–16
Interaction with CYP2D6	Yes	No	Yes
Circulating active metabolites	*O*-desmethyl	None	None

found no additional binding sites ($Ki > 1\ \mu M$) for either drug. Likewise, extensive in vitro testing of milnacipran (**2**) on a battery of >40 receptors resulted in consistently low binding numbers ($Ki > 10\ \mu M$), providing further evidence of remarkable selectivity.

A comparison of clinical parameters for all three drugs offers additional understanding of relative similarities and differences (Table 14.3). As far as dosing, duloxetine (**3**) offers the advantages of lower daily doses and the possibility of once-a-day therapy; it also has a longer elimination half-life than the other two. Milnacipran (**2**), however, does not have significant interactions with CYP2D6, and may offer an added advantage for use by renally impaired patients due to the co-existence of two elimination routes (renal and hepatic). Finally, both milnacipran (**2**) and duloxetine (**3**) have improved bioavailabilities compared to venlafaxine (**1**).

There have been no published studies comparing the SSNRIs among themselves, although it has been suggested that the efficacy and tolerability of venlafaxine (**1**) and duloxetine (**3**) were globally comparable. However, comparison studies between each SSNRI and one or more SSRIs suggest that the level of efficacy of the three SSNRIs is similar. Further studies to provide comparative data among the three drugs are required before any firm conclusions can be made as to their relative efficacy.

Regarding side-effect profiles, all three SSNRIs are generally well tolerated, most adverse events occurring early in treatment, with a mild to moderate severity and a tendency to decrease or disappear with continued treatment. Venlafaxine (**1**) seems to be the least well-tolerated SNRI, combining a higher level of serotonergic adverse events (nausea, sexual dysfunction, withdrawal problems) with dose-dependent hypertension. In contrast, milnacipran (**2**) and duloxetine (**3**) appear better tolerated and essentially devoid of cardiovascular toxicity.

14.2 SYNTHESIS OF VENLAFAXINE HYDROCHLORIDE (Basappa et al., 2004; Chavan et al., 2004; Yardley et al., 1990)

The synthesis of venlafaxine (**1**) as reported by the original inventors is straightforward (Scheme 14.1). As the drug molecule is commercialized as a racemate, the synthetic

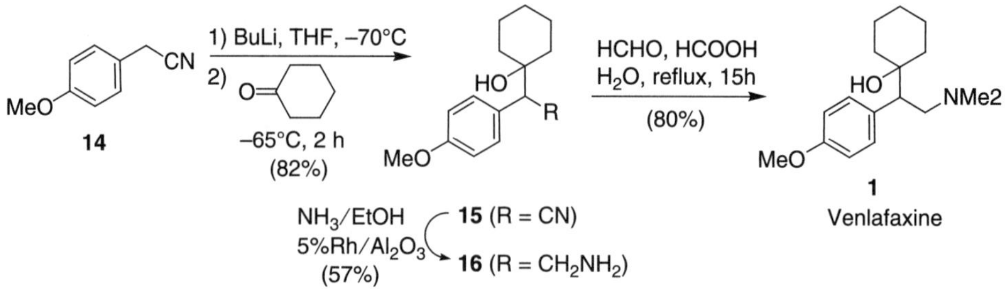

Scheme 14.1.

route does not require a stereoselective approach. Thus, the anion of commercially available nitrile **14** was condensed with cyclohexanone to provide intermediate nitrile adduct **15**, which was hydrogenated to amine **16** using rhodium on alumina catalysis. The synthesis concluded with installation of the required dimethyl amine moiety via standard reductive alkylation of **16**, affording venlafaxine (**1**) in 35% overall yield.

An improved modification of this protocol has been described, which provides the desired molecule in three steps and 68% overall yield (Scheme 14.2). The route was essentially the same as in Scheme 14.3, except that the intermediate nitrile **15** was converted to the bicyclic aminal **17**, which was then further alkylated to provide the final target. This method was reported as being amenable for large-scale synthesis.

A recent literature report described a "green" procedure for the condensation of arylacetonitriles with cyclic ketones using phase-transfer catalysis. This process was applied to the synthesis of venlafaxine, which was realized in overall 30% yield in two steps from commercially available **14**. The condensation step was run in aqueous sodium hydroxide in the presence of tetrabutylammonium sulfate, to provide quantitative yield of intermediate **15**. Hydrogenation in a formalin–methanol mixture provided the final product venlafaxine (**1**) in 30% overall yield. This protocol did not necessitate intermediate purification steps, making it attractive from the commercial standpoint.

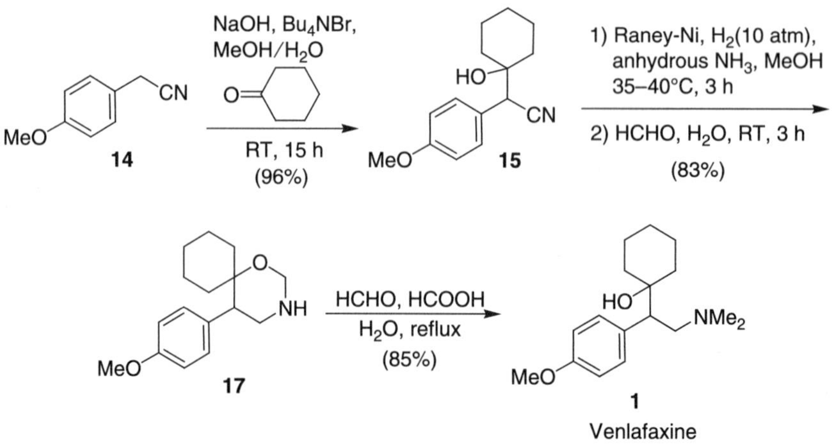

Scheme 14.2.

14.3 SYNTHESIS OF MILNACIPRAN (Bonnaud et al., 1986, 1987; Mouzin et al., 1978; Shuto et al., 1995; Zhang and Eaton, 2002)

The original synthesis of milnacipran is described in Scheme 14.3. Given that the commercial drug is a racemate, the synthetic route is simply concerned with relative stereochemistry around the cyclopropane ring. Thus, the carbanion of benzyl nitrile **18** was added to chloromethyloxirane to provide a *cis–trans* mixture of cyclopropanes **19**. Thermal convergence of the mixture **19** to lactone **20** set the stage for nucleophilic attack by potassium phthalimide, which generated acid **21** bearing the desired relative configuration around the cyclopropane ring. Standard diethyl amide formation led to intermediate **22**, which was easily converted to milnacipran (**2**) via aminolysis with aqueous methylamine.

Alternatively, lactone **20** could be converted to bromide **23** in the presence of hydrobromic acid (Scheme 14.4). Subsequent treatment with thionyl chloride generated the doubly functionalized intermediate **24**, which was subsequently treated with diethylamine to install the required amide functionality. Nucleophilic attack with potassium phthalimide on the bromide moiety resulted in the generation of intermediate amide **22**, at which point this route intersected that already described in Scheme 14.3. Treatment with hydrazine provided milnacipran (**2**) in 37% overall yield.

An improved route to milnacipran (**2**) and derivatives is described in Scheme 14.5. In this approach, lactone **20** was opened with lithium diethylamide to provide amide alcohol **25**, which was readily transformed into azide **26**. Hydrogenation on palladium–carbon directly led to the desired target in 86% yield over the three steps.

Finally, an entirely different approach to milnacipran (**2**) was recently reported in the literature (Scheme 14.6). In this case, the general strategy is based on position-selective deprotonation of cyclopropane carboxamides. Thus, cyclopropane amide **27**, which was easily prepared from commercially available cyclopropane carboxylic acid, underwent

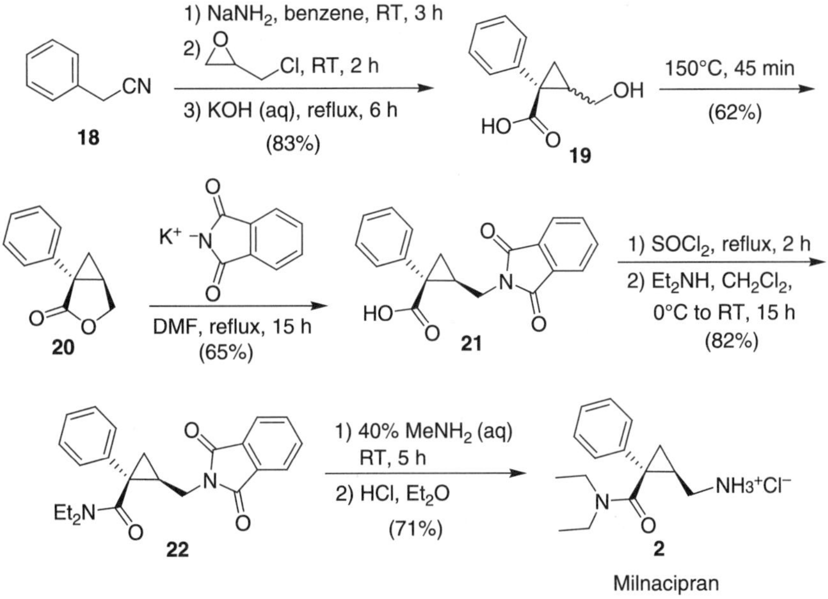

18 → 19: 1) $NaNH_2$, benzene, RT, 3 h; 2) chloromethyloxirane (Cl), RT, 2 h; 3) KOH (aq), reflux, 6 h (83%)

19 → 20: 150°C, 45 min (62%)

20 → 21: K^+ ^-N-phthalimide, DMF, reflux, 15 h (65%)

21 → 22: 1) $SOCl_2$, reflux, 2 h; 2) Et_2NH, CH_2Cl_2, 0°C to RT, 15 h (82%)

22 → 2: 1) 40% $MeNH_2$ (aq) RT, 5 h; 2) HCl, Et_2O (71%)

2 ($NH_3^+Cl^-$) Milnacipran

Scheme 14.3.

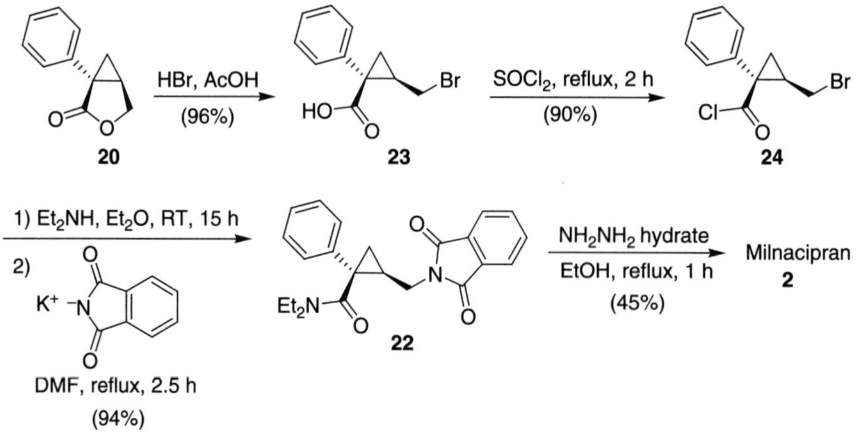

20 → HBr, AcOH (96%) → 23 → $SOCl_2$, reflux, 2 h (90%) → 24 → 1) Et_2NH, Et_2O, RT, 15 h; 2) K^+ ^-N-phthalimide, DMF, reflux, 2.5 h (94%) → 22 → NH_2NH_2 hydrate, EtOH, reflux, 1 h (45%) → Milnacipran 2

Scheme 14.4.

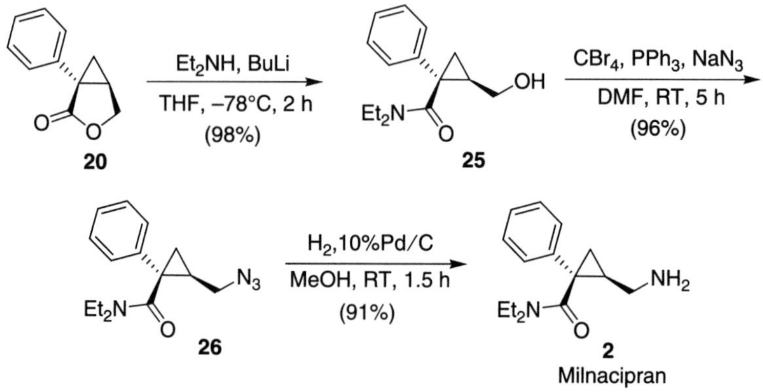

20 → Et_2NH, BuLi, THF, –78°C, 2 h (98%) → 25 → CBr_4, PPh_3, NaN_3, DMF, RT, 5 h (96%) → 26 → H_2,10%Pd/C, MeOH, RT, 1.5 h (91%) → 2 Milnacipran

Scheme 14.5.

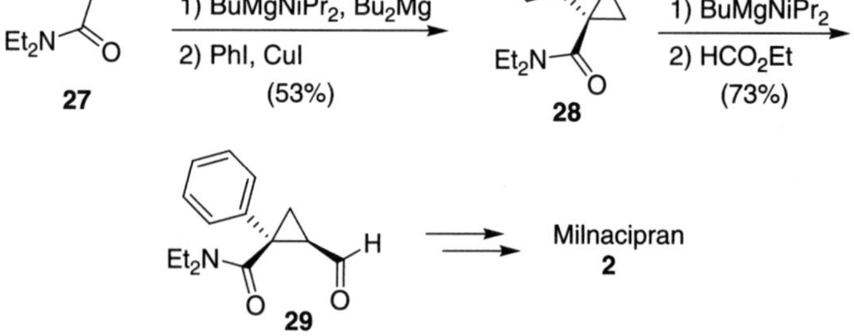

27 → 1) $BuMgNiPr_2$, Bu_2Mg; 2) PhI, CuI (53%) → 28 → 1) $BuMgNiPr_2$; 2) HCO_2Et (73%) → 29 → → Milnacipran 2

Scheme 14.6.

α-metallation with butylmagnesium diisopropylamide. The resulting anion was phenylated with iodobenzene to generate intermediate amide **28**. Further metallation, occurring now at the β position, provided an amido-Grignard reagent that easily reacted with ethyl formate to give aldehyde **29**, a known precursor of milnacipran (**2**).

14.4 SYNTHESIS OF DULOXETINE (Berglund, 1994; Borghese, 2003; Bymaster et al., 2003; Deeter et al., 1990; Kamal et al., 2003; Liu et al., 2000, 2005; Noyori et al., 2001; Ratovelomanana-Vidal et al., 2003; Robertson et al., 1991; Wheeler and Kuo, 1995)

The original synthesis of duloxetine (**3**) is relatively straightforward, involving a four-step sequence from readily available 2-acetylthiophene **30** (Scheme 14.7). Understandably, the main synthetic challenge stems from the presence of a chiral center, because duloxetine (**3**) is marketed as the (*S*)-enantiomer as shown. Thus, a Mannich reaction between **30** and dimethylamine generated ketone amine **31**, which was then reduced to provide intermediate racemic alcohol amine **32**. The desired optically active (*S*)-alcohol **32a** was accessed via resolution of racemate **32** with (*S*)-(+)-mandelic acid, which provided the necessary substrate for etherification with 1-fluoronaphthalene to afford optically active amine **33**. Finally, *N*-demethylation with 2,2,2-trichloroethyl chloroformate and cleavage of the intermediate carbamate with zinc powder and formic acid led to the desired target duloxetine (**3**).

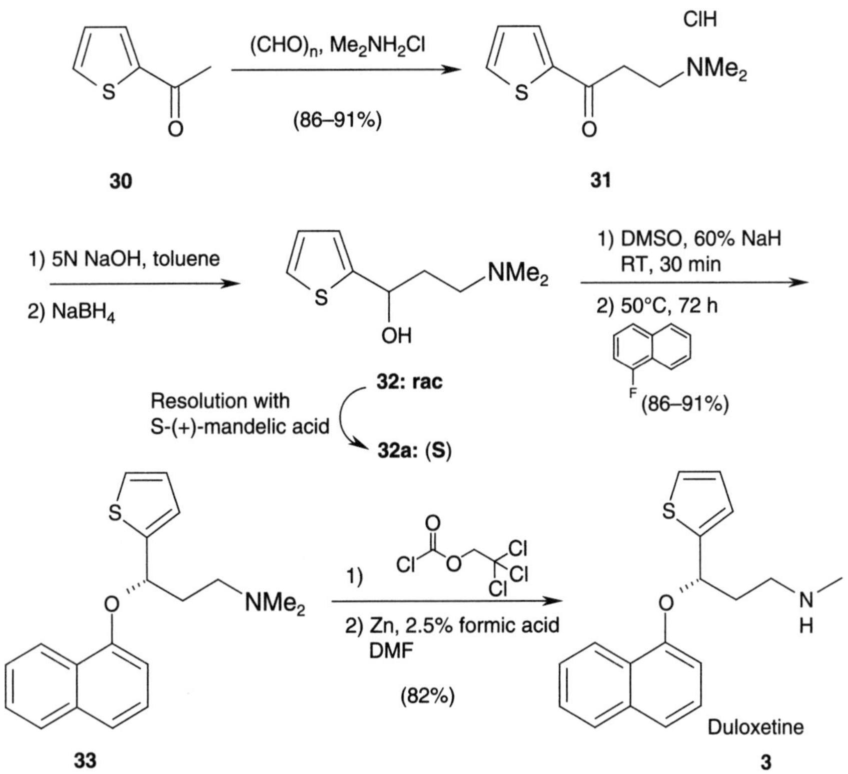

Scheme 14.7.

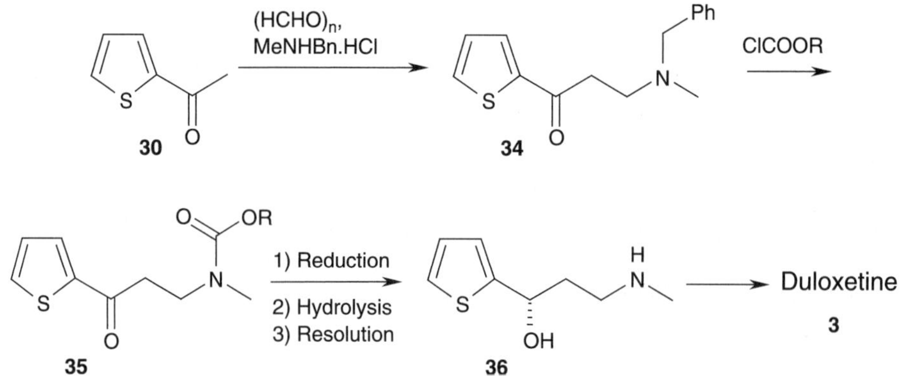

Scheme 14.8.

A subsequent process, still reliant on salt resolution, is described in Scheme 14.8. In this case, β-keto amine **34** was converted to carbamate **35**, which was then transformed into the desired amino alcohol intermediate **36**. The resolution was performed in three distinct stages: (1) initial resolution of the racemate, (2) racemization of the (*R*)-enriched mixture, and (3) second-order asymmetrically induced crystallization of the (*S*)-salt.

Subsequent synthetic literature around duloxetine (**3**) is focused around the description of stereoselective approaches to the critical alcohol intermediate **36**, or similar derivatives.

The first report in this regard described a method for direct formation of the desired optically active (*S*)-alcohol **32a**, via enantioselective reduction with a chiral amine complex of lithium aluminum hydride (Scheme 14.9). Therefore, the necessary chiral hydride complex **38** was preformed in toluene at low temperature from chiral amino alcohol **37**. The resulting hydride solution was then immediately combined with ketone **31** to afford the desired (*S*)-alcohol **32a** in excellent yield and enantiomeric excess. In addition to providing a more efficient route to the desired drug molecule, this work also led to the establishment of the absolute configuration of duloxetine (**3**) as (*S*).

A subsequent report provided a modified approach to duloxetine (**3**, Scheme 14.10). In this case, the crucial stereoselective reduction was achieved utilizing catalytic

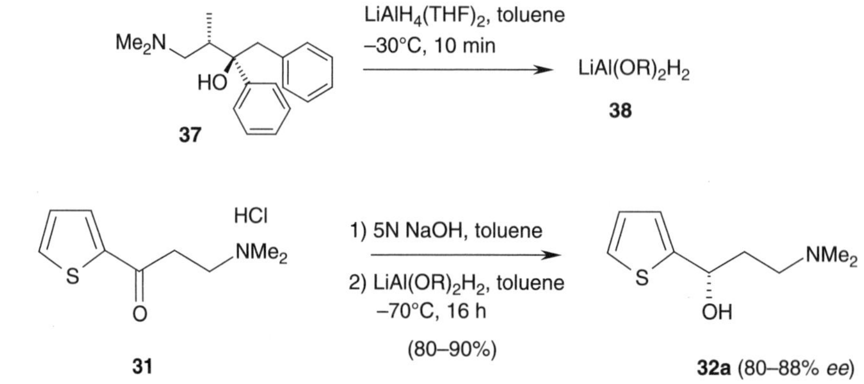

Scheme 14.9.

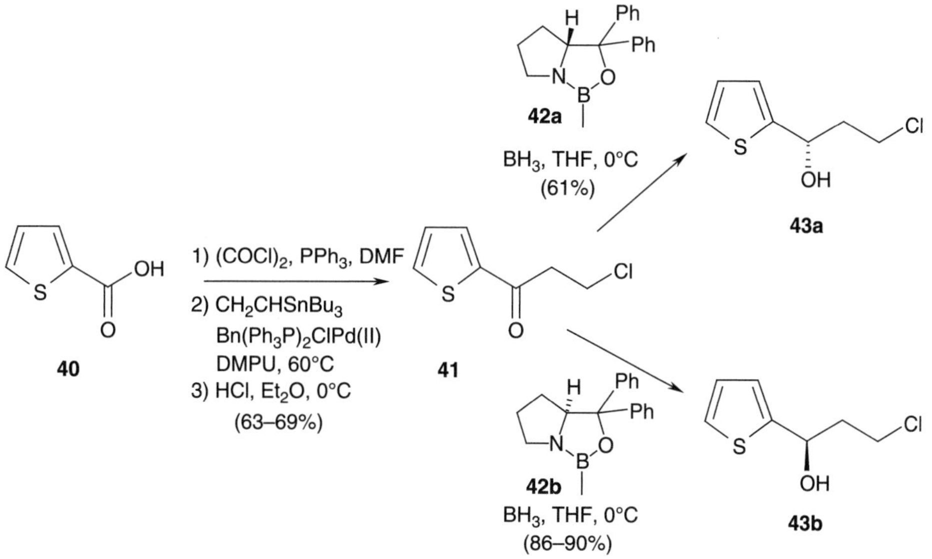

Scheme 14.10.

oxazaborolidines **42**. The necessary racemic ketone intermediate **41** was prepared from thiophene carboxylic acid **40** in three steps. Thus, acid chloride formation, followed by Stille coupling with tributyl(vinyl)tin and addition of hydrogen chloride across the newly formed double bond afforded the desired ketone **41**. Reduction utilizing oxazaborolidines **42** provided access to both enantiomeric chloroalcohols **43a** and **43b**.

Both intermediates **43a** and **43b** were converted to the final molecule duloxetine (**3**), as described in Scheme 14.11. Therefore, route A involved direct transformation of the (*S*)-chloroalcohol **43a** into the corresponding iodide, followed by amination and etherification. In contrast, route B consisted of Mitsunobu inversion of (*R*)-chloroalcohol **43b**

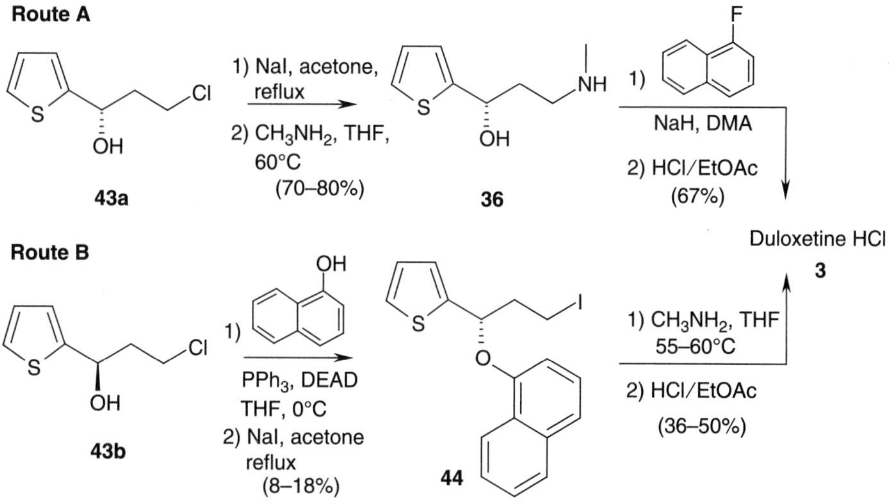

Scheme 14.11.

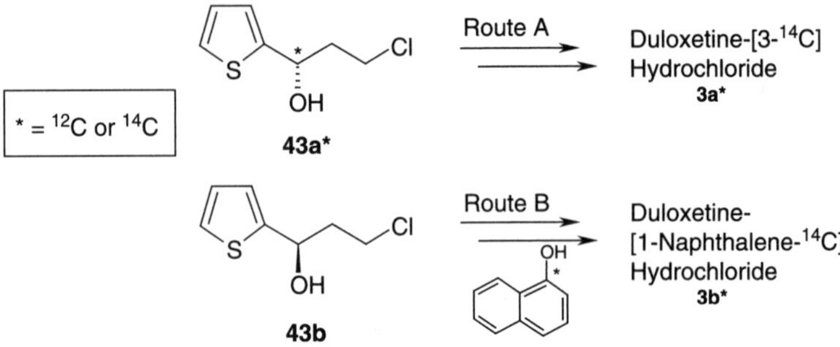

Scheme 14.12.

followed by iodination to provide intermediate (*S*)-ether iodide **44**, which then underwent amination to the final product duloxetine (**3**).

This synthetic flexibility was put to use in the preparation of ^{14}C-labeled isotopomers of duloxetine (**3**) for further biological studies (Scheme 14.12). Labeled thiophene carboxylic acid **40*** was readily converted to radioactive intermediate **43a***, which was then converted to duloxetine-[3-^{14}C] hydrochloride (**3a***) via route A. In contrast, **43b** was submitted to route B, utilizing labeled [1-^{14}C]-naphthalene, to provide duloxetine-[1-naphthalene-^{14}C] hydrochloride (**3b***).

Kinetic resolution technology has also been applied to the duloxetine problem (Scheme 14.13). In this case, chloroketone **41** was converted to racemic alcohol **43** using sodium borohydride. The racemate was then treated with vinyl butanoate in hexanes, in the presence of catalytic immobilized *Candida antarctica* Lipase B (CALB). The reaction was stopped after reaching 50% conversion, leading to the isolation of the desired (*S*)-chloroalcohol **43a**, as well as the (*R*)-ester **45** in good yields and excellent enantiomeric excesses. Chloroalcohol **43a** was converted to duloxetine (**3**) via the

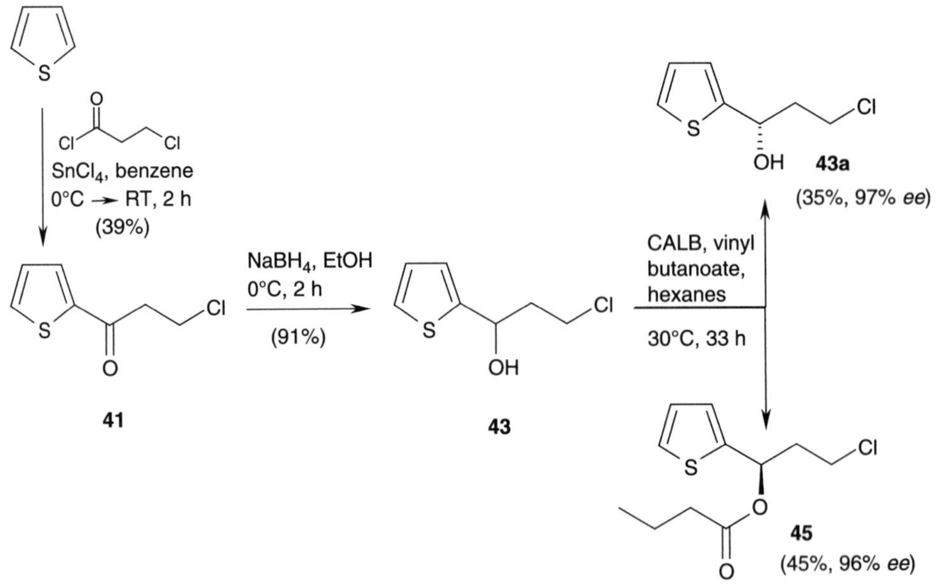

Scheme 14.13.

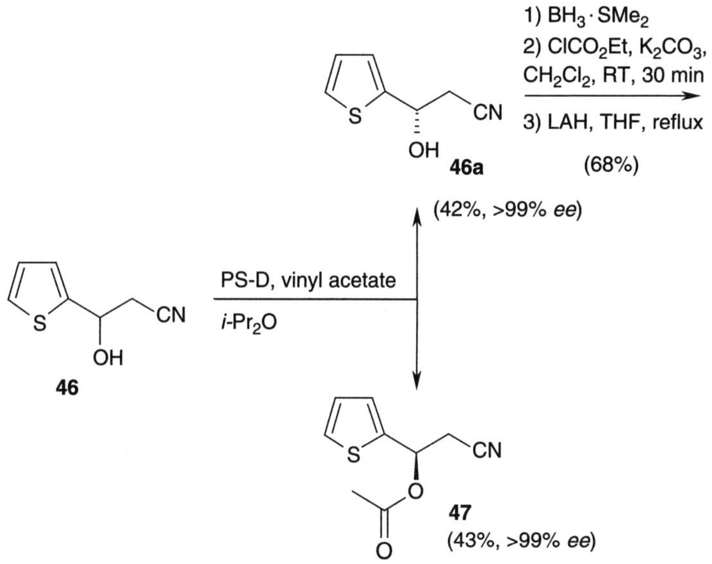

Scheme 14.14.

three-step sequence already described in Scheme 14.11 (iodination, amine displacement, and etherification) in 20% overall yield.

Similarly, lipase-catalyzed kinetic resolution has also been applied to intermediate nitrile alcohol **46** (Scheme 14.14). Best results were obtained by using immobilized *Pseudomonas cepacia* (PS-D) in diisopropyl ether, leading to excellent yield and enantiomeric excess of the desired (*S*)-alcohol **46a**, along with (*R*)-nitrile ester **47**. Reduction of **46a** with borane-dimethylsulfide complex, followed by conversion to the corresponding carbamate and subsequent lithium aluminum hydride reduction gave rise to the desired (*S*)-aminoalcohol intermediate **36**, a known precursor of duloxetine (**3**).

Enantioselective catalytic hydrogenation has also been applied to the preparation of optically active duloxetine intermediates (Scheme 14.15). In one such report, β-keto amine **31** was converted to amino alcohol **32a** in excellent enantiomeric excess without

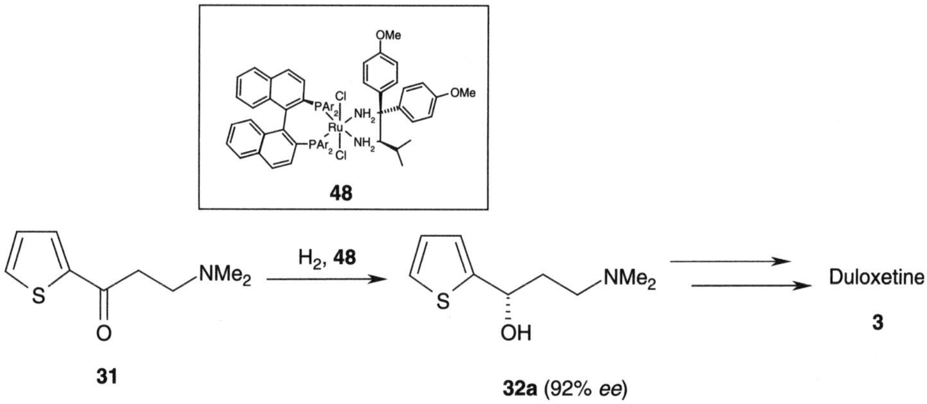

Scheme 14.15.

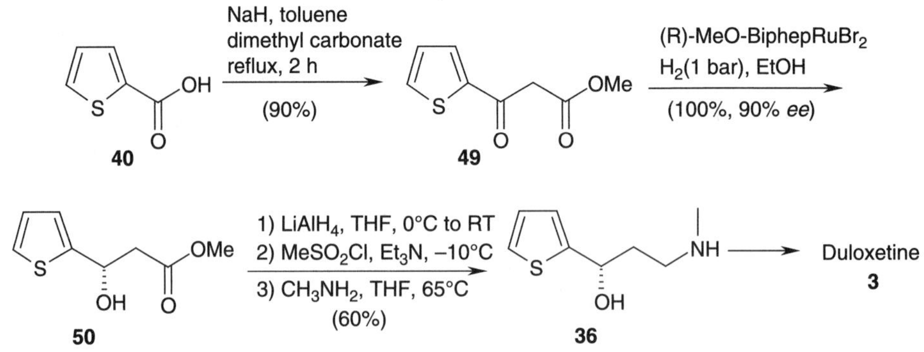

Scheme 14.16.

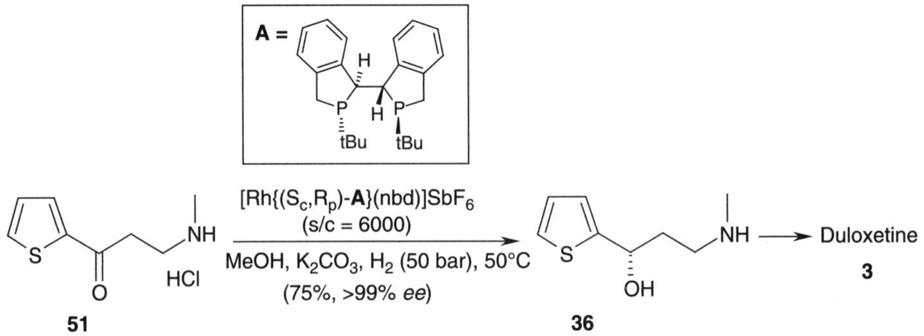

Scheme 14.17.

affecting the thiophene ring, catalyzed by chiral diphosphine/diamine ruthenium complex **48**. The alcohol product was then converted to duloxetine (**3**) utilizing literature procedures already described.

In a related report, ruthenium-catalyzed enantioselective hydrogenation of β-keto esters was utilized to prepare the crucial alcohol intermediate **36** (Scheme 14.16). The required β-keto ester **49** was readily prepared from commercial thiophene carboxylic acid **40**. Hydrogenation of **49** then led to the desired (*S*)-alcohol **50** in quantitative yield and 90% enantiomeric excess, catalyzed by a chiral diphosphine–ruthenium complex generated *in situ*. Catalyst–substrate ratios used were as low as 1/20,000, rendering this approach amenable to industrial application. Alcohol **50** was then converted to known intermediate **36** in three steps and 60% overall yield.

More recently a rhodium-catalyzed enantioselective synthesis of duloxetine (**3**) has been reported (Scheme 14.17). In this work, readily available amino ketone **51** was converted to (*S*)-aminoalcohol **36** in 75% yield and greater than 99% *ee*. The intermediate alcohol was subsequently converted into duloxetine (**3**) in a single step via standard etherification.

REFERENCES

Basappa, Kavitha, C. V., and Rangappa, K. S. (2004). *Bioorg. Med. Chem. Lett.*, 14: 3279–3281.

Berglund, R. (1994). US Patent 5362886.

Bonnaud, B., Mouzin, G., Cousse, H., and Patoiseau, J. P. (1986). EP 200638 A1.

Bonnaud, B., Cousse, H., Mouzin, G., Briley, M., Stenger, A., Fauran, F., and Couzinier, J. P. (1987). *J. Med. Chem.*, 30: 318–325.

Borghese, A. (2003). WO 062219.

Briley, M. (1998). "Milnacipran, A Well-Tolerated Specific Serotonin and Norepinephrine Reuptake Inhibiting Antidepressant," *CNS Drug Reviews*, 4: 137–148.

Bymaster, F., Beedle, E., Findlay, J., Gallagher, P., Krushinski, J., Mitchell, S., Robertson, D., Thompson, D., Wallace, L., and Wong, D. (2003). *Bioorg. Med. Chem. Lett.*, 13: 4477–4480.

Chavan, S. P., Khobragade, D. A., Kamat, S. K., Sivadasan, L., Balakrishnan, K., Ravindranathan, T., Gurjar, M., and Kalkote, U. R. (2004). *Tetrahedron Lett.*, 45: 7291–7295.

Cymbalta US Package Insert and European Summary of Product Characteristics.

Deeter, J., Frazier, J., Staten, G., Staszak, M., and Weigel, L. (1990). *Tetrahedron Lett.*, 31: 7101–7104.

Farvolden, P., Kennedy, S. H., and Lam, R. W. (2003). *Expert Opin. Investig. Drugs*, 12: 65–86.

Ferrier, I. N. (1999). *J. Clin. Psychiatry*, 60(Suppl 6): 10–14.

Gorman, J. M. and Kent, J. M. (1999). *J. Clin. Psychiatry*, 60(Suppl 4): 33–38.

Jain, R. (2004). *Prim Care Companion J. Clin. Psychiatry*, 6(Suppl 1): 7–11.

Kamal, A., Ramesh Khanna, G. B., Ramu, R., and Krishnaji, T. (2003). *Tetrahedron Lett.*, 44: 4783–4787.

Kirwin, J. L. and Goren J. L. (2005). *Pharmacotherapy*, 25: 396–410.

Koch, S., Hemrick-Luecke, S. K., and Thompson, L. K. (2003). *Neuropharmacology*, 45: 935–944.

Lantz, R. J., Gillepsie, T. A., Rash, T. J., Kuo, F., Skinner, M., Kuan, H-Y., and Knadler, M. P. (2003). "Metabolism, Excretion, and Pharmacokinetics of Duloxetine in Healthy Human Subjects," *Drug Metab. Dispos.*, 31: 1142–1150.

Liu, H., Hoff, B. H., and Anthonsen, T. (2000). *Chirality*, 12: 26–29.

Liu, D., Gao, W., Wang, C., and Zhang, X. (2005). *Angew. Chem. Int. Ed.*, 44: 1687–1689.

Mouzin, G., Cousse, H., and Bonnaud, B. (1978). *Synthesis*, 4: 304–305.

Noyori, R., Koizumi, M., Ishii, D., and Ohkuma, T. (2001). *Pure Appl. Chem.*, 73: 227–232.

Ratovelomanana-Vidal, V., Girard, C., Touati, R., Tranchier, J. P., Ben Hassine, B., and Genêt, J. P. (2003). *Adv. Synth. Catal.*, 345: 261–274.

Robertson, D., Wong, D., and Krushinski, J. (1991). US Patent 5023269.

Shuto, S., Takada, H., Mochizuki, D., Tsujita, R., Hase, Y., Ono, S., Shibuya, N., and Matsuda, A. (1995). *J. Med. Chem.*, 38: 2964–2968.

Stahl, S. M. (1998). *J. Clin. Psychiatry*, 59(Suppl 4): 5–14.

Stahl, S. M., Grady, M. M., Moret, C., and Briley, M. (2005). "SNRIs: Their Pharmacology, Clinical Efficacy, and Tolerability in Comparison with Other Classes of Antidepressants," *CNS Spectrums*, 10: 732–747.

Thomson Micromedex® Healthcare Series, DrugPoint® Summary for Venlafaxine Hydrochloride, and Duloxetine Hydrochloride.

Yardley, J. P., Morris Husbands, G. E., Stack, G., Butch, J., Bicksler, J., Moyer, J. A., Muth, E. A., Andree, T., Fletcher, H., James, M. N. G., and Sielecki, A. R. (1990). *J. Med. Chem.*, 33: 2899–2905.

Walter, M. M. (2005). "Monoamine Reuptake Inhibitors: Highlights of Recent Research Developments," *Drug Development Research*, 65: 97–118.

Westenberg, H. G. (1999). *J. Clin. Psychiatry*, 60(Suppl 17): 4–8.

Wheeler, W. and Kuo, F. (1995). *J. Label. Compd. Radiopharm.*, 36: 213–223.

Wong, D. T. and Bymaster, F. P. (2002). "Dual 5-HT and Noradrenaline Uptake Inhibitor Class of Antidepressants–Potential for Greater Efficacy of Just Hype?," *Progress in Drug Research*, 58: 170–222.

Zhang, M.-X. and Eaton, P. (2002). *Angew. Chem. Int. Ed.*, 41: 2169–2170.

15

GABA$_A$ RECEPTOR AGONISTS FOR INSOMNIA: ZOLPIDEM (AMBIEN®), ZALEPLON (SONATA®), ESZOPICLONE (ESTORRA®, LUNESTA®), AND INDIPLON

Peter R. Guzzo

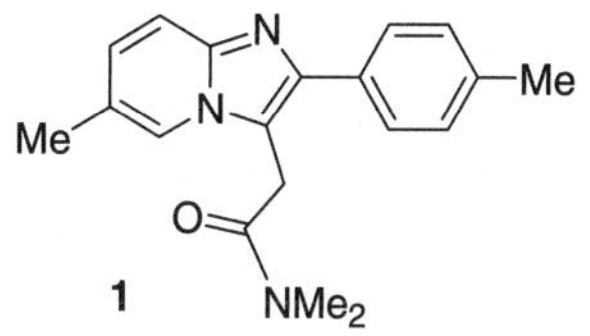

1

USAN: Zolpidem tartrate
Trade name: Ambien®
Sanofi-Aventis
Launched: 1999
M.W. 307.39

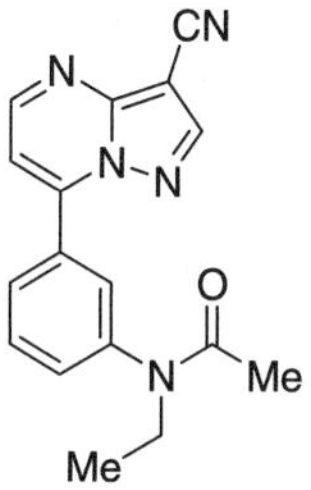

2

USAN: Zaleplon
Trade name: Sonata®
Wyeth
Launched: 1999
M.W. 305.33

The Art of Drug Synthesis. Edited by Douglas S. Johnson and Jie Jack Li

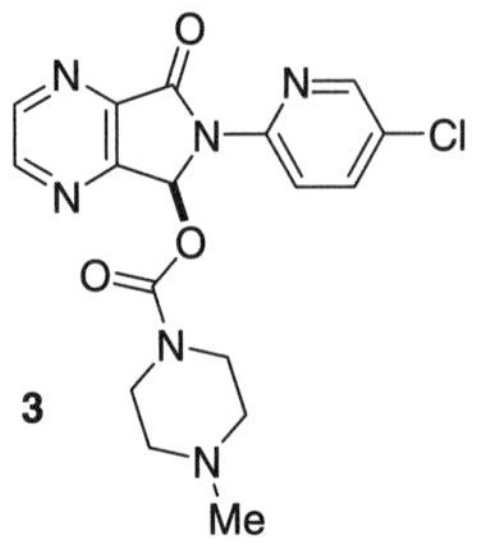

3

USAN: Eszopiclone
Trade name: Estorra®, Lunesta®
Sepracor
Launched: 2005
M.W. 388.81

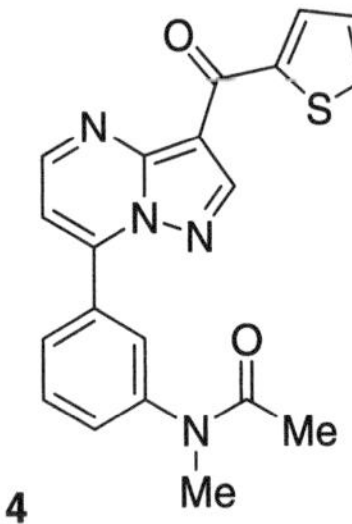

4

USAN: Indiplon
Neurocrine Biosciences
Pre-registration in 2006
M.W. 376.43

15.1 INTRODUCTION

Approximately one-third of human life is spent sleeping, and although some people may try to find ways to reduce this time, sleep is essential for our general health. Insomnia is the inability to obtain enough quality sleep. It is a common sleep disorder that is assigned to any of the following symptoms: difficulty falling asleep, waking often, waking earlier than desired, and waking feeling tired. The National Sleep Foundation has reported that approximately 35% of respondents experienced symptoms of insomnia, with a higher incidence seen in the older population, and approximately 15% experience chronic insomnia symptoms (National Sleep Foundation). Insomnia is classified as transient, short-term, or chronic. Transient or short-term insomnia is generally caused by stress or a disruption in normal circadian rhythms. Chronic insomnia is present for a month or longer and is further subdivided into (1) primary insomnia, where there is no evidence of medical or psychiatric disorders, or (2) secondary insomnia, where there is an underlying medical disorder with insomnia as a consequence (Monti, 2004; Monti and Monti, 2005; Richards, 2005). Insomnia is often a symptom of another emotional or medical disorder, but pharmacological treatment can be beneficial for all forms of insomnia. However, with secondary insomnia, the coexisting medical or psychiatric condition must be treated to address the primary medical condition, with the insomnia treatment given as an adjunct measure. Persistent insomnia can lead to reduced wake time performance and is a strong risk factor for other serious problems such as major depression, overall poor health, and societal problems. It is a source of large economic burden.

Treatments for insomnia include pharmacological and nonpharmacological techniques (e.g., stimulus control, relaxation methods). Pharmacological options have steadily progressed from traditional folk remedies to the chloral hydrate and barbiturates used from around 1910–1960, to benzodiazepines through the 1960s and 1970s, to the nonbenzodiazepine gamma amino butyric acid (GABA) A receptor agonists introduced since the mid-1980s to present; these last form the subject of this chapter. Barbiturates were initially popular; however, it was discovered that their use could lead to dependency and other serious side-effects. Benzodiazepines were hailed as safer alternatives to the barbiturates and became the dominant sedatives. However, the benzodiazepines are not only active as sedatives, but also possess muscle-relaxant, anticonvulsant, and anxiolytic properties. The benzodiazepines do exhibit several undesirable side-effects such as residual drowsiness, daytime tiredness, depression, effects on respiration, disruption of normal sleep architecture, memory impairment, as well as a tendency for tolerance and drug dependency upon prolonged use (Sanger, 2004). The nonbenzodiazepine $GABA_A$ agonists demonstrate comparable sleep promoting effects to the benzodiazepines but have a better safety profile and, to date, little evidence of tolerance or drug dependency. When the nonbenzodiazepine agents zaleplon, zolpidem, and zopiclone are compared with conventional benzodiazepines, fewer clinically important interactions with other drugs have been reported (Hesse et al., 2003). These observations may be explained by the difference in metabolism by CYP450 isoforms. For example, benzodiazepines triazolam and midazolam are metabolized nearly exclusively by CYP3A4; however, the nonbenzodiazepines are metabolized by several CYP450 isoforms in addition to CYP3A4.

The barbiturates, benzodiazepines, and nonbenzodiazepines all owe their sedative properties to acting as $GABA_A$ receptor agonists at modulatory or allosteric sites. Gamma amino butyric acid (GABA) is the major inhibitory neurotransmitter in the central nervous system, and three types of receptors, A–C, have been characterized. $GABA_A$ receptors, sometimes referred to as benzodiazepine receptors, are the most abundant of the inhibitory neurotransmitter receptors. They are pentameric membrane proteins and are ligand gated chloride ion channels that can be modulated by multiple binding sites. Currently, seven $GABA_A$ subunits with multiple isoforms and at least eight $GABA_A$ receptor subtypes comprised of various subunits are known (Möhler et al., 2004; Sieghart et al., 2005). More selective interaction with $GABA_A$ receptor subtypes has led to more selective pharmacological activity. The increasingly selective treatments for insomnia have mirrored the increased understanding of the $GABA_A$ receptor, its function, and the role of the diversity of receptor subtypes. For example, the benzodiazepine class of drugs is now known to bind nonselectively at several different $GABA_A$ subunits, including α_1, α_2, α_3, and α_5, increasing GABA release and dampening neuronal activity. The lack of selectivity of the benzodiazepines leads to multiple effects such as sedative, anxiolytic, muscle-relaxant, and amnesic properties. The nonbenzodiazepines, however, bind preferentially to α_1 subunits of the $GABA_A$ receptor complex, leading to more selective sedation effects. The nonbenzodiazepines generally compare favorably with the benzodiazepines by displaying reduced rebound insomnia, tolerance, and withdrawal symptoms. All the nonbenzodiazepines approved for the treatment of insomnia in the United States have a schedule IV drug classification. The pharmacokinetics and pharmacodynamics for three of the agents in this chapter, zaleplon, zolpidem, and zopiclone, have been recently reviewed (Drover, 2004).

15.2 SYNTHESIS OF ZOLPIDEM

Zolpidem (**1**) is an effective hypnotic agent indicated for the short-term treatment of insomnia. Zolpidem interacts with the $GABA_A$ receptor, and its pharmacological effect is blocked by the benzodiazepine-receptor antagonist flumazenil (Sanger and Depoortere, 1998). Zolpidem displaces benzodiazepines more selectively from the cerebellum than the hippocampus or spinal cord, consistent with preferential interaction with the $\alpha_1 GABA_A$ receptor subtype (sometimes referred to as the benzodiazepine ω_1 receptor). Studies with recombinant GABA$_A$ receptors also show preferential binding of zolpidem to receptors that contain the α_1 subunit.

In clinical trials, zolpidem shortened sleep latency, improved the quality of sleep, and accelerated the restoration of normal sleep patterns (Lee, 2004). In insomniac patients it increased the amount of slow wave, restorative sleep as seen in normal sleepers. Zolpidem has high oral bioavailability (70%), a short duration of action ($t_{1/2} = 2$ h), and is relatively highly bound to plasma proteins (92%). The recommended dose is generally 10 mg/day as needed. Zolpidem is extensively metabolized, mainly by CYP3A4 but also by CYP1A2 and CYP2C9, and its major metabolites do not appear to have pharmacological activity. It has minimal daytime residual effects, and a low risk for tolerance and abuse. The safety profile showed a low incidence of adverse events, close to that observed with placebo. The medicine is available in over 80 countries.

The synthesis of zolpidem began with an alkylation/condensation reaction of aminopyridine **5** and bromide **6** to give imidazopyridine **7** (Scheme 15.1). Mannich-type reaction with formaldehyde and dimethylamine provided **8**. Treatment of **8** with methyliodide to form the quaternary salt **9**, followed by reaction with sodium cyanide, gave **10**. Acidic hydrolysis followed by reaction of the resultant acid **11** with carbonyldiimidazole (CDI) and dimethylamine afforded zolpidem (**1**) in 46% overall yield (George et al., 1991; Rossey and Long, 1988).

A more efficient and convergent industrial-scale synthesis that avoids toxic methyl iodide and sodium cyanide was developed (Scheme 15.2). Condensation of *N,N*-dimethyl-2,2-dimethoxyacetamide with imidazopyridine **7** under acidic conditions afforded hydroxy derivative **12**. Conversion of the hydroxyl group to a chloride with thionyl chloride followed by reductive removal of the chloride with sodium borohydride provided zolpidem.

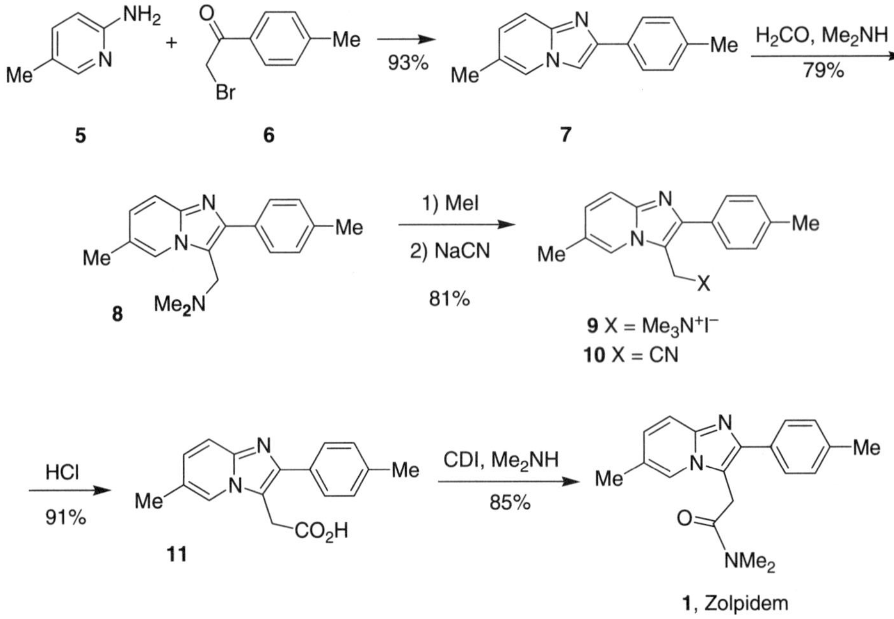

Scheme 15.1. Synthesis of zolpidem.

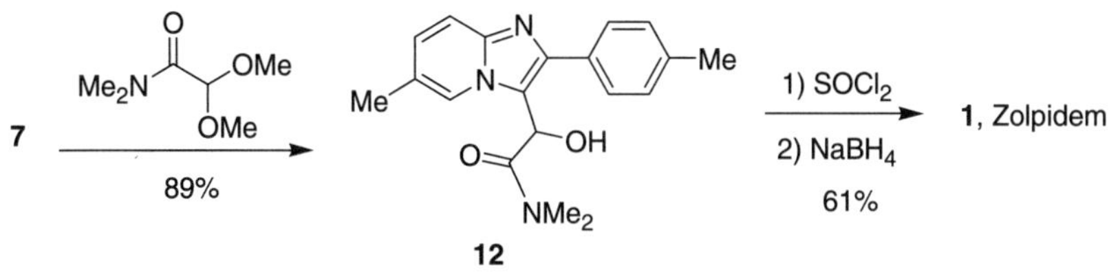

Scheme 15.2. Industrial-scale synthesis.

15.3 SYNTHESIS OF ZALEPLON

Zaleplon (**2**) selectively binds to the benzodiazepine ω_1 receptor subtype of the $GABA_A$ receptor complex (Barbera and Shapiro, 2005). It is indicated for the short-term treatment of insomnia and is classified as a schedule IV controlled substance. The drug has moderate oral bioavailability (30%), is rapidly absorbed, and has a short half-life of approximately 1 h. The recommended dose is generally between 5 and 10 mg. Consistent with its pharmacokinetic profile, zaleplon is effective in promoting sleep initiation, can be used in the middle of the night with little residual daytime effects, but is generally less effective with promoting sleep maintenance. The adverse effects associated with zaleplon are shorter in duration and less in magnitude when compared with benzodiazepines. Zaleplon is extensively metabolized to three major inactive metabolites. It is metabolized primarily by the aldehyde oxidase pathway and the cytochrome P450 CYP3A4 pathway. Product labeling suggests that an alternative hypnotic be used if the patient is taking a CYP3A4 inducer, such as rifampicin, or a CYP3A4 or aldehyde oxidase inhibitor, such as cimetidine, due to significant alterations in concentrations of zaleplon.

The synthesis of zaleplon (Scheme 15.3) starts by condensing acetophenone derivative **13** with dimethylformamide dimethylacetal (DMF-DMA) to provide enamine **14** (Mealy et al., 1996; Dusza et al., 1987). A subsequent alkylation with ethyl iodide in the presence of sodium hydride gave **15**. Zaleplon (**2**) is assembled by condensing aminopyrazole **16** with enamine **15** in refluxing acetic acid.

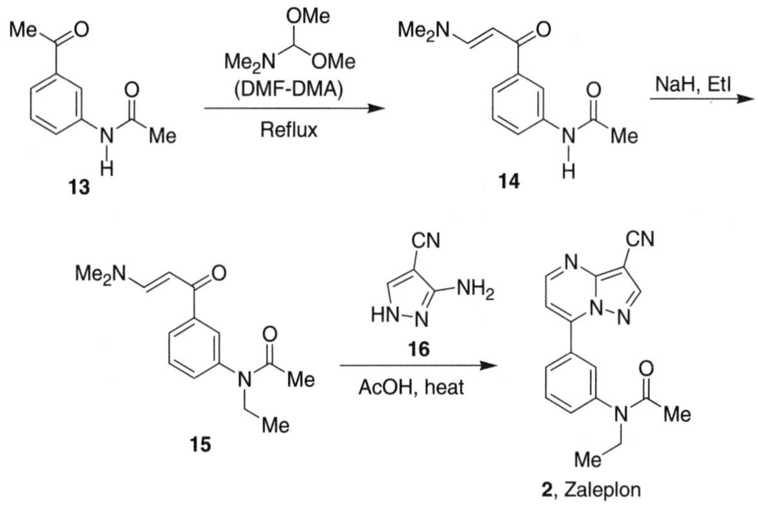

Scheme 15.3. Synthesis of zaleplon.

15.4 SYNTHESIS OF ESZOPICLONE

Racemic zopiclone was originally discovered and marketed by Rhone-Poulenc Rorer, which became Aventis after a merger (now part of Sanofi-Aventis after yet another recent merger). Eszopiclone (**3**), the (*S*)-enantiomer of zopiclone, was launched by Sepracor in 2005 as Lunesta® (Cotrel and Roussch, 2001, 2002). Eszopiclone binds at the GABA$_A$ receptor with approximately 50-fold higher binding affinity than the (*R*)-enantiomer and is the important enantiomer for the hypnotic effects (Culy et al., 2003). Currently, there is less information on the pharmacokinetics of eszopiclone compared with what is known for zopiclone. Racemic zopiclone is rapidly absorbed, has an oral bioavailability of approximately 80%, and a half-life of about 5 h. Zopiclone is extensively metabolized predominantly by CYP3A4 but also by CYP2C8 and CYP2B6 to two major metabolites: the *N*-oxide and *N*-desmethyl compounds. The desmethyl metabolite of eszopiclone does not show sedative effects but has displayed anxiolytic effects in animal models; the anxiolytic effects are likely not due to interaction at the GABA$_A$ receptor. Following the oral administration of racemic zopiclone, the enantiomers display different pharmacokinetics. The (*S*)-enantiomer exhibits significantly higher maximal concentrations (C_{max}), exposure levels (AUC_∞), longer half-life, and reduced clearance rates. The half-life ($t_{1/2}$) of eszopiclone was just over 5 h in healthy humans. A 2.5-mg dose of eszopiclone was found to be equivalent to a 5-mg dose of zopiclone, with few significant differences observed for any pharmacokinetic parameter.

Eszopiclone has been approved for the treatment of patients who experience difficulty falling asleep, poor sleep maintenance, and for long-term treatment of insomnia. Clinical trials have shown that eszopiclone improved sleep onset, sleep maintenance, total sleep time, sleep quality, and daytime functioning compared with placebo. Improved wake time alertness, concentration, and sense of well-being were reported. Eszopiclone was well tolerated, with only mild adverse events reported. There was no evidence of drug–drug interactions, tolerance, residual drowsiness or treatment-related rebound insomnia. The recommended dose to improve sleep onset and maintenance is generally between 1 and 3 mg.

The racemic synthesis of eszopiclone is shown in Scheme 15.4 (Cotrel et al., 1975; Jeanmart et al., 1978). The treatment of pyrazine anhydride (**17**) with 2-amino-5-chloropyridine (**18**) gave amide **19** in good yield. Treatment of amide **19** with refluxing thionyl chloride produced imide **20**. Mono-reduction of the imide with KBH_4 afforded alcohol **21**. Acylation of alcohol **21** with **22** using NaH in DMF produced racemic zopiclone (**23**). Several methods have been described for the preparation of the (*S*)-zopiclone (Culy et al., 2003). Chiral resolution of **23** by diastereomeric salt formation and recrystallization has been described with both malic acid (Blaschke et al., 1993) and *O,O′*-dibenzoyltartaric acid (Cotrel et al., 1992). Another conceptually different approach involving an enzymatic resolution of carbonate intermediates has also been described (Solares et al., 2002; Palomo et al., 2003). An immobilized form of lipase B from *Candida antarctica* (Chirazyme-L2) catalyzed the hydrolysis of carbonate **24** to give optically active carbonate **25** with the correct absolute (*S*)-stereochemistry required for the synthesis of eszopiclone and racemic **21**. The enzyme preparation can be recycled ten times without any loss to the activity of the catalyst or enantioselectivity of the reaction. Also, racemic **21** can be recycled to improve the efficiency of the asymmetric synthesis.

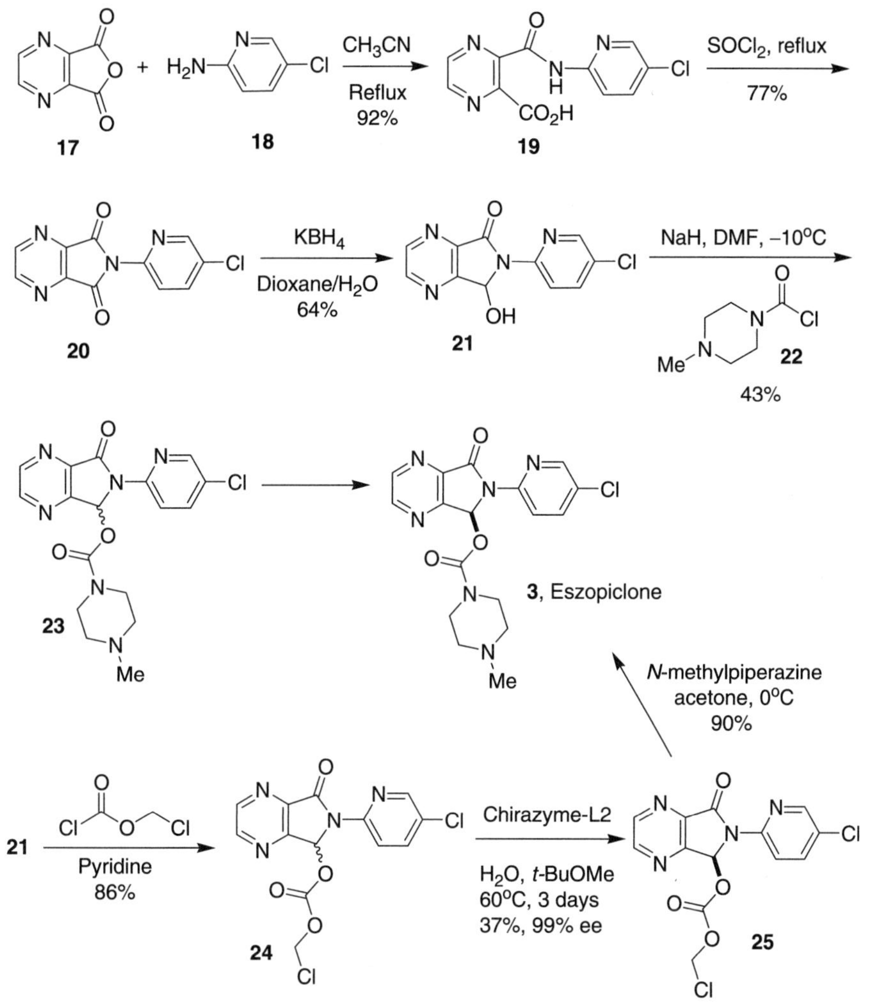

Scheme 15.4. Synthesis of eszopiclone.

15.5 SYNTHESIS OF INDIPLON

Indiplon (**4**) is a nonbenzodiazepine sedative, which was in preregistration for the treatment of insomnia in 2006. Indiplon is a potent $GABA_A$ receptor agonist with selectivity for the α_1 subunit. The compound originated at American Cyanamid and was subsequently developed by Neurocrine Biosciences.

Indiplon is rapidly absorbed (1–2 h) and eliminated ($t_{1/2} = 1.3$ h). In-vitro studies on indiplon show two major metabolites: *N*-demethylation due to CYP3A4/5 and *N*-deacetylation by carboxylesterases. Its short half-life has enabled the development of two dosing paradigms with different formulations: (1) an immediate release formulation to improve sleep initiation and for dosing in the middle of the night, and (2) a modified release formulation, which provides for immediate release and sustained release of the drug to help with sleep initiation, duration, maintenance, and sleep quality (Neubauer,

2005; Sorbera et al., 2003). In clinical trials with both formulations, indiplon has demonstrated efficacy in sleep onset and duration compared with placebo. There was no evidence of next-day sedation effects, except for a modest effect in elderly subjects that was observed with the modified release version at the highest study dose. Indiplon immediate-release formulation could be taken in the middle of the night to help patients return to sleep without residual morning effects. Indiplon given at two doses, 10 mg and 20 mg, did not show evidence of changes to respiratory function, withdrawal symptoms, or rebound insomnia in patients.

Several syntheses of indiplon have been described and two routes are shown in Scheme 15.5 (Sorbera et al., 2003; Dusza et al., 2002). Treatment of acetophenone **26** with refluxing dimethylformamide dimethylacetal (DMF-DMA) provided enamide **27**. Alkylation of the amide with methyl iodide using NaH in DMF afforded **28**. β-Ketonitrile **29** was treated with DMF-DMA to give enamide **30**. Cyclization with aminoguanidine produced aminopyrazole **31**. The condensation of enamide **28** with aminopyrazole **31** in acetic acid furnished indiplon (**4**). Alternatively, enamine **28**

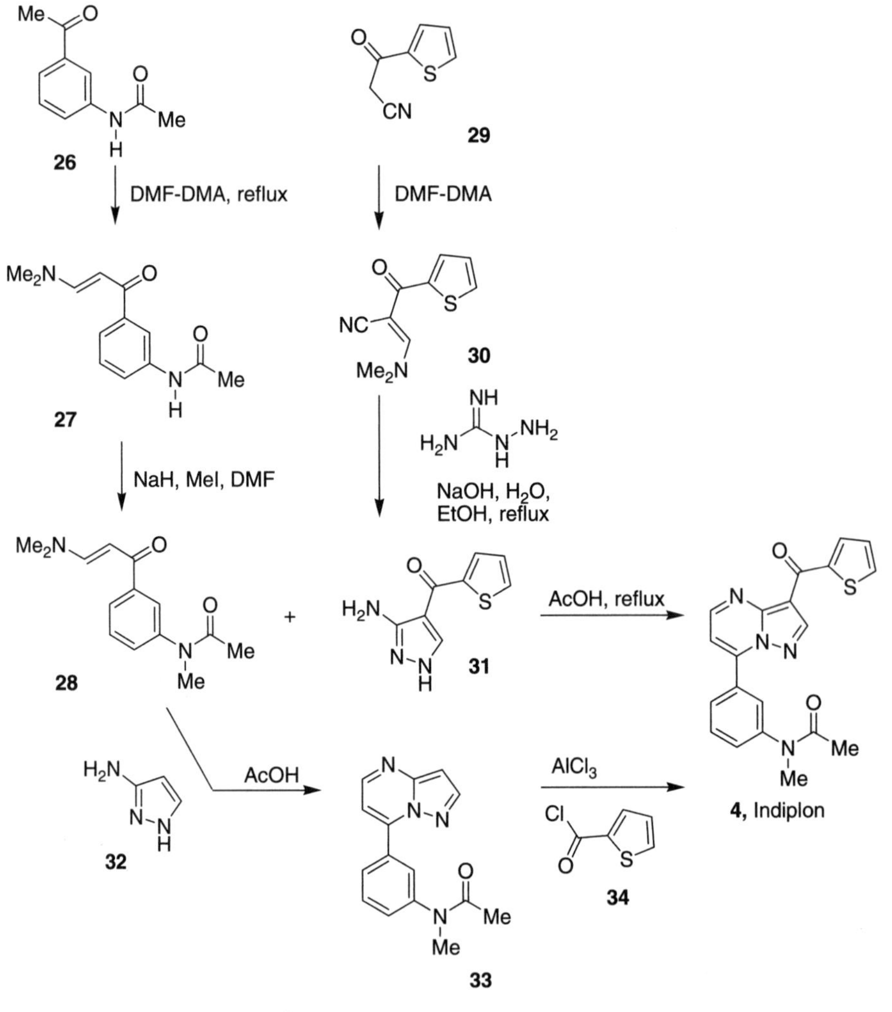

Scheme 15.5. Syntheses of indiplon.

could be condensed with a simpler aminopyrazole **32** to afford the pyrazolopyrimidine **33**. Friedel–Crafts acylation of **33** with acid chloride **34** furnished indiplon (**4**).

REFERENCES

Barbera, J. and Shapiro, C. (2005). *Drug Safety*, 28: 301–318.

Blaschke, G., Hempel, G., and Müller, W. E. (1993). *Chirality*, 5: 419–421.

Cotrel, C., Jeanmart, C., and Messer, M. N. (1975). US 3,862,149. (to Rhone-Poulenc S.A.).

Cotrel, C. and Roussel, G. (1992). EP 495,717 (to Rhone-Poulenc Rorer SA).

Cotrel, C. and Roussel, G. (2001). US 6,319,926 (to Sepracor Inc.).

Cotrel, C. and Roussel, G. (2002). US 6,444,673 (to Sepracor Inc.).

Culy, C., Castañer, J., and Bayés, M. (2003). *Drugs Fut.*, 28: 640–646.

Drover, D. R. (2004). *Clin. Pharmacokinet.*, 43: 227–238.

Dusza, J. P., Tomcufcik, A. S., and Albright, J. D. (1987). US 4,626,538 (to American Cyanamide, Co.).

Dusza, J. P., Tomcufcik, A. S., Albright, J. D., and Beer, B. (2002). US 6,399,621 (assigned to American Cyanamid Company).

George, P., Rossey, G., Depoortere, H., Mompon, B., Allen, J., and Wick, A. (1991). *Il Farmaco*, 46: 277–288.

Hesse, L. M., von Moltke, L. I., and Greenblatt, D. J. (2003). *CNS Drugs*, 17: 513–532.

Jeanmart, C., Cotrel, C., and Janot, M. M.-M. (1978). *Chimie Organique*, 377–378.

Lee, Y.-J. (2004). *CNS Drugs*, 18: 17–23.

Mealy, N. and Castañer, J. (1996). *Drugs Fut.*, 21: 37–39.

Möhler, H. and Rudolph, U. (2004). *Drug Disc. Today: Ther. Strat.*, 1: 117–123.

Monti, J. M. (2004). *Curr. Med. Chem.: CNS Agents*, 4: 119–137.

Monti, J. M. and Monti, D. (2005). *Expert Opin. Ther. Patents*, 15: 1353–1359.

National Sleep Foundation, Washington DC, USA; see www.sleepfoundation.org

Neubauer, D. N. (2005). *Expert Opin. Invest. Drugs*, 14: 1269–1276.

Palomo, J. M., Mateo, C., Fernández-Lorente, G., Solares, L. F., Diaz, M., Sánchez, V. M., Bayod, M., Gotor, V., Guisan, J. M., and Fernandez-Lafuente, R. (2003). *Tetrahedron: Asymmetry*, 14: 429–438.

Richards, C. (2005). *Drug Discovery Today*, 10: 611–614.

Rossey, G. and Long, D. (1988). US 4,794,185 (to Synthelabo).

Sanger, D. J. (2004). *CNS Drugs*, 18: 9–15.

Sanger, D. J. and Depoortere, H. (1998). *CNS Drug Rev.*, 4: 323–340.

Sieghart, W. and Ernst, M. (2005). *Curr. Med. Chem.: CNS Agents*, 5: 217–242.

Solares, L. F., Diaz, M., Brieva, R., Sánchez, V. M., Bayod, M., and Gotor, V. (2002). *Tetrahedron: Asymmetry*, 13: 2577–2582.

Sorbera, L. A., Castañer, J., and Martin, L. (2003). *Drugs Fut.*, 28: 739–746.

16

$\alpha_2\delta$ LIGANDS: NEURONTIN® (GABAPENTIN) AND LYRICA® (PREGABALIN)

Po-Wai Yuen

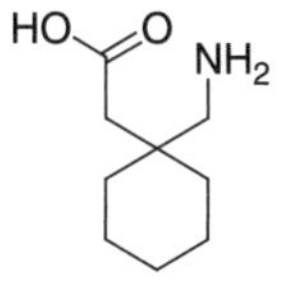

1

USAN: Gabapentin
Trade name: Neurontin®
Pfizer Inc
Launched: 1993
M.W. 171.24

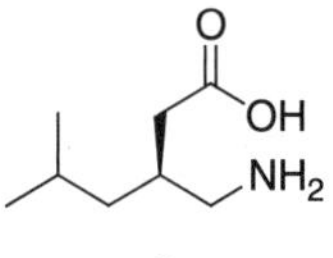

2

USAN: Pregabalin
Trade name: Lyrica®
Pfizer Inc
Launched: 2004
M.W. 159.23

16.1 INTRODUCTION

Drug discovery is a process full of unexpected twists and turns. The discoveries of gabapentin (**1**) and pregabalin (**2**) are unquestionably great examples that illustrate the high degree of serendipity around this process. Satzinger and co-workers (Satzinger et al., 1976) initiated the GABA project in 1973 to look for an epilepsy treatment. Their research was based on γ-aminobutyric acid (GABA), an inhibitory neurotransmitter that seems to have a role in controlling convulsion states in rats. When GABA levels in the brain drop

The Art of Drug Synthesis. Edited by Douglas S. Johnson and Jie Jack Li

below a threshold level, seizures appear (Karlsson et al., 1974), and when its levels increase during a convulsion the seizures stop (Hayashi, 1958). One would think the most straightforward approach for treating seizures would be to administer GABA to a convulsing subject. Unfortunately, GABA does not enter the central nervous system (CNS) with systemic administration. To solve this problem, Satzinger's group decided to prepare lipophilic GABA analogs with the hope that these compounds could access the CNS via passive diffusion and mimic the pharmacology of GABA. In 1975, Satzinger and Hartenstien reported the preparation of gabapentin (**1**) (Satzinger et al., 1976) and showed its activity in several animal models of epilepsy. Although gabapentin and its analogs have much improved lipophilicity compared to GABA, it was discovered much later that they did not cross the blood–brain barrier by passive diffusion as Satzinger had originally envisioned. It was determined that the *L*-type amino acid transporter provided the uptake mechanism to deliver these compounds into the CNS (Su et al., 1995). In 1993, gabapentin received United States approval as an add-on treatment for seizures under the trade name of Neurontin®. Later, it was discovered that gabapentin also showed activity in animal models of neuropathic pain. Eventually, in 2002, Neurontin® received U.S. approval for the treatment of post-herpetic neuralgia. With Neurontin® on the market, several other open label studies were conducted that suggested it might also have utility against anxiety and panic disorders; however, in both of these instances, high doses of gabapentin were required.

The story of the discovery of pregabalin (**2**) began in 1991, when Silverman and Taylor published a paper on the anticonvulsant effect of 3-alkyl GABA analogs (Silverman et al., 1991). Out of this collection of compounds, 3-isobutyl GABA stood out as the most active analog in the series, which protected mice from seizures induced by corneal electroshock.

Yuen et al. synthesized the two enantiomers of 3-isobutyl GABA enantioselectively in late 1991 (Yuen et al., 1994). The two compounds were then tested in the [^{3}H]-gabapentin binding assay. It was found that the (*S*)-enantiomer (pregabalin (**2**)) was 17 times more potent than the (*R*)-enantiomer in this assay, with an IC_{50} of 37 nM, which was twice as potent as gabapentin (**1**). In the maximal electroshock model, pregabalin was four-fold more potent than gabapentin, with an oral ED_{50} of 20 mg/kg. The (*R*)-enantiomer was not active in this model, even at a high dose of 300 mg/kg (Taylor et al., 1993).

The GABA pharmacophore is embedded in the structures of gabapentin and pregabalin; therefore, it is inviting to speculate that these molecules might interact with gabaergic targets in the CNS. However, after nearly 30 years of careful study, a consensus has emerged that these compounds are unlike other gabaergic agents (Fig. 16.1). Unlike baclofen (**3**), gabapentin (**1**) and pregabalin (**2**) do not directly interact with GABA receptors.

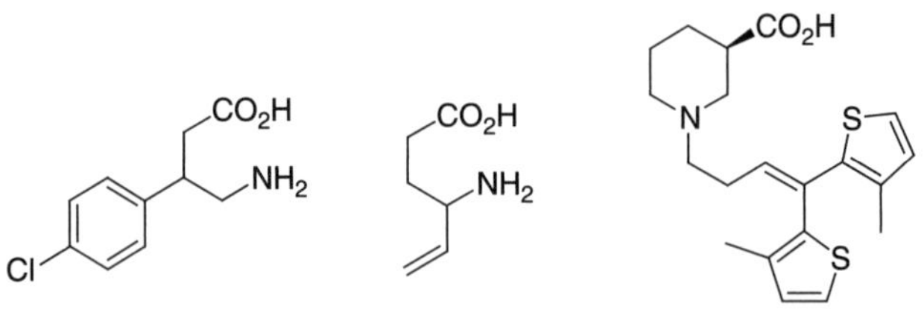

Figure 16.1. Marketed GABAergic agents.

They are different from vigabatrin (**4**) in that they are not inhibitors of GABA transaminase, and, in contrast to tiagabine (**5**), these molecules do not directly inhibit GABA reuptake (Bryans and Wustrow, 1999; Taylor, 1994).

In the early 1990s, researchers at the former Parke–Davis Neuroscience Research Unit in Cambridge, England, began conducting binding studies of [^{3}H]-gabapentin to brain tissues. They found that gabapentin and pregabalin bind heavily to synapse-rich areas of the brain, especially in the cortex. Brown and Gee carried out protein fractionation studies of pig brains with [^{3}H]-gabapentin and found a purified protein that had potent affinity for both gabapentin and pregabalin. This protein was sequenced and it was discovered to be the $\alpha_2\delta$ subunit of voltage-gated calcium channels (Gee et al., 1996). Dooley further demonstrated that pregabalin interacts with the calcium channels by showing that it inhibits norepinephrine (NE) release from rat cortical slices in a dose-dependent manner (Dooley et al., 2000).

Based on the biochemical, neurochemical, and behavioral pharmacology observations, a working hypothesis was developed: gabapentin- and pregabalin-like compounds with *L*-amino acid transporter activity cross the blood–brain barrier via an active transport (Su et al., 1995) and, once inside the CNS, these molecules interact with the $\alpha_2\delta$ subunit of voltage-gated calcium channels, thereby modulating calcium flux. This leads to decreased neurotransmitter release, which can be measured as a behavioral endpoint (Brown et al., 1996).

Although gabapentin (**1**) and pregabalin (**2**) are quite similar structurally, there is an interesting difference in the oral absorption profile of these compounds, which provides a clear advantage in favor of pregabalin. Increasing oral doses of gabapentin causes nonlinear increases in plasma levels such that, at higher doses, a two-fold increase in the dose increases steady-state plasma concentrations by approximately one-third. This dose nonlinearity may be due to saturation of intestinal transport by gabapentin at high doses. For pregabalin, there is a much more linear dose versus blood level relationship. Therefore, pregabalin can achieve therapeutically efficacious blood levels at lower doses than those required with gabapentin.

There are several unique physical properties of pregabalin that distinguish it from most CNS drugs. It has excellent aqueous solubility, which leads to rapid dissolution. With excellent oral bioavailability, micromolar plasma drug levels can be achieved easily. The parent drug compound is solely responsible for its clinical efficacy and it does not interact with the cytochrome P450 enzymes. Pregabalin is also cleared exclusively by renal elimination without going through metabolism (Radulovic et al., 1996).

Pregabalin has gone through one of the most extensive clinical development programs within the Pfizer organization. The list of successfully completed double-blind clinical studies includes partial seizures (add-on treatment), diabetic neuropathy, post-herpetic neuralgia, generalized anxiety disorder, social anxiety disorder, fibromyalgia pain (and sleep), and tooth extraction pain. In 2005, pregabalin received U.S. approval for the treatment of diabetic neuropathy, post-herpetic neuralgia, and as an add-on treatment for partial seizures under the trade name of Lyrica®.

16.2 SYNTHESIS OF GABAPENTIN

There are several methods documented in the literature for the synthesis of gabapentin (**1**). Satzinger, Hartenstein, Herrmann, and Heldt disclosed the original synthesis in a 1976 Goedecke A.-G. patent (Satzinger et al., 1976). The key intermediate 1,1-cyclohexanediacetic acid (**7**) was prepared according to a process developed by van Wessem and Sakal (van Wessem and Sakal, 1960). Cyclohexanone was treated with ethyl cyanoacetate in the presence of ammonia in ethanol to give the Guareschi salt **6**, which was then hydrolyzed with concomitant decarboxylation to give 1,1-cyclohexanediacetic acid (**7**). The diacid **7** was dehydrated with acetic anhydride to provide anhydride **8**, which was further treated with methanol to give monomethyl ester **9**. Acid **9** was subjected to a Curtius-type rearrangement to afford isocyanate **10**. Both the isocyanate and the methyl ester functional groups of compound **10** were hydrolyzed with aqueous hydrochloric acid to give gabapentin hydrochloride (**11**). This hydrochloride salt was treated with anion exchange chromatography to give pure gabapentin (**1**) after crystallization from a mixed solvent of ethanol and ether (Scheme 16.1).

Hartenstein and Satzinger reported another synthesis of gabapentin in 1977 (Hartenstein and Satzinger, 1977). In this alternative synthesis, these researchers treated the anhydride intermediate **8** in their original synthesis with hydroxylamine to give the *N*-hydroxyimide **12**. It was then treated with benzenesulfonyl chloride to generate the *N*-benzenesulfonyloxyimide **13**. Intermediate **13** underwent a Lossen rearrangement to give the carbamate-ester **14**. Subsequent acid hydrolysis of compound **14** produced gabapentin hydrochloride (**11**), which was again converted to gabapentin (**1**) via anion exchange chromatography (Scheme 16.2).

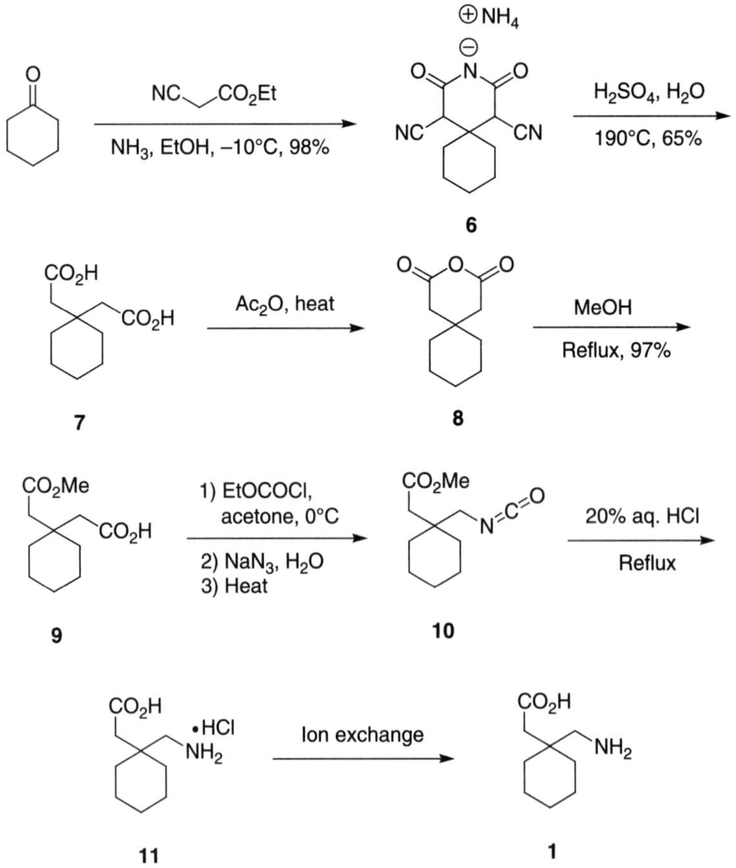

NC‿CO_2Et; NH_3, EtOH, −10°C, 98% → **6** ($\oplus NH_4$, $\ominus$N, NC, CN) → H_2SO_4, H_2O; 190°C, 65% →

7 (CO_2H, CO_2H) → Ac_2O, heat → **8** → MeOH; Reflux, 97% →

9 (CO_2Me, CO_2H) → 1) EtOCOCl, acetone, 0°C; 2) NaN_3, H_2O; 3) Heat → **10** (CO_2Me, N=C=O) → 20% aq. HCl; Reflux →

11 (CO_2H, •HCl NH_2) → Ion exchange → **1** (CO_2H, NH_2)

Scheme 16.1. The original Goedecke synthesis of gabapentin (**1**).

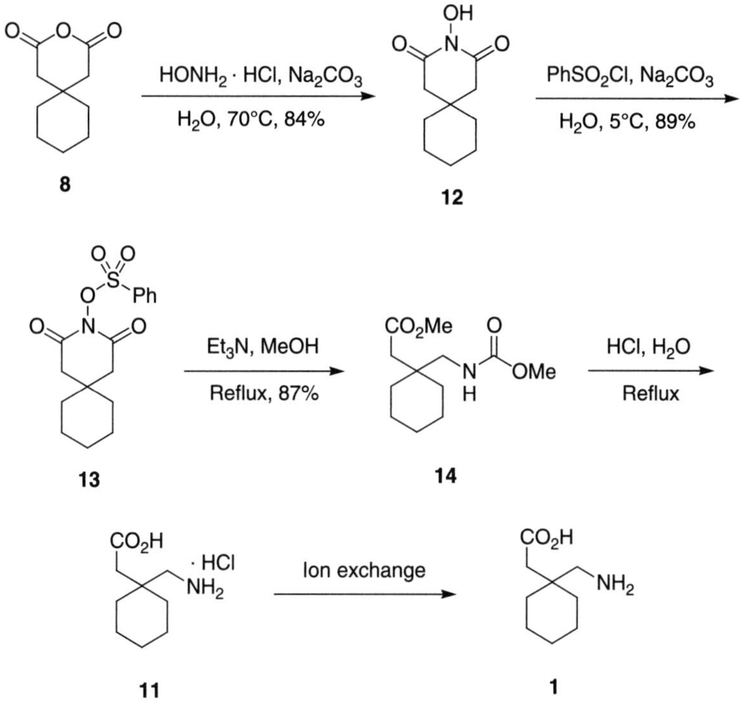

Scheme 16.2. The Hartenstein synthesis of gabapentin (**1**).

Mettler and co-workers reported several synthetic routes to prepare gabapentin in 1991 (Griffiths et al. 1991). Their first synthesis of gabapentin hydrochloride started with a Knoevenagel condensation between cyclohexanone and diethyl malonate to form the cyclohexylidene-malonate **15**. This intermediate was treated with HCN at pH 11 to give the Michael adduct **16**. Malonate **16** was then hydrogenated over a nickel catalyst to produce spirolactam **17**, which was hydrolyzed with aqueous hydrochloric acid with concomitant decarboxylation to give gabapentin hydrochloride (**11**) in good yield (Scheme 16.3).

Mettler et al. found that their original procedure was not very convenient for large-scale production of the malonate intermediate **16**. Safety precautions required to handle excess solid potassium cyanide were both difficult and expensive. To compound the problem even further, a large amount of TiO_2-pyridine complex was generated in the first step of this process. This solid waste material required special purification treatment before its final disposal.

Mettler and colleagues reported an alternative synthesis of malonate **16** in the same paper (Griffiths et al., 1991) in which they condensed cyclohexanone with ethyl cyanoacetate instead of diethyl malonate in the Knoevenagel reaction to give ethyl cyano(cyclohexylidene)-acetate (**18**). In the presence of a catalytic amount of sodium cyanide, the Michael addition of HCN to cyanoacetate **18** proceeded in good yield at room temperature to generate the dicyanoester **19**. Intermediate **19** was selectively converted to malonate **16** with pressurized HCl treatment in ethanol (Scheme 16.4).

Mettler and co-workers also developed another procedure to produce gabapentin (**1**) as the free amino acid (Griffiths et al., 1991). Cyanoacetate intermediate **18** was treated with

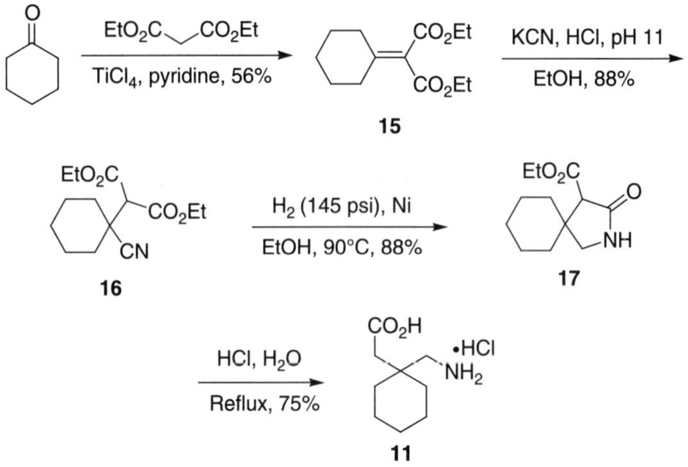

Scheme 16.3. Mettler's initial synthesis of gabapentin hydrochloride (**11**).

sodium cyanide in a refluxing ethanol–water mixture to give (1-cyanocyclohexyl)acetonitrile (**20**). Nitrile **20** underwent a Pinner reaction with benzyl alcohol saturated with hydrogen chloride gas to give the benzyl ester **21**. Ester **21** was hydrogenated over rhodium on carbon catalyst to give gabapentin (**1**) in 27% yield. However, this synthesis is not very attractive considering the cost of benzyl alcohol and the low yield of the final hydrogenation step to produce gabapentin (**1**) (Scheme 16.5).

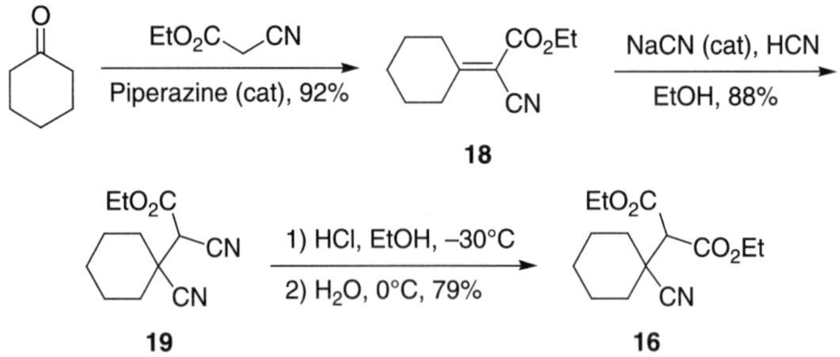

Scheme 16.4. Improved synthesis of malonate intermediate **16**.

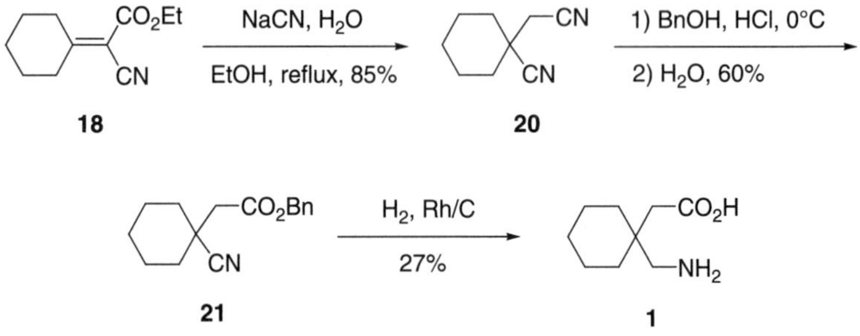

Scheme 16.5. Mettler's synthesis of gabapentin (**1**).

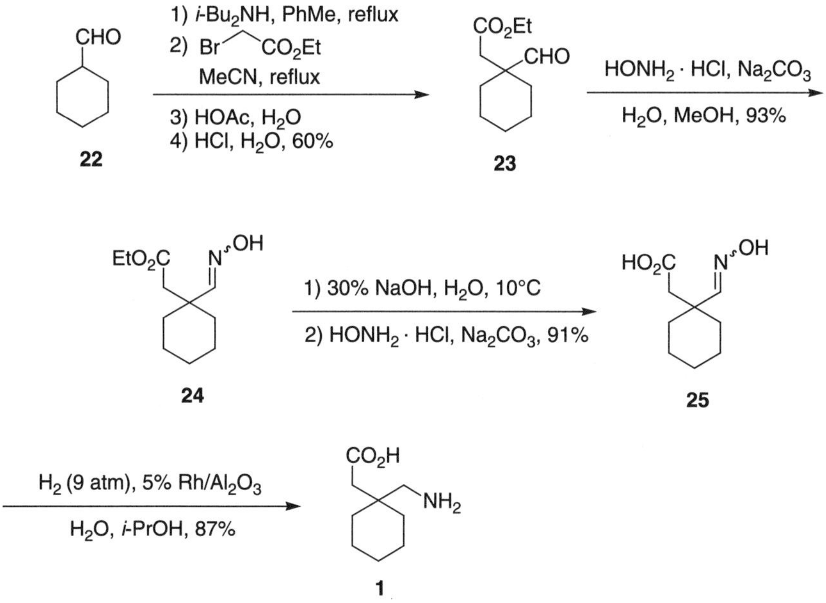

Scheme 16.6. Rossi's synthesis of gabapentin (**1**).

Rossi and Vecchio disclosed a synthesis of gabapentin (**1**) in their 2002 patent application (Rossi and Vecchio, 2002). As shown in Scheme 16.6, cyclohexanecarboxaldehyde (**22**) was converted into the corresponding enamine and then alkylated with ethyl bromoacetate followed by hydrolysis of the iminium salt to give aldehyde **23**. This aldehyde was treated with hydroxylamine to form oxime **24**. During hydrolysis of ester **24** with sodium hydroxide to form acid **25**, approximately 25% of the oxime was also hydrolyzed back to the aldehyde. Therefore, further treatment of the mixture with hydroxylamine at pH 5 was necessary to regenerate oxime **25**. Gabapentin (**1**) was then obtained in good yield by catalytic hydrogenation of oxime **25** in the presence of 5% rhodium on aluminum oxide catalyst.

In a 2002 patent application, Cannata, Micoli, and Corcella, of Zambon Group S.P.A., described another process to synthesize gabapentin (**1**) (Cannata et al., 2002). The key intermediate for their synthesis was 1,1-cyclohexanediacetic acid monoamide (**26**). Amide **26** was subjected to the Hofmann rearrangement conditions with sodium hydroxide and sodium hypochlorite to give gabapentin hydrochloride (**11**), which was then converted to gabapentin (**1**) by anion exchange chromatography (Scheme 16.7).

In 2003, Velardi and Fornaroli, of Procos S.P.A., reported their work on the synthesis of gabapentin (**1**) (Velardi and Fornaroli, 2003). Their approach is outlined in Scheme 16.8.

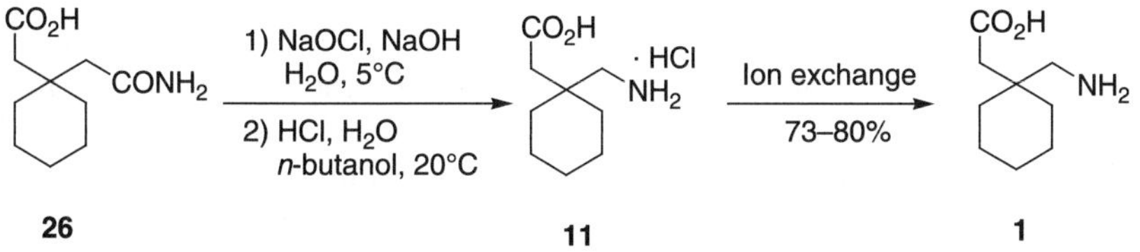

Scheme 16.7. The Zambon synthesis of gabapentin (**1**).

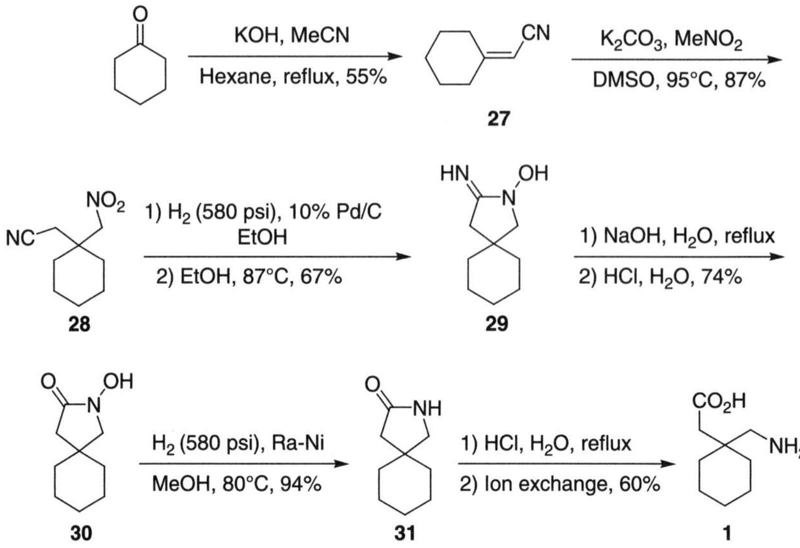

Scheme 16.8. The Procos synthesis of gabapentin (**1**).

Cyclohexanone was first condensed with acetonitrile to form cyclohexylidineacetonitrile (**27**), which was treated with nitromethane in the presence of potassium carbonate to give the Michael adduct **28**. The nitro compound **28** was then hydrogenated with 10% palladium on carbon with subsequent ring closure to produce hydroxypyrrolidinimine **29**. This intermediate was then converted into the hydroxypyrrolidinone **30** by base hydrolysis. Hydrogenation of hydroxypyrrolidinone **30** using Raney nickel as catalyst gave pyrrolidinone **31**. Under hydrochloric acid refluxing conditions, pyrrolidinone **31** was converted to gabapentin hydrochloride, which yielded gabapentin (**1**) after application of anion exchange chromatography.

The gabapentin lactam (**31**) is an important intermediate for the preparation of gabapentin (**1**). In a small-scale synthesis, Parsons and co-workers (Bryans et al., 2003; Cagnoli et al., 2003) devised a new approach to prepare the lactam **31** in 2003. The key step of this synthesis involved a radical cyclization reaction to form spirocyclic lactam **34** as a single diastereomer. 1-Cyclohexene-1-carboxaldehyde was condensed with benzoylhydrazine to form hydrazone **32**. The hydrazone was then reduced with dimethylamine-borane/methanesulfonic acid and *N*-acylated with trichloroacetyl chloride to give hydrazide **33**. Trichloromethylamide **33** was then subjected to the 5-*exo*-trig halogen atom transfer radical cyclization mediated by copper(I) chloride and TMEDA to form trichlorospirolactam **34** in good yield. Hydro-de-chlorination and hydrazide cleavage took place together smoothly when the trichloro-spirolactam **34** was treated with excess wet Raney nickel to give gabapentin lactam **31** (Scheme 16.9).

In a laboratory-scale synthesis, Li and co-workers (Li et al., 2004) of the Zhejiang University of Technology, utilized a Meldrum's acid derivative **35** as the starting point of their synthesis. Michael addition of nitromethane to the cyclohexylidene-Meldrum's acid **35** gave the nitro compound **36**. In a one-pot procedure, intermediate **36** was converted to gabapentin hydrochloride (**11**) by catalytic hydrogenation of the nitro group using platinum as the catalyst followed by acid hydrolysis and subsequent decarboxylation of the Meldrum's acid moiety to complete the transformation (Scheme 16.10).

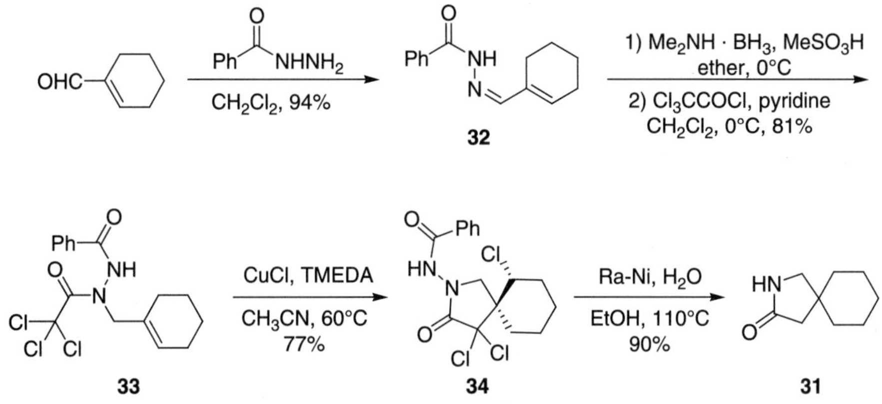

Scheme 16.9. The Parsons synthesis of gabapentin lactam (**31**).

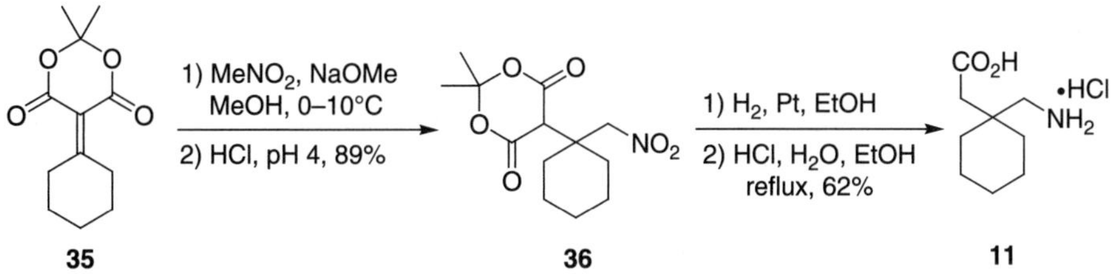

Scheme 16.10. The Zhejiang University synthesis of gabapentin hydrochloride (**11**).

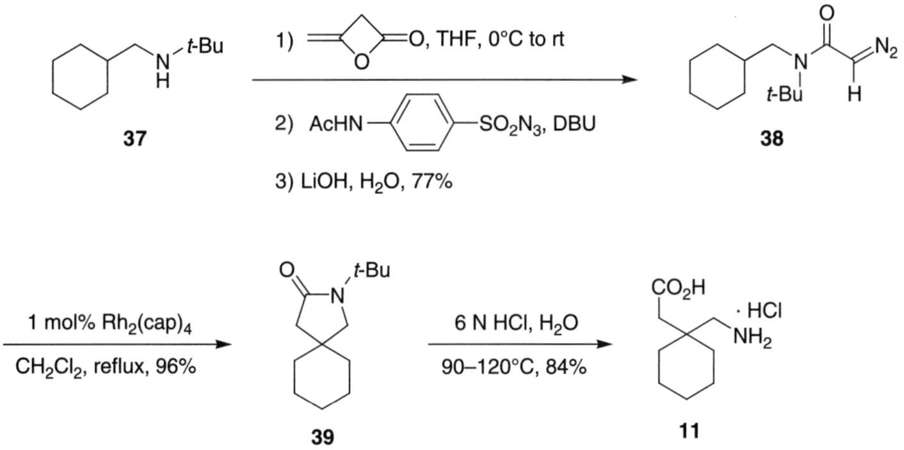

Scheme 16.11. The Chengdu Institute synthesis of gabapentin hydrochloride (**11**).

In 2003, Hu and co-workers, of the Chengdu Institute of Organic Chemistry, investigated a transition metal catalyzed C–H insertion reaction to prepare gabapentin hydrochloride (**11**) (Chen et al., 2003, 2005). The α-diazoacetamide **38** was prepared by reacting the *N-tert*-butylamine **37** with diketene followed by 4-acetamidobenzenesulfonyl azide/DBU treatment. The α-diazoacetamide **38** was then subjected to an intramolecular C–H insertion reaction catalyzed by 1 mol% of $Rh_2(cap)_4$ to form *N-tert*-butyl gabapentin lactam **39** in good yield. This lactam was then hydrolyzed under refluxing aqueous hydrochloric acid to give gabapentin hydrochloride (**11**) (Scheme 16.11).

16.3 SYNTHESIS OF PREGABALIN

($\pm$)-3-Isobutyl GABA was first synthesized by Silverman and Andruszkiewicz in 1989 (Andruszkiewicz and Silverman, 1989). It first involved a conjugate addition of nitromethane to ethyl 5-methyl-hexenoate (**40**) to give nitroester **41**, which was hydrogenated to give lactam **42**. Hydrolysis of the lactam under acidic conditions gave racemic 3-isobutyl GABA (**43**) in 58% overall yield (Scheme 16.12).

The first enantioselective synthesis of pregabalin (**2**) was developed by Yuen and co-workers at Pfizer using Evans' chiral oxazolidinone chemistry (Yuen et al., 1994). 4-Methylpentanoic acid (**44**) was converted into the corresponding acid chloride and then coupled with Evans' chiral auxiliary. The resulting acyloxazolidinone **45** was then alkylated with benzyl bromoacetate to give the benzyl ester intermediate **46**. The chiral auxiliary was removed by peroxide treatment followed by a modified bisulfite work-up at pH 7 to give the corresponding acid intermediate. Under the usual acidic bisulfite work-up conditions, a significant amount of diacid by-product was formed due to an undesired hydrolysis of the benzyl ester. Borane reduction of the resulting acid intermediate gave alcohol **47** in good yield. The alcohol was then converted to the corresponding azide **48** under standard conditions. With catalytic hydrogenation, the benzyl group was removed and the azide was simultaneously reduced to the amine to give pregabalin (**2**) in 99.5% *ee* and 65% chemical yield (Scheme 16.13).

While trying to scale-up the synthesis to kilogram quantities, Pfizer process chemists encountered unexpected difficulties. A significant amount of lactone was formed after the borane reduction step, leading to a 10% overall yield. By replacing benzyl bromoacetate with *t*-butyl bromoacetate as the alkylating agent, this problem was solved and the overall yield was increased to 33% (Hoekstra et al., 1997). However, the cost of goods was too high to be practical. Therefore, other synthetic routes to produce pregabalin (**2**) were considered.

The first approach investigated was the *L*-leucine approach (Hoekstra et al., 1997), as shown in Scheme 16.14, in which *L*-leucine was converted to the bromoester **49**. The bromide was displaced with excess diethyl sodiomalonate to give triester **50** in good yield. The *t*-butyl ester was deprotected by formic acid treatment and the resulting acid

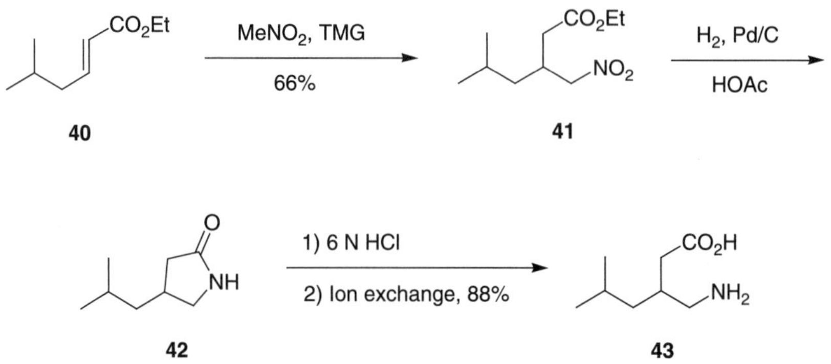

CO_2Et **40** → ($MeNO_2$, TMG; 66%) → CO_2Et, NO_2 **41** → (H_2, Pd/C; HOAc) → O, NH **42** → (1) 6 N HCl; 2) Ion exchange, 88%) → CO_2H, NH_2 **43**

Scheme 16.12. The original Silverman synthesis of ($\pm$)-3-isobutyl GABA (**43**).

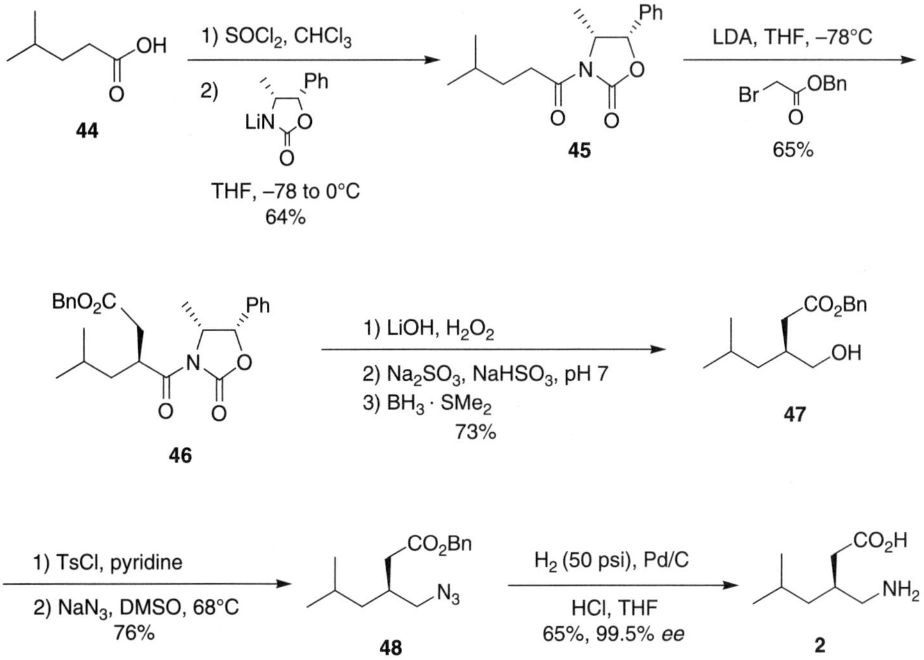

Scheme 16.13. The original Pfizer synthesis of pregabalin (**2**).

was reduced with borane to the corresponding alcohol. The diester alcohol was treated with aqueous hydrochloric acid to effect hydrolysis of the esters followed by decarboxylation and lactone formation to give **51**. Lactone **51** was opened up to the iodoester with trimethylsilyl iodide in ethanol and the iodide was further converted to azidoester **52**.

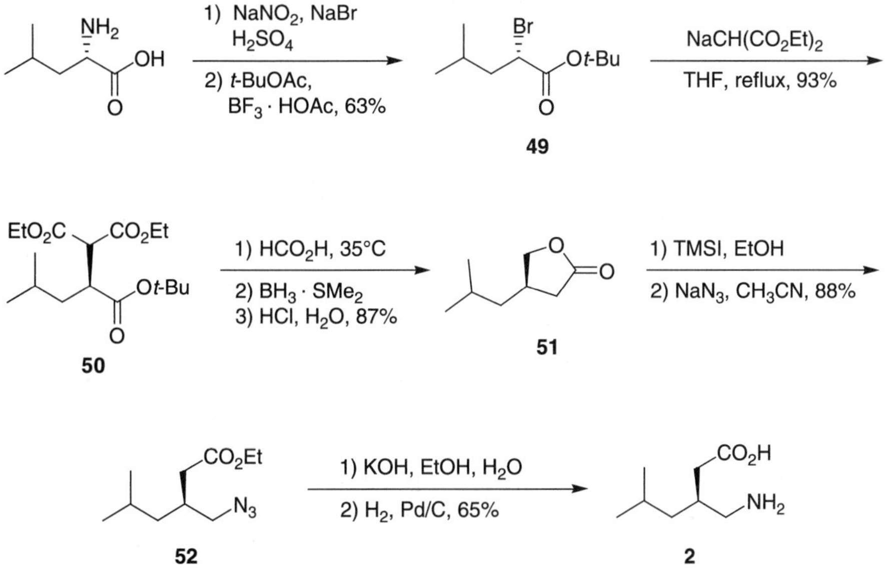

Scheme 16.14. The *L*-leucine approach to pregabalin (**2**).

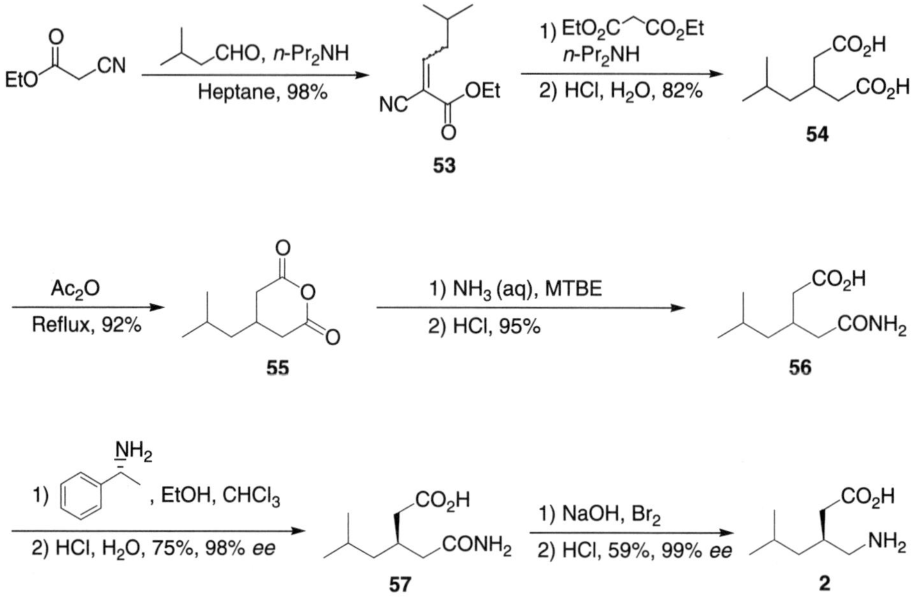

Scheme 16.15. The γ-isobutylglutaric acid approach to pregabalin (**2**).

Saponification followed by catalytic hydrogenation of the azidoester gave pregabalin (**2**) in 29% overall yield. With the use of trimethylsilyl iodide in this process, the cost of goods was still too high to be practical.

The γ-isobutylglutaric acid approach (Hoekstra et al., 1997; Huckabee and Sobieray, 1996) was the next approach investigated by Huckabee and Sobieray at Pfizer (Scheme 16.15). Ethyl cyanoacetate was first condensed with isovaleraldehyde to give the unsaturated ester **53**. Michael addition of diethyl malonate followed by acid-induced decarboxylation gave glutaric acid **54**, which was then converted to the amide acid **56** via the anhydride intermediate **55**. Acid **56** can be resolved with (*R*)-methylbenzylamine to give the chiral acid **57** in 98% *ee*. Hofmann rearrangement of amide **57** gave pregabalin (**2**) in good yield. The antipode of acid **57** from the resolution step can be hydrolyzed back to glutaric acid **54** and recycled.

Pregabalin (**2**) can also be prepared from diethyl malonate as shown in Scheme 16.16 (Grote et al., 1996; Hoekstra et al., 1997). Diethyl malonate was condensed with isovaleraldehyde followed by Michael addition of potassium cyanide to the unsaturated diester **58** to give β-cyanodiester **59**. The diester was then saponified; the resulting malonic acid was decarboxylated and the cyano acid hydrogenated using a nickel catalyst to give (±)-3-isobutyl GABA (**43**). The key step of this process was a mandelic acid resolution of racemic 3-isobutyl GABA (**43**) to give pregabalin (**2**). Although the (*R*)-enantiomer generated as a by-product in this late-stage resolution cannot be recycled efficiently, this process is still the most cost-effective process to produce pregabalin (**2**) thus far.

Many of the synthetic routes derived for pregabalin (**2**) before 2003 have relied on classical chiral resolution of either an intermediate or the final racemic 3-isobutyl GABA (**43**). In 2003, Jacobsen and co-workers published their work on a novel enantioselective synthesis of pregabalin (**2**) (Sammis and Jacobsen, 2003). The Harvard group was able to promote conjugate addition of hydrogen cyanide into the unsaturated imide

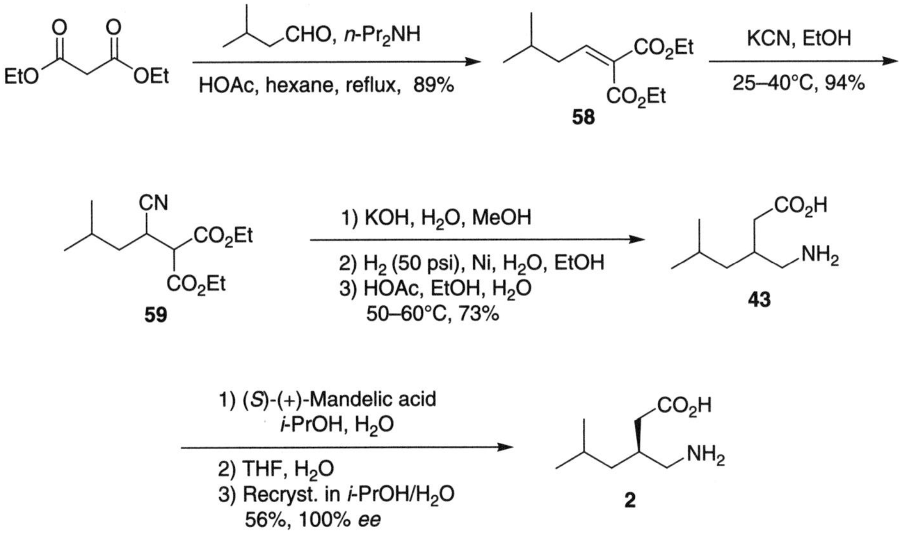

Scheme 16.16. The malonate approach to pregabalin (**2**).

60 using 10 mol% of the (*R*,*R*)-aluminum salen catalyst **61** to give the (*S*)-cyanoimide **62** in 96% *ee* and 93% chemical yield. Imide **62** was hydrolyzed to give cyanoacid **63**. The cyanoacid was then hydrogenated in the presence of platinum oxide to give pregabalin hydrochloride (**64**), as seen in Scheme 16.17, in 84% overall yield from the unsaturated imide **60**. However, considering the cost of TMSCN and the high-pressure requirement of the hydrogenation reaction, it was deemed commercially impractical.

Kissel, Ramsden, and other researchers at Pfizer and Chirotech jointly published a novel chiral synthesis of pregabalin (**2**) in 2003 based on asymmetric hydrogenation (Burk et al., 2001, 2003). Their synthesis started with the condensation of isobutyraldehyde with acrylonitrile under Baylis–Hillman conditions to give allylic alcohol **65**. This alcohol was activated as the carbonate **66** and subjected to palladium-catalyzed carbonylation conditions to give cyanoester **67**. The ester **67** was hydrolyzed and converted to

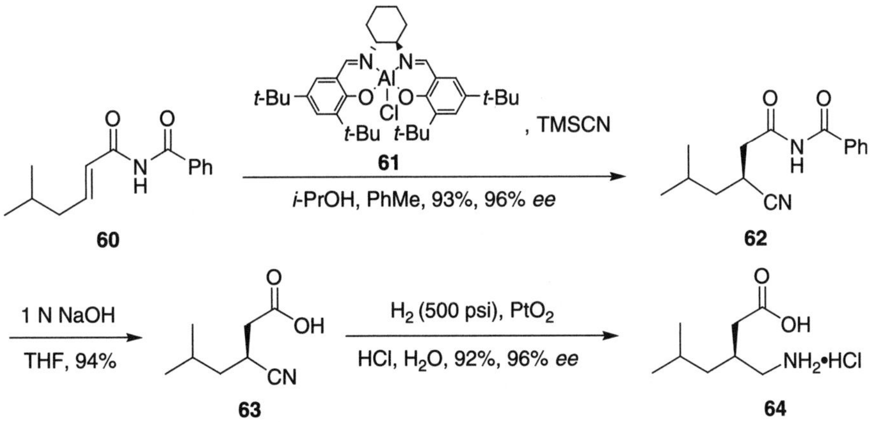

Scheme 16.17. The Jacobsen synthesis of pregabalin hydrochloride (**64**).

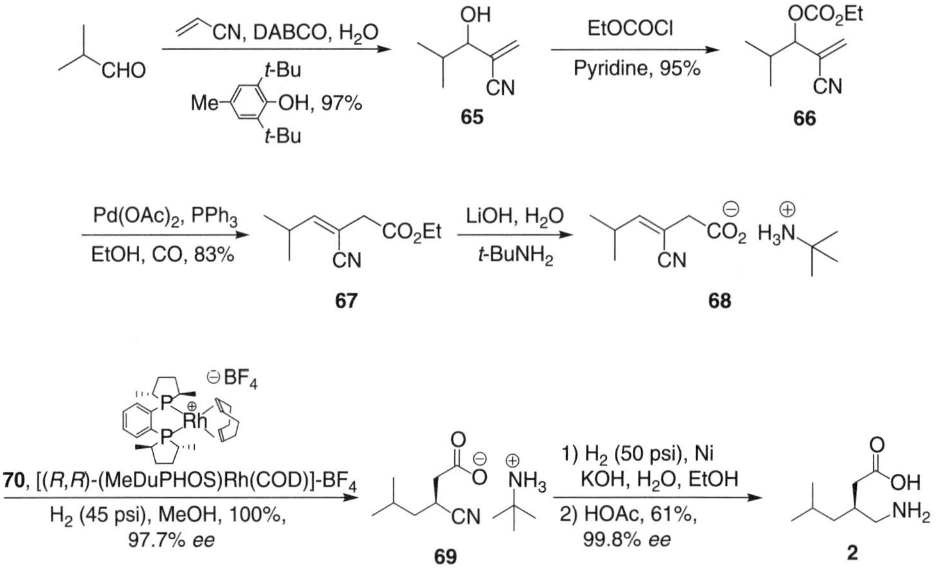

Scheme 16.18. The Pfizer/Chirotech asymmetric hydrogenation route to pregabalin (**2**).

the corresponding *t*-butyl ammonium salt **68**, which was hydrogenated with the (*R*,*R*)-methylDuPHOS rhodium catalyst **70** to give the chiral cyanoacid ammonium salt **69** in 98% *ee*. Hydrogenation of nitrile **69** over nickel gave pregabalin (**2**) in greater than 99% *ee* (Scheme 16.18).

Recently, Hu and co-workers, at Pfizer, filed a patent application on a process to prepare pregabalin (**2**) via enzymatic resolution (Hu et al., 2005). The Pfizer researchers investigated a large number of commercially available hydrolases to catalyze the hydrolysis of β-cyanodiester **59**, an intermediate from the malonate route (Scheme 16.16), to form (3*S*)-3-cyano-2-ethoxcarbonyl-5-methylhexanoic acid potassium salt (**71**)

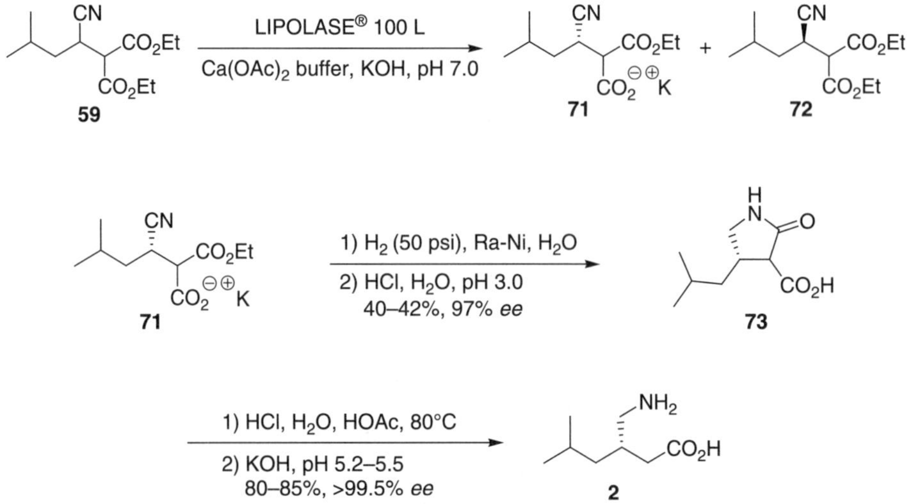

Scheme 16.19. The Pfizer enzymatic resolution route to pregabalin (**2**).

enantioselectively. With this enzyme-screening study, they determined that LIPOLASE® 100L, type EX, was the most effective hydrolase for the large-scale preparation of monoester **71**. Hydrogenation of nitrile **71** over Raney nickel followed by acid treatment gave the pyrrolidinone-3-carboxylic acid **73** in 97% *ee*. Acid **73** was then decarboxylated and hydrolyzed with hydrochloric acid to give crystalline pregabalin (**2**) in greater than 99.5% *ee* (Scheme 16.19). The (*R*)-3-cyano-2-ethoxycarbonyl-5-methylhexanoic acid ethyl ester (**72**) left over from the LIPOLASE-catalyzed hydrolysis reaction can be recycled back to the β-cyanodiester **59** by sodium ethoxide in ethanol treatment, providing a 50% savings in cost of goods over the malonate approach described above.

Although gabapentin (**1**) and pregabalin (**2**) are quite simple structurally, they have challenged chemists to design commercially viable processes for their bulk production. Most importantly, these two compounds have proven to be effective drugs for the treatment of a number of CNS conditions, thereby opening up a novel area of investigation in the field of calcium channel research.

REFERENCES

Andruszkiewicz, R. and Silverman, R. B. (1989). *Synthesis*, 12: 953–955.

Brown, J. P., Boden, P., Singh, L., and Gee, N. S. (1996). *Review in Contemporary Pharmacotherapy*, 7: 203–214.

Bryans, J. S. and Wustrow, D. J. (1999). *Med. Res. Rev.*, 19: 149–177.

Bryans, J. S., Chessum, N. E. A., Huther, N., Parsons, A. F., and Ghelfi, F. (2003). *Tetrahedron*, 59: 6221–6231.

Burk, M. J., Goel, O. P., Hoekstra, M. S., Mich, T. F., Mulhern, T. A., and Ramsden, J. A. (2001). WO 01/55090.

Burk, M. J., de Koning, P. D., Grote, T. M., Hoekstra, M. S., Hoge, G., Jennings, R. A., Kissel, W. S., Le, T. V., Lennon, I. C., Mulhern, T. A., Ramsden, J. A., and Wade, R. A. (2003). *J. Org. Chem.*, 68: 5731–5734.

Cagnoli, R., Ghelfi, F., Pagnoni, U. M., Parsons, A. F., and Schenetti, L. (2003). *Tetrahedron*, 59: 9951–9960.

Cannata, V., Nicoli, A., and Corcella, F. (2002). WO 02/34709.

Chen, Z., Chen, Z., Jiang, Y., and Hu, W. (2003). *Synlett.*, 13: 1965–1966.

Chen, Z., Chen, Z., Jiang, Y., and Hu, W. (2005). *Tetrahedron*, 61: 1579–1586.

Dooley, D. J., Donovan, C. M., and Pugsley, T. A. (2000). *J. Pharmacol. Exp. Ther.*, 295: 1086–1093.

Gee, N. S., Brown, J. P., Dissanayake, V. U. K., Offord, J., Thurlow, R., and Woodruff, G. N. (1996). *J. Biol. Chem.*, 271: 5768–5776.

Griffiths, G., Mettler, H., Mills, L. S., and Previdoli, F. (1991). *Helv. Chim. Acta*, 74: 309–314.

Grote, T. M., Huckabee, B. K., Mulhern, T. A., Sobieray, D. M., and Titus, R. D. (1996). WO 96/40617.

Hartenstein, J. and Satzinger, G. (1977). DE 2611690.

Hayashi, T. (1958). *Nature*, 182: 1076–1077.

Hoekstra, M. S., Sobieray, D. M., Schwindt, M. A., Mulhern, T. A., Grote, T. M., Huckabee, B. K., Hendrickson, V. S., Franklin, L. C., Granger, E. J., and Karrick, G. L. (1997). *Org. Proc. Res. Dev.*, 1: 26–38.

Hu, S., Martinez, C. A., Tao, J., Tully, W. E., Kelleher, P., and Dumond, Y. (2005). US 2005/0283023.

Huckabee, B. K. and Sobieray, D. M. (1996). WO 96/38405.

Karlsson, A., Fonnum, F., Malthe-Sorensson, D., and Storm-Mathisen, J. (1974). *Biochem. Pharmacol.*, 23: 3053–3061.

Li, J.-H., Li, Z.-G., and Chen, Q.-G. (2004). *J. Chem. Res.*, 758–759.

Radulovic, L. L., Busch, J. A., Windsor, B. L., McNally, W. P., Sinz, M. W., and Bockbrader, H. N. (1996). *Pharm. Res.*, 13: S480.

Rossi, P. and Vecchio, E. (2002). WO 02/074727.

Sammis, G. M. and Jacobsen, E. N. (2003). *J. Am. Chem. Soc.*, 125: 4442–4443.

Satzinger, G., Hartenstein, J., Herrmann, M., and Heldt, W. (1976). DE 2460891.

Silverman, R. B., Andruszkiewicz, R., Nanavati, S. M., Taylor, C. P., and Vartanian, M. G. (1991). *J. Med. Chem.*, 34: 2295–2298.

Su, T.-Z., Lunney, E., Campbell, G., and Oxender, D. L. (1995). *J. Neurochem.*, 64: 2125–2131.

Taylor, C. P. (1994). *Neurology*, 44: S10–S16.

Taylor, C. P., Vartanian, M. G., Yuen, P.-W., Bigge, C. F., Suman-Chauhan, N., and Hill, D. R. (1993). *Epilepsy Research*, 14: 11–15.

van Wessem, G. C. and Sakal, E. L. (1960). US 2960441.

Velardi, F. and Fornaroli, M. (2003). US 2003/0009055.

Yuen, P.-W., Kanter, G. D., Taylor, C. P., and Vartanian, M. G. (1994). *Bioorg. Med. Chem. Lett.*, 4: 823–826.

17

APPROVED TREATMENTS FOR ATTENTION DEFICIT HYPERACTIVITY DISORDER: AMPHETAMINE (ADDERALL®), METHYLPHENIDATE (RITALIN®), AND ATOMOXETINE (STRATERRA®)

David L. Gray

NH_2

1

USAN: Amphetamine
Trade name:
Adderall® (Shire, 1996)
M.W. 135.2

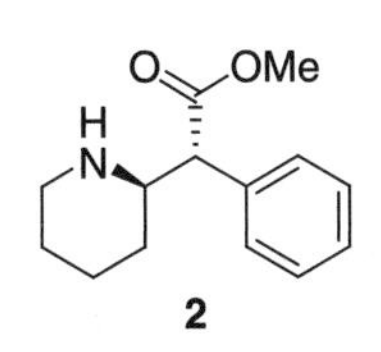

2

USAN: Methylphenidate
Trade name:
Ritalin® (Novartis, 1955)
Concerta® (Johnson and Johnson, 2000)
Focalin® (Celgene/Novartis, 2001)
Daytrana® (Shire, 2006)
M.W. 233.3

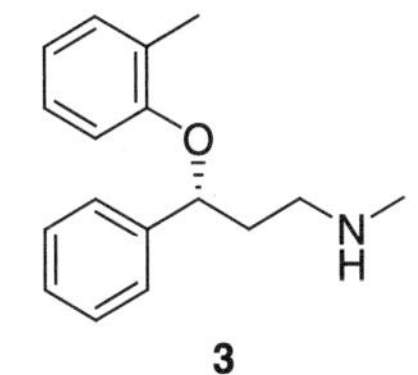

3

USAN: Atomoxetine
Trade name:
Straterra® (Eli Lilly, 2002)
M.W. 255.3

The Art of Drug Synthesis. Edited by Douglas S. Johnson and Jie Jack Li

17.1 INTRODUCTION

Attention deficit hyperactivity disorder (ADHD) is one of the more common psychiatric conditions afflicting young children (Biederman and Faraone, 2005; Doyle, 2004). It is estimated that somewhere between 3 and 7% of school-age children worldwide may suffer from ADHD and that about half of those children will continue to deal with the disorder into adulthood (Barkley, 1977; Faraone et al., 2003; Swanson et al., 1998). The disease strikes both sexes and can be difficult to diagnose with certainty because each patient presents the major symptoms to varying degrees and in different ways. The ADHD diagnosis is a subjective observation made by a trained clinician who reviews the history of a patient looking for a set of behaviors and patterns as outlined in the DSM IV (Barkley, 2002). The latest revision of the criteria that must be present for an ADHD diagnosis include measures of inattention, hyperactivity, difficulty with impulse control, inappropriate outbursts, and difficulty in school (Faraone et al., 2000). The fact that ADHD children do not have physical symptoms of the disease beyond their behavior and school performance has led to a great deal of controversy around ADHD diagnosis and treatment (Faraone et al., 2005). Current diagnostic tools often consist of questionnaires to parents and teachers; however, recent advances in brain imaging techniques (PET and fMRI) (Dougherty et al., 1999; Lou et al., 1989; Sowell et al., 2003; Vaidya and Gabrieli, 1999) have provided some links between the disorder and brain morphology that might have future utility in diagnosis. Although the presenting symptoms may seem to be more of a nuisance to others than an actual danger to the child, studies have shown that children with untreated ADHD are much more likely to end up in serious accidents, commit crime, lose their job, or abuse drugs (Fisher et al., 2002; Rasmussen and Gillberg, 2000). These trends are postulated to arise from deficits in controlling impulse and from impaired decision-making. When ADHD persists into adulthood, the symptomology tends to lessen with age, although the same unfortunate trends in rates of imprisonment, drug abuse, motor vehicle accidents, and unemployment make adult ADHD extremely costly both to the individual sufferer and to society (Discala et al., 1998; Mannuza et al., 1997; Steinhausen et al., 2003). The current arsenal of therapies that have proven efficacy in large-scale clinical trials include (1) behavioral therapy and psychiatric visits, (2) stimulant medications, and (3) nonstimulant medications (Faraone, 2003).

17.1.1 Stimulants Versus Nonstimulants

A large percentage of ADHD patients show major improvements in disease symptoms when treated with stimulant medications. To many, the idea of giving a psychomotor stimulant to a hyperactive individual seems counterintuitive. The confusion arises from the "stimulant" class to which these medications belong. In the early days of psychopharmacology, compounds were classified by their measurable actions on typical human subjects or laboratory animals. Compounds that stimulated the CNS in a manner similar to adrenaline (**4**, Fig. 17.1) (for example, increases in locomotor activity, loss of appetite, restlessness) were classified as stimulants. Stimulants exert their CNS effects by increasing the synaptic concentration of the neurotransmitter dopamine (**5**) via one of two pathways—either by directly stimulating the release of dopamine, or by blocking the function of the dopamine reuptake transporter (DAT) (McMillen, 1983). The DAT transporter is the major mechanism for removing extracellular dopamine and bringing it back into the neuron for future release, thereby regulating the concentration of dopamine in the synapse.

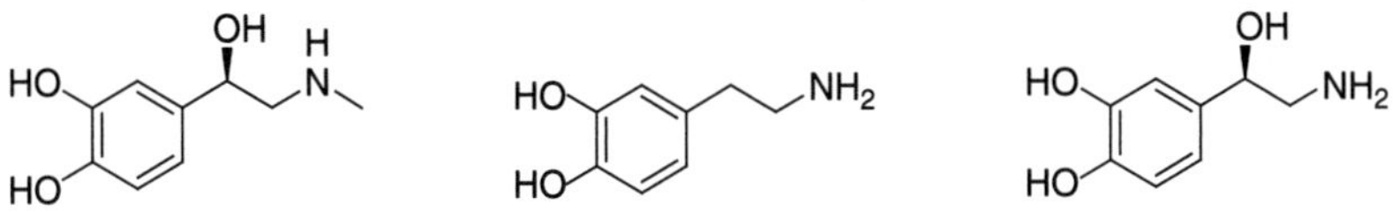

4, adrenaline 5, dopamine 6, norepinephrine, noradrenaline

Figure 17.1. Chemical structures of neurotransmitters.

Although stimulants act like adrenaline on most people—making them feel restless and energized—in the ADHD population, these substances generally have a rapid and beneficial effect on the major symptoms of the disorder, with treated patients becoming calmer, more focused on tasks, and displaying increased performance on school tests. This observation led to a hypothesis that ADHD symptoms might be related to a chronically reduced level of available dopamine in cortical brain regions that are associated with cognition, impulse control, and executive functioning. This has been called the "dopamine hypothesis of ADHD" (Volkow et al., 1998, 2001). Dopamine is the key trigger of the brain's reward system, and elevation of dopamine in the mesolimbic region leads to the sensations of pleasure and well-being. As a result, most stimulants have a very high potential for abuse, because a rapid rise in mesolimbic dopamine concentrations (above a certain threshold level) will lead the patient or animal to seek more of whatever produced the initial sensation (Wilens et al., 2003). Most stimulants, including amphetamine (**1**) and methylphenidate (**2**), have been classified by the U.S. DEA as Schedule II regulated substances, which places tight controls on the prescription and dispensing of these drugs. Although the potential for abuse is a concern, numerous studies have demonstrated that ADHD patients do not normally misuse these drugs, and seem to actually have a lower rate of substance abuse compared to the untreated ADHD population (Faraone and Wilens, 2003; Gawin et al., 1985). It is thought that, when taken as prescribed, current stimulant therapies do not release mesolimbic dopamine quickly enough to activate reward-seeking behaviors in ADHD patients. Pharmaceutical companies have tailored formulations around the specific human pharmacokinetics of amphetamine (**1**) and methylphenidate (**2**) in order to maximize the therapeutic benefit to ADHD patients while minimizing side-effect risk. This topic will be discussed with the individual drugs in the next sections.

In late 2002, the U.S. FDA approved Eli Lilly's atomoxetine (Straterra®, **3**) as the first new chemical entity approved for the treatment of ADHD since 1975. This compound was designed to mimic some of the actions of the stimulants by inhibiting the reuptake of norepinephrine (**6**, Fig. 17.1), a neurotransmitter that is structurally and physiologically related to dopamine (**5**), and which is thought to play a major role in attention, cognition, and executive reasoning via actions in the forebrain. Blocking norepinephrine reuptake also leads to an increase in the synaptic dopamine concentration in the prefrontal cortex. In clinical trials, Straterra® was shown to have efficacy on ADHD symptoms while having no abuse liability, and this drug is not regulated by the DEA (Michelson et al., 2003).

As of 2006, there are several branded medications approved for the treatment of ADHD; however, there are only three chemicals that make up the primary active ingredients in these drugs: the (*S*)-enantiomer of amphetamine (**1**), the 2(*R*),2′(*R*)-enantiomer of methylphenidate (**2**), and the (*R*)-enantiomer of atomoxetine (**3**). An older approved ADHD drug, pemoline (Cylert®), was withdrawn from the market in 2005 due to reported liver toxicity. The concern over the abuse potential of stimulants has led researchers to devise a number of novel formulations that take advantage of pharmacokinetics to minimize the effect of the drugs on reward pathways. Adderall®, Adderall®XR, and Vyvanse® (a prodrug in registration) are formulations of amphetamine (**1**), and Concerta®, Ritalin®, Focalin®, and Daytrana® are formulations of methylphenidate (**2**).

All three ADHD-approved chemical entities have at least one chiral center, a feature that has led to a number of interesting syntheses of these compounds over the years. Amphetamine (**1**) and methylphenidate (**2**) were discovered before the modern era of asymmetric and enantioselective synthesis, and are sold as racemic, single-enantiomer, and enantio-enriched formulations. Atomoxetine (**3**), first presented in a 1977 Eli Lilly patent, was developed as a single-enantiomer drug (Molloy and Schmiegel, 1977).

17.2 SYNTHESIS OF AMPHETAMINE

Amphetamine (**1**) is a very simple phenethylamine, described in the chemical literature as early as 1887 (Edeleano, 1887). Smith, Kline and French (now GSK) filed a patent on the synthesis and use of amphetamine in 1930 (Nabenhauer, 1930), and the enantiomers were assigned in 1932 (Leithe, 1932; V-Braun and Friehmelt, 1933). Not surprisingly, early access to chiral material relied on classical crystallization-based resolutions (Gillingham, 1962; Nabenhaur, 1942). The early, racemic syntheses of amphetamine fall into four major classifications according to the method used to make the C–N bond:

- reductive amination of a ketone (Scheme 17.1),
- direct displacement of a leaving group by an amine (Scheme 17.3)
- nitro alkane addition followed by reduction of the nitro group, (Scheme 17.4) and
- metal-promoted amination of an unsaturated carbon compound.

The reductive amination of methyl benzyl ketone (**7**) was one of the earliest methods used in amphetamine synthesis and continues to be an option in manufacturing because of its simple elegance and the availability of cheap starting materials. According to the Smith-Kline patent (1930) the oxime **8** forms as a mixture of isomers upon exposure of the ketone to hydroxylamine hydrochloride under mildly basic conditions (Nabenhauer, 1930). Reduction of the oxime can be accomplished using a variety of reducing agents. The initial report employed sodium in methanol for converting **8** to the target amphetamine **1**. A number of variations on this theme have been put forward over the years (Foreman et al., 1969; Hartung and Munch, 1931; Kabalka and Guindi, 1990; Micovic et al., 1996). Several groups had interest in the metabolic fates of the individual enantiomers of **1**, as they had markedly different pharmacology when administered as single enantiomers. Isotopically labeled amphetamine was a powerful tool for studying the in vivo metabolism of this compound, and the reduction

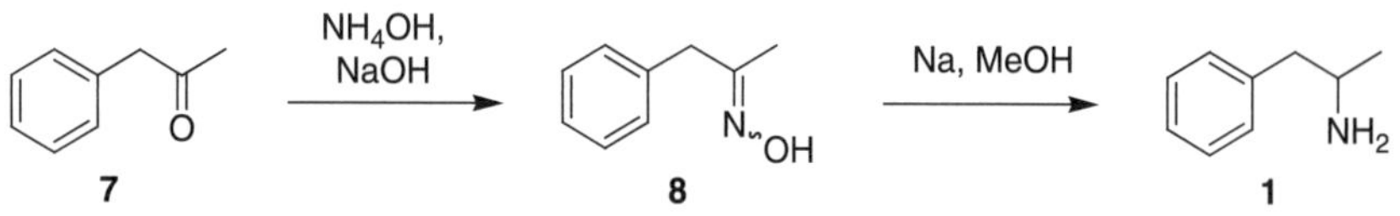

NH_4OH, NaOH
Na, MeOH
O
N–OH
NH_2
7 8 1

Scheme 17.1. Amphetamine synthesis via reductive amination.

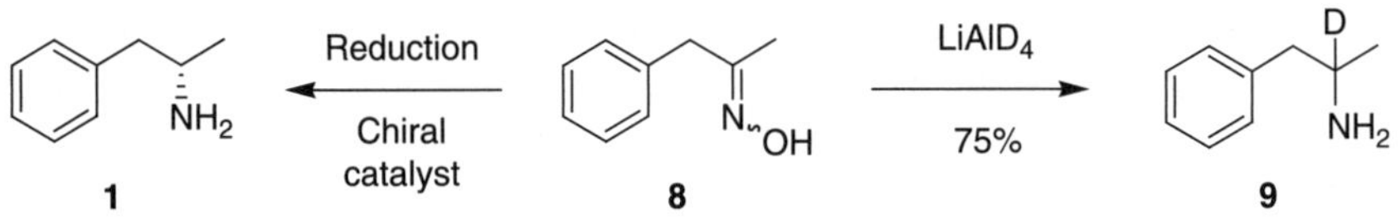

Reduction
Chiral catalyst
$LiAlD_4$
75%

Scheme 17.2. Synthesis of labeled amphetamine via reductive amination.

of oxime **8** offered a convenient way to introduce deuterium at the methine carbon of **1**. In 1969, Foreman demonstrated that treatment of **8** with $LiAlD_4$ in ether (Scheme 17.2) proceeded in up to 75% yield to afford the labeled product **9** (Foreman et al., 1969). Using tools like **9**, it was eventually determined that the *l*-isomer of amphetamine is rapidly metabolized in humans and that most of the CNS effects of amphetamine can be attributed to the *d*-isomer (Sadler, 1993).

At least one group attempted to synthesize enantio-enriched **1** via asymmetric reduction of **8** with modest success (Krasik and Alper, 1992). The classical resolution of racemic **1** (via the bis-tartrate salt, for example) (Nabenhaur, 1942) remains as the simplest and most economical method for preparing chiral amphetamine (**1**) on a large scale.

In 1946, Hass investigated the biological effects of the so-called "pressor" amines, including amphetamine (**1**) (Hass et al., 1946). His reported synthesis is an early example of a second method for forming the C–N bond within amphetamine—the direct displacement of a leaving group by ammonia (Scheme 17.3). This route began with the Markovnikov addition of HCl across the double bond of allyl benzene (**10**) to make chloride **11**, which was sealed inside an iron pipe with ammonia in MeOH and heated to 160°C for 9 h. This manipulation resulted in the aminolysis of the chloride, giving amphetamine (**1**) in 51% yield after recrystallization.

A few years later, Hass reported an alternative synthesis of racemic amphetamine, which exemplifies the use of a nitroalkane as the source of the nitrogen atom (Scheme 17.4) (Hass et al., 1950). In this route, calcium-hydroxide-promoted Henry reaction of nitroethane and benzaldehyde (**12**) afforded an 86% yield of the nitro alkene **13**. Simultaneous hydrogenation

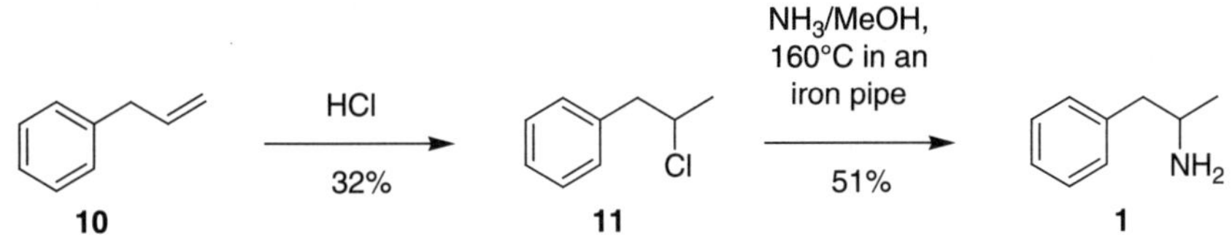

HCl, 32%
NH_3/MeOH, 160°C in an iron pipe, 51%

Scheme 17.3. Amphetamine synthesis via direct aminolysis of a leaving group.

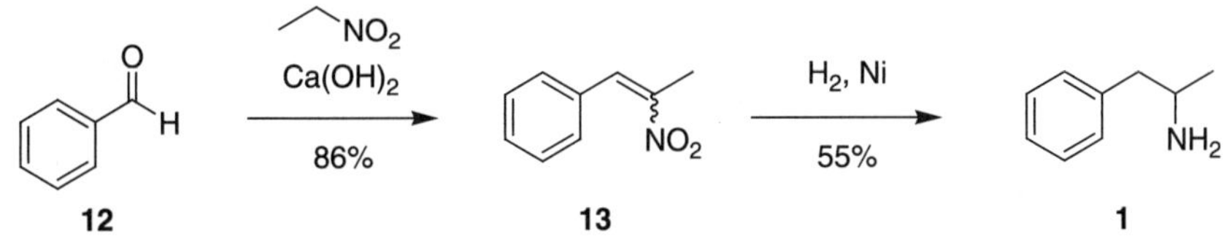

NO_2, $Ca(OH)_2$, 86%
H_2, Ni, 55%

Scheme 17.4. Amphetamine synthesis via addition of a nitroalkane and reduction.

of the olefin and nitro groups over nickel(0) then yielded 55% of the racemic amphetamine **1**. Interest in amphetamine synthesis led several groups to investigate methods for the chiral reduction of **13**, including chemical and chemoenzymatic approaches (Fuji, 1992; Hartung et al., 2000; Iwai, 1965; Mori et al., 1990).

17.2.1 Pharmacokinetic Properties of *d*- and *l*-Amphetamine

In 1942, Nabenhauer reported the large-scale synthesis of amphetamine along with the resolution of its enantiomers via a crystallization that employed potassium bis-tartrate salts as the resolving agent (Nabenhaur, 1942). The availability of the individual enantiomers eventually led to the unraveling of a complex story surrounding the pharmacology and pharmacokinetics (PK) of the individual amphetamine enantiomers. Adderall® is the brand name for an approved ADHD medication composed of four amphetamine salts mixed in a 1 : 1 : 1 : 1 ratio—racemic amphetamine aspartate, racemic amphetamine sulfate, *d*-amphetamine saccharate, and *d*-amphetamine sulfate. The combination of two parts racemate and two parts *d*-isomer affords a pill that contains a 3 : 1 ratio of enantiomers favoring the more active *d*-amphetamine isomer (**1**). Adderall® is also available in an extended-release formulation branded Adderall®XR (Couch et al., 2005).

It is commonly held that dopamine-driven reward-seeking pathways are activated when dopamine concentrations are elevated rapidly above a certain threshold. Thus, there is both a magnitude and a temporal component to the action of substances that produce a "high." The major drugs of abuse rapidly cross the blood–brain barrier and elevate dopamine, and it is this pharmacodynamic "quick-surge" phenomenon that makes them so addictive. Current ADHD formulations of amphetamine are designed to provide several distinct benefits to the patient. First, *d*-amphetamine has a moderate half-life in humans (about 12 h) (Brunton et al., 2006). Through formulation, the duration of action of this drug can be extended through a typical day with one (Adderall®XR) or two (Adderall®) doses of the drug. The second benefit has to do with providing a pharmacokinetic profile where the drug concentration slowly builds to a therapeutic range and then maintains that concentration without major fluctuation for a number of hours (through the typical school day, for instance). The effect of this slow and even PK profile is to provide therapeutic elevation of synaptic dopamine while minimizing the type of rapid surge in dopamine that is thought to activate reward-seeking behaviors or "highs." Formulators have taken advantage of several novel methods to achieve the desired PK profiles and paid more than the usual attention to the shape and slope of the PK curves (Fig. 17.2) (Physicians Desk Reference, 2006). Adderall®XR makes use of increasingly common technology to encase the drug inside biodegradable polymer-coated beads. These polymer coatings degrade at different rates and release the drug over an extended period of time.

17.2.2 Chiral Synthesis of Amphetamine

In 1973, Barfknecht and co-workers, published an asymmetric synthesis of amphetamine and amphetamine analogs beginning with methyl benzyl ketone (**7**, Scheme 17.5) (Nichols et al., 1973). Reductive amination was carried out using chiral α-methyl-benzylamine (for example, *S*-**14**), which is readily available as both enantiomers. The initial imine formation was driven to completion by removing H_2O with a Dean–Stark apparatus. The resulting imine was reduced with Raney nickel at 50 psi and the product isolated and recrystallized as the HCl salt of amine **15** in good overall yield. Hydrogenolysis of the *N*-benzyl group over 10% Pd/C gave (*d*)-amphetamine (**1**) of high optical purity. A few years later, Repke

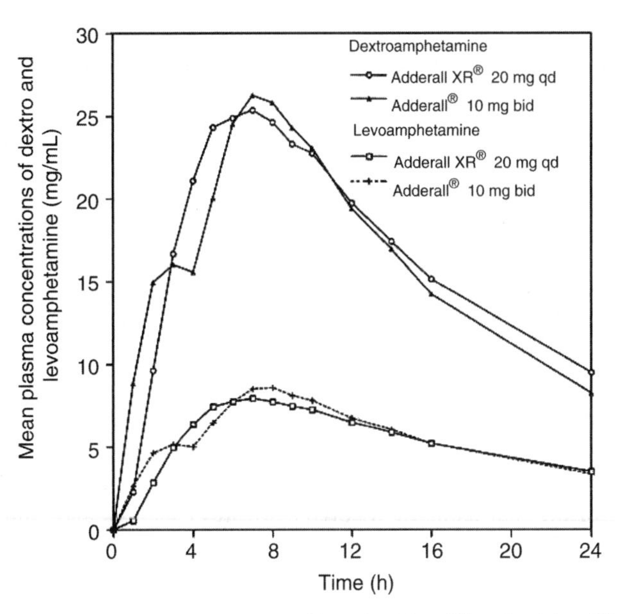

Figure 17.2. Pharmacokinetic profiles of Adderall® and Adderall®XR.

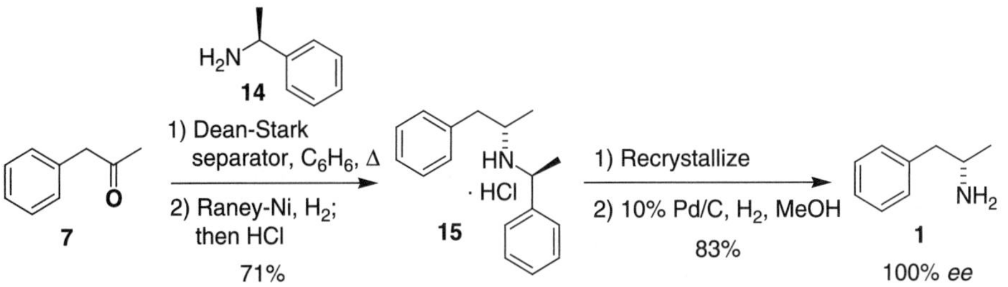

Scheme 17.5. Asymmetric synthesis of *d*-amphetamine.

put forward a synthesis of chiral amphetamine starting from *d*-phenylalanine (Repke et al., 1978), which was notable among several approaches to enantiopure phenethylamines **1** (Bell et al., 1995; Rozwadowska, 1993; Shi et al., 2004; Wagner et al., 2003).

17.3 SYNTHESIS OF METHYLPHENIDATE

Racemic *threo*-methylphenidate was approved for the treatment of fatigue, nausea, and depression in 1955 under the brand name Ritalin®, and was first used in children in 1958 with an approval for "hyperkinetic disorder" (ADHD) in 1960. Methylphenidate has an even shorter half-life than amphetamine, and its use in treating ADHD was limited by the fact that schoolchildren would need to visit a nurse during the day to take a second dose of this scheduled drug (in order to maintain efficacy throughout the entire school day). In humans, injecting methylphenidate produces effects similar to intravenous cocaine, but oral methylphenidate is adsorbed very slowly from the gut into the blood and takes an unusually long time (estimated 2.5 h) to reach a peak concentration in the brain (Brunton et al., 2006). This slow oral adsorption is advantageous for patients as discussed earlier, because it leads to a gentle rise in synaptic dopamine concentration that is not pleasurable and does not activate reward-seeking behavior.

Much like amphetamine, the biological activity of methylphenidate has been known for many decades (Meier et al., 1954). Methylphenidate was first synthesized in 1944 by Pannizon, who published a four-step synthesis of **2** and several related molecules from readily available starting materials and filed a U.S. patent (Novartis) detailing the synthesis and medical uses of the series in that same year (Panizzon, 1944; Hartmann and Panizzon, 1950). In Pannizon's original route (Scheme 17.6), benzylnitrile (**16**) was reacted with 2-chloropyridine (**17**) and sodium amide to afford pyridyl nitrile **18** in 70% yield after recrystallization. Hydrolysis of the nitrile under strongly acidic conditions led to the amide **19**, which was esterified with acidic methanol to deliver methyl ester **20** in good yield. Hydrogenation of the pyridine ring over platinum in aqueous acetic acid unmasked the piperidine, providing methylphenidate (**2**) as an 80 : 20 mixture of *erythro* and *threo* racemates. Using classical techniques, Shafi'ee and Hite were able to establish the relative stereochemistry of the four isomeric methylphenidates (Shafi'ee and Hite, 1969).

Early pharmacologic profiling of the individual diastereoisomers of **2** led first to the determination that the interesting biological activity resided in the *threo* racemate, and finally to the discovery that the *d-threo* enantiomer (*R,R* isomer) was the most potent inhibitor of the dopamine reuptake transporter (Shafi'ee and Hite, 1969; Srinivas et al., 1992, 1993). In 1960, a Novartis patent disclosed a process for taking the predominately *erythro* mixture (80 : 20 mixture of esters **20** that arose from a modest *erythro* preference in Pannizon's Pt-based pyridine hydrogenation), and epimerizing the mixture using 50% aqueous KOH to afford a new mixture of diastereoisomeric acids **21** that not only favored the desired *threo* racemate, but from which pure *threo* racemic acid (**22**) could be precipitated upon neutralizing the base (Scheme 17.7) (Rometsch, 1960).

Novartis also disclosed the purification of the *erythro* racemate (**23**) and the isolation of a single *threo* enantiomer through successive crystallizations. This original synthesis has remained a major method for preparing **2** for almost 50 years, with only minor modifications being made to the original reaction conditions (Deutsch et al., 1996). After all, the route comprised only four steps, the starting materials and reagents were inexpensive, and it used no chromatography. In 1998, Winkler published a novel four-step synthesis of the

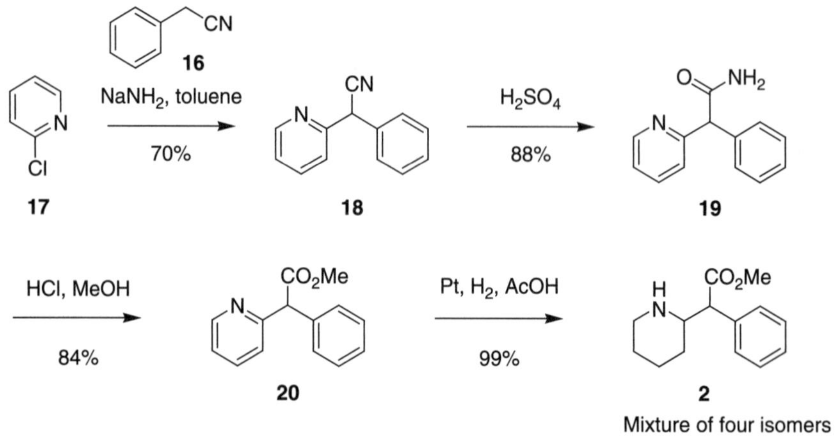

Scheme 17.6. First synthesis of methylphenidate.

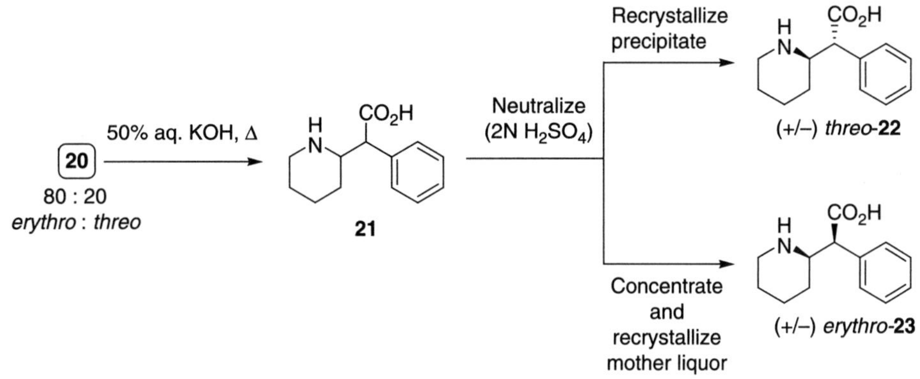

Scheme 17.7. Separation of (±) *threo* and *erythro* ritalinic acids **22** and **23**.

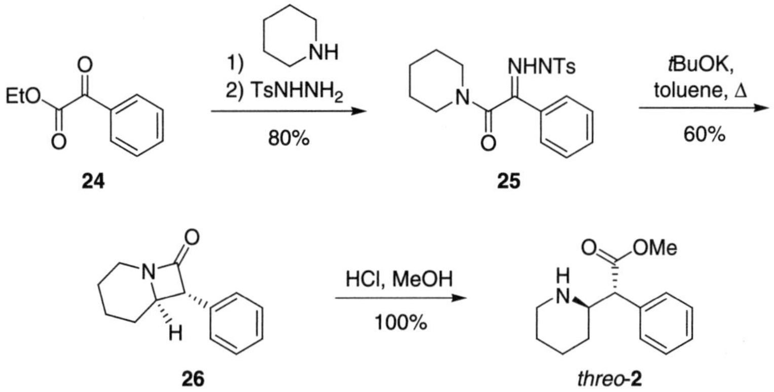

Scheme 17.8. Winkler synthesis of *threo*-methylphenidate.

threo racemate featuring the formation and opening of a β-lactam (Axten et al., 1998). The preparation commences with the reaction of phenylglyoxylate **24** (Scheme 17.8) and piperidine to obtain the corresponding piperidine amide, which was directly converted to tosylhydrazone **25** by the condensation of tosylhydrazine with the more electrophilic benzylic carbonyl. Thermally initiated decomposition of the tosylhydrazone led to the formation of a β-lactam (**26**), which equilibrates via the enamide to a 6 : 1 mixture of diastereoisomers under the basic reaction conditions. After separating away the minor diastereoisomer, exposure of **26** to acidic methanol causes lactam opening and esterification, to deliver racemic *threo*-methylphenidate (**2**) in quantitative yield with no loss of stereochemical integrity from **26**.

17.3.1 Methylphenidate Formulations

In 2001, Celgene obtained FDA approval to re-launch the single enantiomer dexmethylphenidate (*d-threo* **2**, formerly Dexedrine®) for ADHD under the brand name Focalin®, and subsequently sold the product to Novartis. *d-threo* Methylphenidate has a 2.2 h

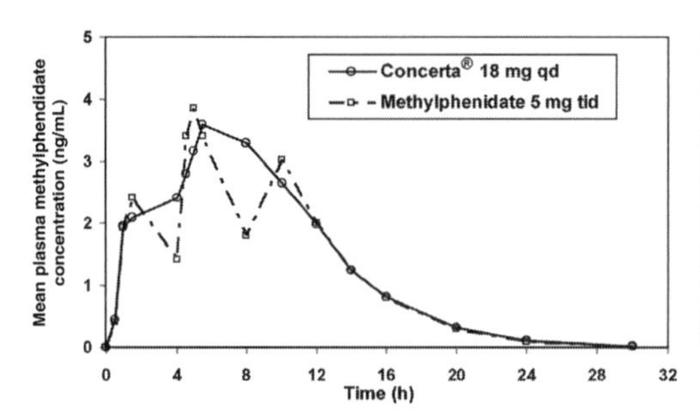

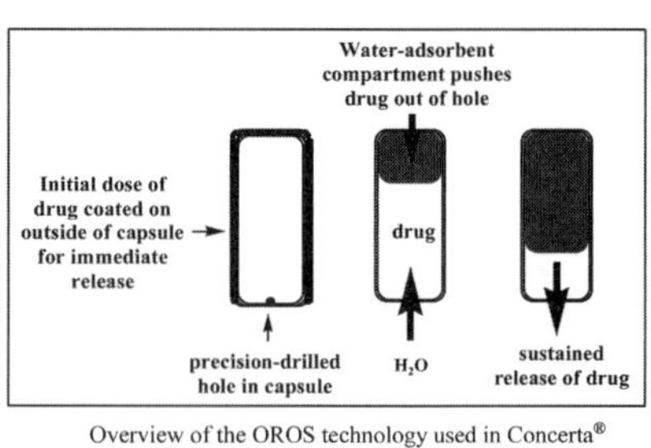

Figure 17.3. Pharmacokinetic profile and OROS technology of Concerta®.

half-life in humans—too short for convenient once- or twice-daily dosing using conventional formulations (Brunton et al., 2006). Similar to the approach with amphetamine, formulators have turned to slow-release polymer-coated preparations of methylphenidate to lengthen its duration of action to approximate a typical school day. Novartis received approval in 2005 to market an extended-release formulation of dexmethylphenidate (**2**) called Focalin®XR for children, adolescents, and adults. Johnson and Johnson launched a very unusual once-daily methylphenidate product called Concerta® in 2002, which uses an actual device—a novel osmotic pump based on Elan's OROS technology (Fig. 17.3)—to release the drug over a specified period of time. The capsule releases an immediate burst of 30% of its contents, and the remaining 70% is released slowly throughout the day. After releasing the drug, the tiny capsule is passed through the intestine unchanged and excreted with the stool. Figure 17.3 depicts the plasma drug concentrations of thrice-daily methylphenidate and Concerta® versus time, together with an overview of the osmotic push drug release system (Swanson et al., 2003). In 2006, Shire obtained FDA approval to market the Daytrana® transdermal methylphenidate patch, representing an additional method for dosing this compound in ADHD patients. Along with formulating for optimum pharmacokinetics with products like Concerta® and Daytrana®, pharmaceutical scientists continue to explore formulations that make it more difficult to extract or otherwise misuse the active ingredients in these medications.

17.3.2 Chiral Synthesis of Methylphenidate

As interest in single-enantiomer drugs grew during the 1980s and into the 1990s, several groups reported the synthesis of the most active (2*R*,2′*R*)-enantiomer of methylphenidate. The first, in 1998, came from Perel et al., who published an eight-step synthesis of (*R*,*R*) methylphenidate starting from the chiral pool (*d*-pipercolic acid, **27**, Scheme 17.9) (Thai et al., 1998). This synthesis commenced with Boc protection of the basic amine functionality in **27** followed by BOP (**28**) mediated transformation of the carboxylic acid to its corresponding Weinreb amide derivative **29** in 90% yield for the two steps. This substance (**29**) was reacted with phenyl lithium in ether at −23°C to afford phenyl ketone **30** in 47% yield, along with some unreacted starting material. The ketone was methyleneated under Wittig conditions employing methyl triphenylphosphonium bromide and *t*-BuOK to give disubstituted olefin **31** in excellent yield.

The chiral center on the piperidine ring retained stereochemical integrity to this point in the synthesis, and provided the basis for a relatively modest level (72 : 28 *threo* : *erythro*) of substrate-based stereocontrol in the BH_3·THF/H_2O_2 hydroboration/oxidation of the

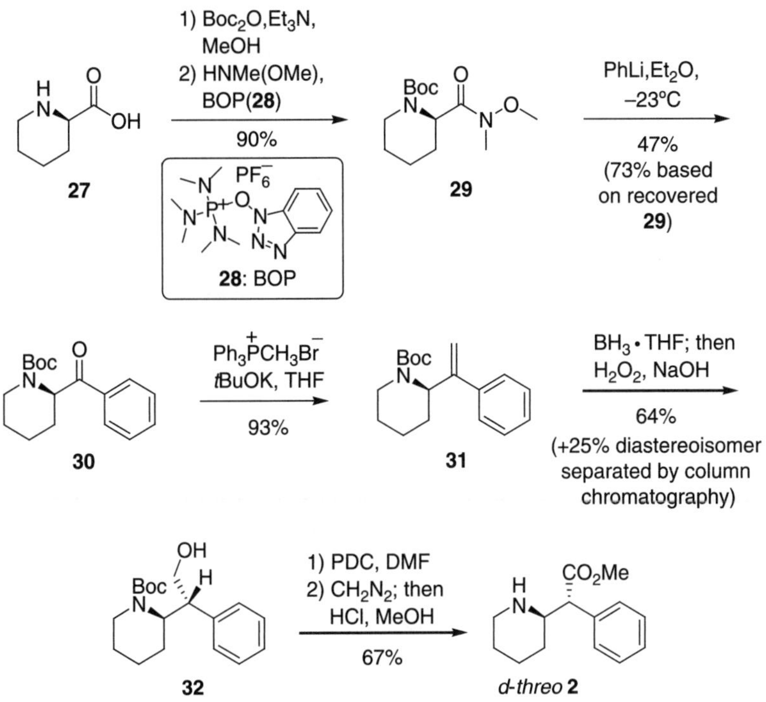

Scheme 17.9. Asymmetric synthesis of *d-threo* methylphenidate from *d*-pipecolic acid.

double bond in **31**. Among several conditions examined, (+)-IPC-BH_2 was the most *threo*-selective hydroboration reagent, affording exclusively the desired isomer **32** after oxidative work-up, albeit in moderate yield (55%). In order to maximize final product throughput, the nonsubstituted borane was used, resulting in two diastereoisomeric products, which were separated by silica gel chromatography. This reaction sequence led to key hydroxy intermediate **32**, which was oxidized to the acid using PDC in DMF and subsequently converted to the methyl ester with diazomethane without any purification at that stage. The last step in this synthesis was the removal of the *N*-Boc protecting group via treatment with methanolic HCl, affording, after recrystallization, optically and chemically pure (2*R*,2′*R*)-methylphenidate (**2**).

A very elegant synthetic approach was reported a year later by Davies et al., leveraging asymmetric C–H activation chemistry to accomplish a one-pot synthesis of *d-threo* methylphenidate (Scheme 17.10) (Davies et al., 1999). *N*-Boc piperidine (**33**) was selectively alkylated by the carbene formed by decomposition of diazoester **34** in a reaction mediated by 25 mol% of chiral Rh (II) catalyst **35**, giving the *N*-Boc protected (2*R*,2′*R*) isomer in a single step. TFA was added to accomplish removal of the Boc group after the C–H insertion reaction was complete, affording (*R*,*R*)-methylphenidate (**2**) with an *ee* of 86% in 52% overall yield.

Much like amphetamine, methylphenidate was originally approved and marketed as a racemate. Novartis and others have filed several patents on the large-scale preparation and resolution of the most potent (*R*,*R*)-isomer (Harris and Zavareh, 1997; Patrick et al., 1987; Prashad, 1998; Prashad and Bin, 2000; Zeitlin and Stirling, 1998). A 1999 Novartis process publication on the synthesis of methylphenidate is illustrative in that, despite

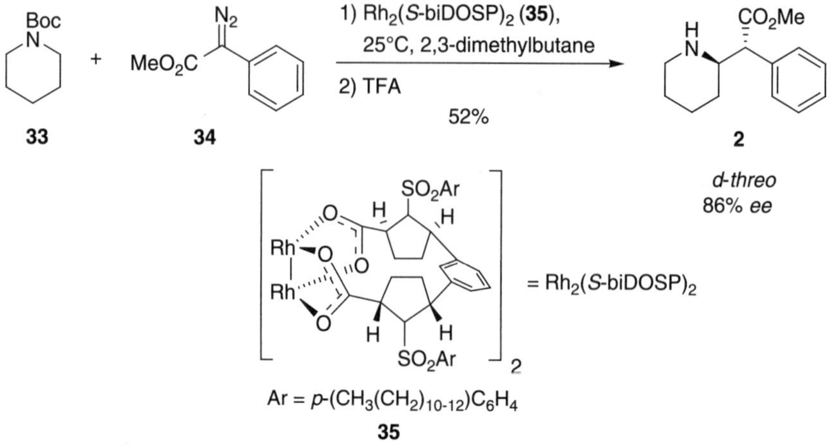

Boc
N
+
MeO_2C
N_2
1) Rh_2(*S*-biDOSP)$_2$ (**35**),
25°C, 2,3-dimethylbutane
2) TFA
52%
H
N
CO_2Me
33
34
2
d-threo
86% *ee*

SO_2Ar
H
H
Rh
O
O
O
O
Rh
O
O
H
H
SO_2Ar
2
= Rh_2(*S*-biDOSP)$_2$
Ar = *p*-$(CH_3(CH_2)_{10\text{-}12})C_6H_4$
35

Scheme 17.10. Synthesis of *d-threo* methylphenidate via asymmetric CH insertion.

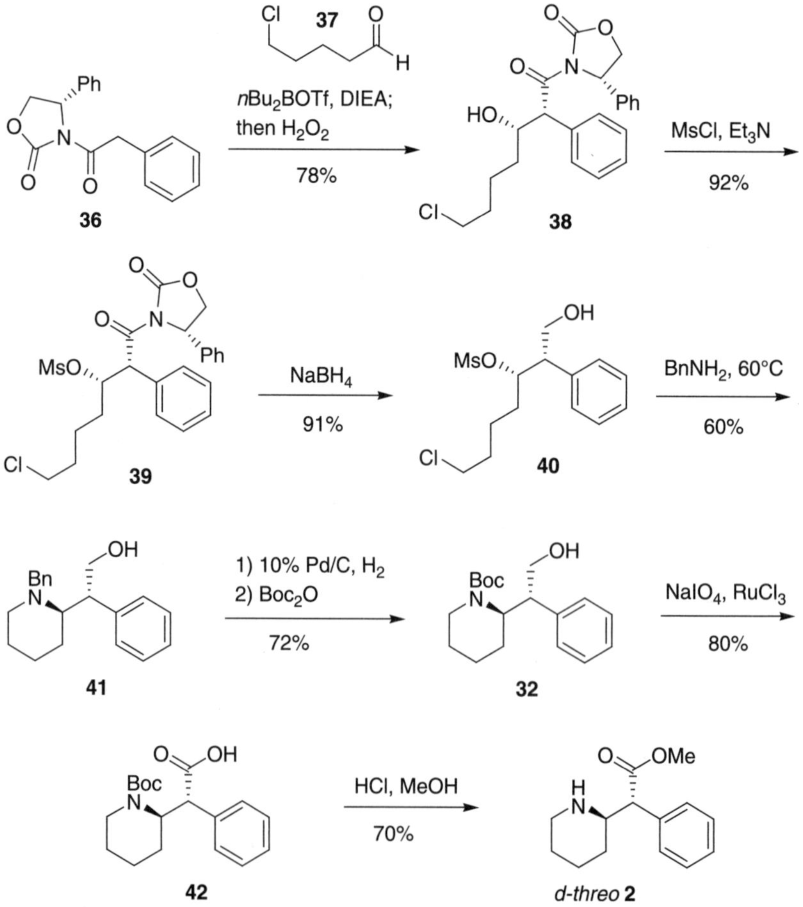

Cl
37
O
H
Ph
O
N
O
O
nBu_2BOTf, DIEA;
then H_2O_2
78%
O
O
O
N
Ph
HO
Cl
MsCl, Et_3N
92%
36
38

O
O
O
N
Ph
MsO
Cl
39
$NaBH_4$
91%
OH
MsO
Cl
40
$BnNH_2$, 60°C
60%

OH
Bn
N
41
1) 10% Pd/C, H_2
2) Boc_2O
72%
OH
Boc
N
32
$NaIO_4$, $RuCl_3$
80%

O
OH
Boc
N
42
HCl, MeOH
70%
O
OMe
H
N
d-threo **2**

Scheme 17.11. Aldol-based asymmetric synthesis of *d-threo* methylphenidate.

having a reasonable large-scale preparation of the *threo* enantiomers in hand, the company invested time and money toward finding a more cost-effective route including exploring new chemical routes as well as novel resolution processes (Prashad et al., 1999). The length of this process route underscores the challenge that current multistereocenter drugs present to the chemical manufacturing groups. This Novartis synthesis begins by acylating the less commonly used (and cheaper) (*R*)-4-phenyl-2-oxazolidinone to afford **36** (Scheme 17.11). Following Evans's work on auxiliary-based aldol chemistry (Evans et al., 1982), condensation with 5-chloro pentanal (**37**) afforded a 78% yield of adduct **38** as the sole diastereoisomer. The next objective was to close the piperidine ring, and this was accomplished by converting the hydroxyl to a mesylate using MsCl (**38** to **39**), reducing the methyl ester to the hydroxy chloromesylate (**39** to **40**), and finally reacting the latter bis-electrophile with benzylamine under basic conditions to afford piperidine **41** in acceptable overall yield.

The benzyl group in **41** was removed via hydrogenation and the unmasked amine immediately reprotected as its *N*-Boc derivative, leading to piperidine **32**. Next, the hydroxyl functionality was oxidized to the acid by the action of $RuCl_3$ and $NaIO_4$, giving acid **42** in 80% yield. Formation of the requisite methyl ester occured upon exposure of acid **42** to acidic MeOH, which conditions also cleaved the carbamate protecting group and left an HCl salt, which was recrystallized to enantiopure **2** with high efficiency.

17.4 SYNTHESIS OF ATOMOXETINE

Atomoxetine (Straterra®, originally tomoxetine or tomoxetin, **3**) was first described and synthesized by chemists at Eli Lilly in the late 1970s and was one of the few compounds that was known to display meaningful selectivity for the norepinephrine reuptake transporter (NET) versus the serotonin reuptake transporter (SERT) and the dopamine reuptake transporter (DAT) (Barnett, 1986; Molloy and Schmiegel, 1997). Atomoxetine was one of several structurally related and commercially successful monoamine reuptake inhibitors that were developed by Lilly for the treatment of various psychiatric disorders (Fig. 17.4). Fluoxetine (**43**) and duloxetine (**44**) have both gained approval in the United States as Prozac® and Cymbalta®, respectively, and nisoxetine (**45**) is widely used as a tool in biology.

Straterra® is the first new chemical entity approved to treat ADHD in over 30 years, and it has major advantages over the stimulant medications in that it has been clinically demonstrated to have no abuse liability. The nonstimulant mechanism of action also appears to come with some caveats in terms of efficacy versus the stimulants. Synthetically, the main challenge within this molecule is installing the chiral center, which is

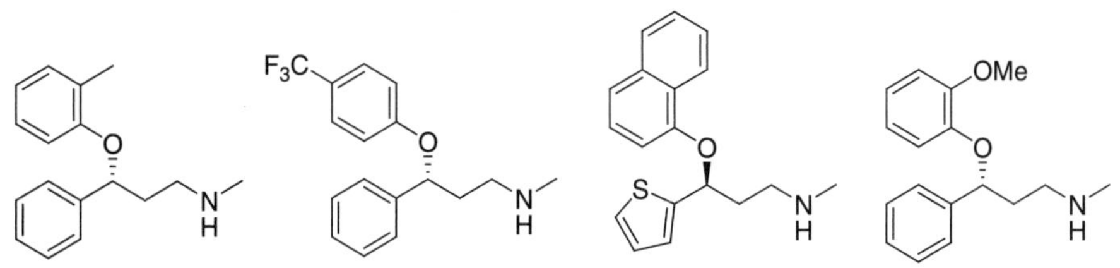

3: atomoxetine, Straterra® **43**: fluoxetine, Prozac® **44**: duloxetine, Cymbalta® **45**: nisoxetine

Figure 17.4. Structurally related monoamine re-uptake inhibitors from Eli Lilly.

unlikely to be derived efficiently from the chiral pool. A number of different approaches toward atomoxetine have successfully dealt with that challenge, including several syntheses that showcase the most common and general methods employed in the pharmaceutical industry for installing a chiral center.

The early literature on the synthesis of atomoxetine includes a number of efforts that deliver a flexible synthesis of the entire "oxetine" group from Lilly, some of which are not ideally suited for atomoxetine (Corey and Reichard, 1989; Koenig and Mitchell, 1994; Mitchell and Koenig, 1995; Molly and Schmiegel, 1977). These publications were followed closely by work from a couple of pioneers in the area of asymmetric, reagent-controlled synthetic methodology who were able to use atomoxetine as a high-profile target for showcasing the power of their methodologies. H.C. Brown's disclosure of an asymmetric route to the Lilly "oxetines" based on the bis-pinene borane IPC_2BCl reducing agent was closely followed by a Sharpless paper detailing the use of titanium isopropoxide/tartrate asymmetric epoxidation methodology to synthesize the same targets (Gao and Sharpless, 1988; Srebnick et al., 1988). The Sharpless and Brown methodologies continue to find broad application in the pharmaceutical industry. Both routes used essentially the same bond construction strategy, which was to set the incorrect stereochemistry of the benzylic hydroxyl group, activate and displace the hydroxyl with *o*-cresol, and finally use a handle on the terminus of the alkyl chain to introduce a methyl amine via displacement of the handle. To showcase the IPC borane methodology, Brown chose to start with chloroketone **46** (Scheme 17.12). This compound was reduced in good yield to the desired (*S*)-benzylic alcohol (**47**) in 94% *ee* using the borane reagent derived from (+)-α-pinene. A single recrystallization of this alcohol from hexanes brought the enantiomeric excess above 99.5%. The second aromatic ring was then appended via Mitsunobu etherification with *o*-cresol (**48**) using DEAD and PPh_3 and this reaction proceeded with clean inversion of configuration, giving ether **49** in 70% yield. The final maneuver was the aminolysis of the primary chloride, which was accomplished using aqueous methylamine at 130°C in a Parr bomb. This final step proceeded in 95% yield to deliver (*R*)-atomoxetine (**3**) with >99% *ee*.

The Sharpless synthesis (Scheme 17.13) begins with the asymmetric epoxidation of cinnamyl alcohol (**50**) using (+)-di-isopropyl tartarate (DIPT) as the chiral ligand for the

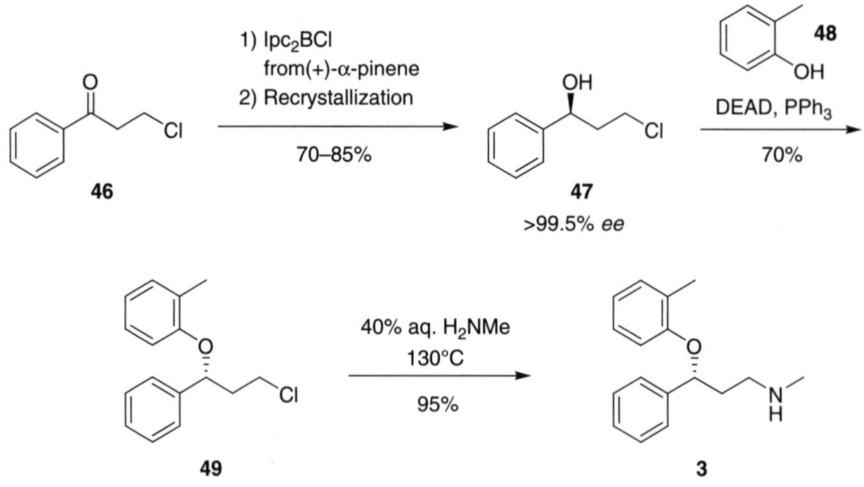

1) Ipc_2BCl from(+)-α-pinene 2) Recrystallization 70–85%; **46**; **47** >99.5% *ee*; **48** DEAD, PPh_3 70%; **49**; 40% aq. H_2NMe 130°C 95%; **3**

Scheme 17.12. Brown synthesis of atomoxetine.

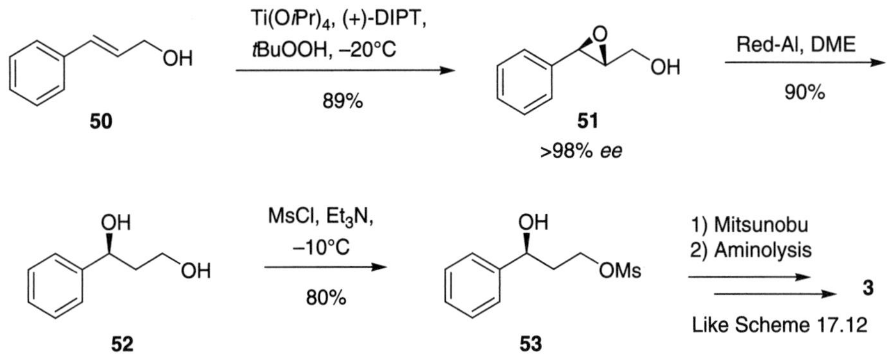

Scheme 17.13. Sharpless synthesis of atomoxetine.

$Ti(OiPr)_4$/TBHP epoxidation system. Under optimized conditions, epoxyalcohol **51** was obtained in 89% chemical yield and >98% *ee*. The epoxide was then reductively opened with complete regioselectivity using Red-Al (sodium bis(2-methoxyethoxy)aluminum hydride) to give the diol **52**. The next maneuver was to selectively monoprotect the hydroxy group, and this was done using MsCl and Et_3N at $-10°C$, leading to mesylate **53** in good yield (Gao and Sharpless, 1988). Having established the chiral center, the synthesis of atomoxetine was finished by Mitsunobu etherification and aminolysis under conditions similar to those reported by Brown.

Several other published approaches to setting the stereocenter at the benzylic carbon in atomoxetine are relevant to current pharmaceutical practice in that they look to enzymes for establishing the requisite chirality (Boaz, 1992; Bracher and Litz, 1996; Chenevert and Fortier, 1991; Fronza et al., 1991; Gu et al., 1993; Kumar et al., 1991; Liu et al., 2000; Schneider and Goeraens, 1992). In one such route, Kumar and colleagues utilized one of the more established enzymatic methodologies to set the atomoxetine chiral center, namely, the bakers yeast reduction of a β-keto ester (**54** to **55**, Scheme 17.14). Having set the required stereochemistry, the nitrogen atom was introduced via amide bond formation, at which point a recrystallization brought the optical purity of amide **56** to >99% *ee*. Amide **56** was then converted to the amine upon treatment with $LiAlH_4$ and

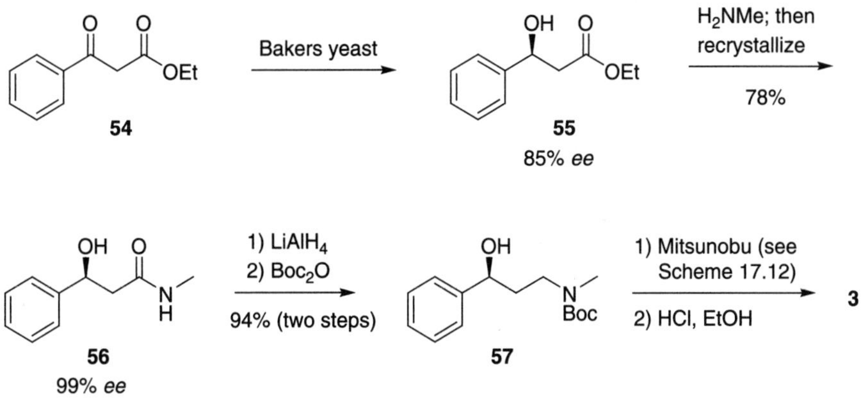

Scheme 17.14. Chemoenzymatic Synthesis of Atomoxetine.

the secondary amine was protected as its *N*-Boc derivative to arrive at intermediate **57** in good yield. Mitsunobu etherification functioned to install the remaining ring and an acidic Boc-deprotection step finished the synthesis of atomoxetine (**3**) (Kumar et al., 1991)

Many of the synthetic routes developed for the preparation of atomoxetine used the Mitsunobu reaction to create the ether linkage in the target (Lapis et al., 2005; Sakuraba and Achiwa, 1991). Although this reaction works well in the lab on a smaller scale, it can be problematic on a manufacturing scale. Chemists at Lilly and elsewhere reported various other methods for the etherification of substrates like **47**, including several procedures that used 2-fluorotoluene as the electrophile in a nucleophilic aromatic substitution (S_NAr) reaction (Heath et al., 2000; Kjell and Lorenz, 2003), and a route featuring a novel α-haloacid coupling (Devin et al., 1997). One of the hallmarks of process chemistry is the balance between length of synthesis, cost of goods, yield, reproducibility, and ease of purification (see Chapter 2). These factors are continually traded off against each other to arrive at a suitably inexpensive and robust process. In the case of atomoxetine, the Lilly process group made the decision to move away from the Mitsunobu chemistry and utilize the more easily scaled S_NAr reaction (Scheme 17.15). A current manufacturing route begins with the familiar chloroketone **46** and proceeds via an oxazaborolidine-catalyzed carbonyl reduction using precatalyst **58** to provide the hydroxy compound **47**. This reaction is nearly quantatative and affords material of 94% *ee*. Next, the primary

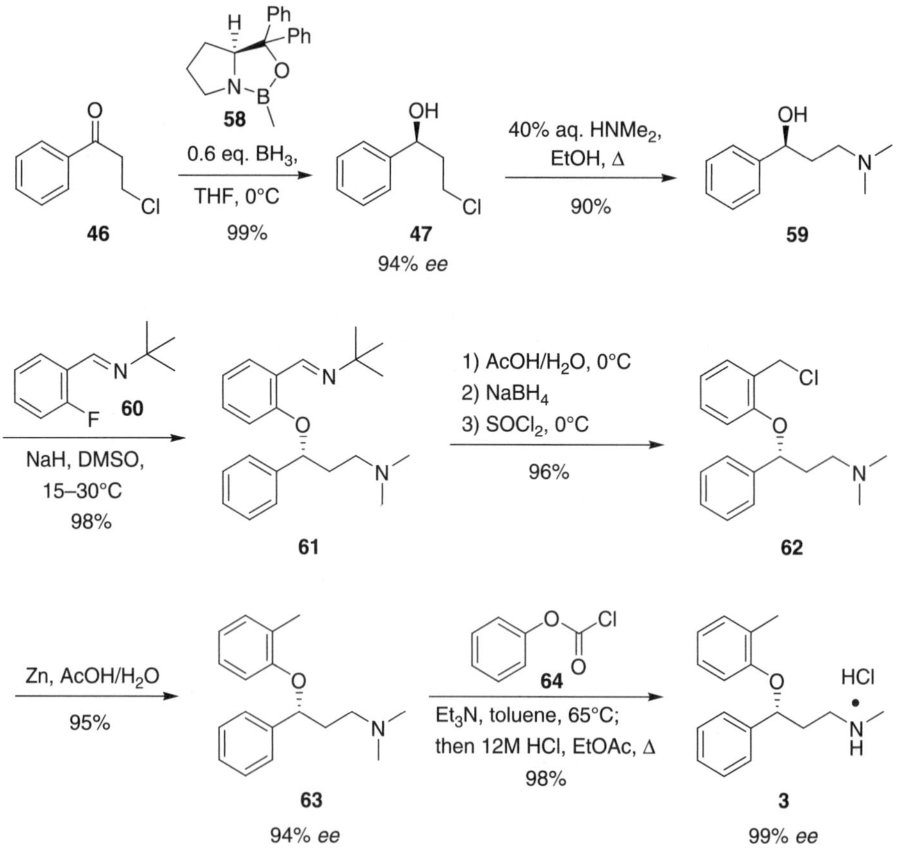

Scheme 17.15. Lilly manufacturing route for atomoxetine.

halide is displaced with dimethylamine in refluxing EtOH, leading to amine **59** in high yield. The key displacement occurs in DMSO near room temperature, using sodium hydride as base and fluoro-2-(*t*-butylamino)benzene (**60**) as the electrophile. Displacement of the activated fluorine results in the formation of the desired ether **61** in 98% yield.

Next, in a four-step sequence, the *t*-butyl imine is removed by first hydrolyzing the imine to the aldehyde, second, reducing the resulting aldehyde to the benzylic alcohol with sodium borohydride, third, converting the benzylic alcohol into a benzylic chloride (**62**) via treatment with $SOCl_2$, and fourth, reductive dehalogenation of the activated chloride using Zn/AcOH. This conversion proceeds in very high overall yield, giving *N,N′*-dimethyl atomoxetine **63** of 94% *ee*. Heating dimethylamino compound **63** at 65°C with Et_3N and phenyl chloroformate (**64**) effectively removes the extra methyl group and the final product is upgraded to >99% *ee* upon its isolation as the atomoxetine HCl salt (**3**). This route adds a few steps when compared with other approaches; however, it does so in exchange for robust chemistry, crystalline intermediates, and reliable chiral purity of the final material.

REFERENCES

Axten, J., Krin, L., Kung, H., and Winkler, J. (1998). *J. Org. Chem.*, 63: 9628–9629.

Barkley, R. (1977). *J. Child Psychol. Psychiatry Allied Discip.*, 18: 137–165.

Barkley, R. (2002). *Clin. Child Fam. Psychol. Rev.*, 5: 89–111.

Barnett, A. (1986). *Drugs Future*, 11: 134.

Bell, F., Cantrell, A., Hoegberg, S., Jaskunas, S., and Johansson, N. (1995). *J. Med. Chem.*, 25: 4929–4936.

Biederman, J. and Faraone, S. (2005). *Lancet*, 366: 237–248.

Boaz, N. (1992). *J. Org. Chem.*, 57: 4289–4292.

Bracher, F. and Litz, T. (1996). *Bioorg. Med. Chem.*, 4: 877–880.

Brunton, L., Lazo, J., and Parker, K. (2006). *Goodman & Gilman's The Pharmacological Basis of Therapeutics*, 11th ed. McGraw-Hill Companies.

Chenevert, R. and Fortier, G. (1991). *Chem. Lett.*, 1603–1604.

Corey, E. and Reichard, G. (1989). *Tetrahedron Lett.*, 30: 5207–5210.

Couch, R., Burnside, B., and Chang, R. (2005). US 6913768 (to Shire Laboratories).

Davies, H., Hansen, T., Hopper, D., and Panaro, S. (1999). *J. Am. Chem. Soc.*, 121: 6509–6510.

Deutsch, H., Shi, Q., Gruszecka-Kowalik, E., and Schweri, M. (1996). *J. Med. Chem.*, 39: 1201–1209.

Devin, P., Heid, R., and Tschaen, D. (1997). *Tetrahedron*, 53: 6739–6746.

DiScala, C., Lescohier, I., Barthel, M., and Li, G. (1998). *Pediatrics*, 102: 1415–1421.

Dougherty, D., Bonab, A., Spencer, T., Rauch, S., Madras, B., and Fischman, A. (1999). *Lancet*, 354: 77–97.

Doyle, R. (2004). *Psychiat. Clin. North America*, 27: 203–214.

Edeleano, L. (1887). *Chem. Ber.*, 20: 616–622.

Evans, D., Ennis, M., and Mathre, D. (1982). *J. Am. Chem. Soc.*, 104: 1737.

Faraone, S. (2003). *Medscape Psychiatry Ment. Health*, 8(2): 1–6.

Faraone, S. and Wilens, T. (2003). *J. Clin. Psychiatry*, 64: 9–13.

Faraone, S., Biederman, J., and Mick, E. (2000). *Am. J. Psychiatry*, 157: 1077–1083.

Faraone, S., Perlis, R., and Doyle, A. (2005). *Biol. Psychiatry*, 57: 1313–1323.

Faraone, S., Sergeant, J., Gillberg, C., and Biederman, J. (2003). *World Psychiatry*, 2: 104–113.

Fischer, M., Barkley, R., Smallfish, L., and Fletcher, K. (2002). *J. Abnorm. Child Psychol.*, 30: 463–475.

Foreman, R., Siegel, F., and Mrtek, R. (1969). *J. Pharm. Sci.*, 58: 189–191.

Fronza, G., Fugante, C., Grassilli, P., and Mele, A. (1991). *J. Org. Chem.*, 56: 6019–6023.

Fuji, M. (1992). *Chem. Lett.*, 6: 933–934.

Gao, Y. and Sharpless, K. (1988). *J. Org. Chem.*, 53: 4081–4084.

Gawin, F., Riodan, C., and Kleber, H. (1985). *Am. J. Alcohol Abuse*, 11: 193–197.

Gillingham, J. (1962). US 3028430 (to Parke Davis).

Gu, J., Li, Z., and Lin, G. (1993). *Tetrahedron*, 49: 5805–5816.

Harris, M. and Zavareh, H. (1997). WO 97/27176, WO 97/32851 (to Medeva Europe).

Hartmann, M. and Panizzon, L. (1950). US 2507631 (to Ciba Pharmaceutical Products – now Novartis) – filed in 1944.

Hartung, C., Breindl, C., Tillak, A., and Beller, M. (2000). *Tetrahedron*, 56: 5157–5162.

Hartung, W. and Munch, J. (1931). *J. Am. Chem. Soc.*, 53: 1875–1879.

Hass, H., Patrick, T., and McBee, E. (1946). *J. Am. Chem. Soc.*, 68: 1009–1011.

Hass, H., Susie, A., and Heider, R. (1950). *J. Org. Chem.*, 15: 8–14.

Heath, P., Ratz, A., and Weigel, L. (2000). WO 00/58262 (to Eli Lilly).

Iwai, I. (1965). *Chem. Pharm. Bull.*, 13: 118–129.

Kabalka, G. and Guindi, L. (1990). *Tetrahedron*, 46: 7443–7457.

Kjell, D. and Lorenz, K. (2003). US 6541668 (to Eli Lilly).

Koenig, T. and Mitchell, D. (1994). *Tetrahedron Lett.*, 35: 1339–1342.

Krasik, P. and Alper, H. (1992). *Tetrahedron: Asymmetry*, 3: 1283–1288.

Kumar, A., Ner, D., and Dike, S. (1991). *Tetrahedron Lett.*, 32: 1901–1904.

Lapis, A., Fatima, A., Martins, J., Costa, V., and Pilli, R. (2005). *Tetrahedron Lett.*, 46: 495–498.

Leithe, W. (1932). *Chem. Ber.*, 65: 660–666.

Liu, H., Hoff, B., and Anthonsen, T. (2000). *J. Chem. Soc., Perkin Trans. 1*, 1767–1769.

Lou, H., Henrikson, L., Bruhn, P., Berner, H., and Neilsen, J. (1989). *Arch. Neurol.*, 46: 48–52.

Mannuza, S., Klein, R., Bessler, A., Malloy, P., and Hynes, M. (1997). *J. Am. Acad. Child Adolesc. Psychiatry*, 36: 1222–1227.

McMillen, B. (1983). *Trends Pharmacol.*, 4: 429–432.

Meier, R., Gross, F., and Tripod, J. (1954). *Klin. Wochenschr.*, 12: 266–270.

Michelson, D., Alder, L., and Spencer, T. (2003). *Biol. Psychiatry*, 53: 112–120.

Micovic, I., Ivanovic, M., Roglic, G., Kiricojevic, V., and Popovic, J. (1996). *J. Chem. Soc. Perkin Trans. I*, 265–269.

Mitchell, D. and Koenig, T. (1995). *Syn. Comm.*, 25: 1231–1238.

Molloy, B. and Schmiegel, K. (1977). US 4018895 (to Eli Lilly).

Mori, A., Ishiyama, I., Akita, H., Suzuki, K., Mitsuoka, T., and Oishi, T. (1990). *Chem. Pharm. Bull.*, 38: 3449–3451.

Nabenhauer, F. (1930). US 1921424 (to Smith, Kline and French).

Nabenhaur, F. (1942). US 2276509 (to Smith, Kline and French).

Nichols, D., Barfknecht, C., and Rusterholz, D. (1973). *J. Med. Chem.*, 16: 480–483.

Panizzon, L. (1944). *Helv. Chim. Acta.*, 27: 1748–1756.

Patrick, K., Caldwell, R., Ferris, R., and Breese, G. (1987). *J. Pharmacology and Experimental Therapeutics*, 241: 152–158.

Physicians Desk Reference, 60th ed. (2006). Thomson PDR, Montvale, NJ.

Prashad, M., Kim, H., Lu, Y., Liu, Y., Har, D., Repic, O., Blacklock, T., and Giannousis, P. (1999). *J. Org. Chem.*, 64: 1750–1753.

Prashad, M. (1998). *Tetrahedron: Asymmetry*, 9: 2133–2136.

Prashad, M. and Bin, H. (2000). US 6162919 (to Novartis).

Rasmussen, P. and Gillberg, C. (2000). *J. Am. Acad. Child Adolesc. Psychiatry*, 39: 1424–1431.

Repke, D., Bates, D., and Ferguson, W. (1978). *J. Pharm. Sci.*, 67: 1167–1168.

Rometsch, R. (1960). US 2957880 (to Ciba Pharmaceutical Products – now Novartis).

Rozwadowska, M. (1993). *Tetrahedron: Asymmetry*, 4: 1619–1621.

Sadler, B. (1993). *Drug Metabolism and Disposition*, 21(4): 717–723.

Sakuraba, S. and Achiwa, K. (1991). *Synlett*, 689–690.

Schneider, M. and Goergens, U. (1992). *Tetrahedron: Asymmetry*, 3: 525–528.

Shafi'ee, A. and Hite, G. (1969). *J. Med. Chem.*, 12: 266–270.

Shi, X., Yao, J., Kang, L., Shen, C., and Yi, F. (2004). *J. Chem. Res.*, 681–683.

Sowell, E., Thompson, P., Welcome, S., Henkenius, A., Toga, A., and Peterson, B. (2003). *Lancet*, 362: 1699–1707.

Srebnick, M., Ramachandran, P., and Brown, H. (1988). *J. Org. Chem.*, 53: 2916–2920.

Srinivas, N., Hubbard, J., Korchinski, E., and Midha, K. (1993). *Pharm. Res.*, 10: 14–21.

Srinivas, N., Hubbard, J., Quinn, D., and Midga, K. (1992). *Clin. Pharmacol. Ther.*, 52: 561–568.

Steinhausen, H., Dreschsler, R., Foldenyi, M., Imhof, K., and Brandeis, D. (2003). *J. Am. Acad. Child Adolesc. Psychiatry*, 42: 1085–1092.

Swanson, J., Gupta, S., Lam, A., Shoulson, I., Lerner, M., Modi, N., Lindemulder, E., and Wigal, S. (2003). *Arch. Gen. Psychiatry*, 60: 204–211.

Swanson, J., Seargeant, J., Taylor, E., Sonuga-Barke, E., Jensen, P., and Cantwell, D. (1998). *Lancet*, 351: 429–433.

Thai, D., Sapko, M., Reiter, C., Bierer, D., and Perel, J. (1998). *J. Med. Chem.*, 41: 591–601.

V-Braun, J. and Friehmelt, E. (1933). *Chem. Ber.*, 66: 684–685.

Vaidya, C. and Gabrieli, J. (1999). *Molecular Psychiatry*, 4: 206–208.

Volkow, D., Wang, G., Fowler, J., Logan, J., Gerasimov, M., Maynard, L., Ding, Y., Gatley, S., Gifford, A., and Franceschi, D. (2001). *J. Neuroscience*, 21: RC121/1–RC121/5.

Volkow, N., Wang, G., Fowler, J., Gatley, S., Logan, J., Ding, Y., and Pappas, N. (1998). *Am. J. Psychiatry*, 155: 1325–1331.

Wagner, J., McElhinny, C., Lewin, A., and Carroll, F. (2003). *Tetrahedron: Asymmetry*, 14: 2119–2125.

Wilens, T., Faraone, S., Biederman, J., and Gunawardene, S. (2003). *Pediatrics*, 111: 179–185.

Zeitlin, A. and Stirling, D. (1998). US 5733756 (to Celgene).

INDEX

The Art of Drug Synthesis. Edited by Douglas S. Johnson and Jie Jack Li